RECEPTORS

RECEPTORS

MODELS FOR BINDING, TRAFFICKING, AND SIGNALING

Douglas A. Lauffenburger
University of Illinois

Jennifer J. Linderman
University of Michigan

New York   Oxford
OXFORD UNIVERSITY PRESS

Oxford University Press

Oxford   New York
Athens   Auckland   Bangkok   Bombay
Calcutta   Cape Town   Dar es Salaam   Delhi
Florence   Hong Kong   Istanbul   Karachi
Kuala Lumpur   Madras   Madrid   Melbourne
Mexico City   Nairobi   Paris   Singapore
Taipei   Tokyo   Toronto

and associated companies in
Berlin   Ibadan

Library of Congress Cataloging-in-Publication Data
Lauffenburger, Douglas A.
Receptors: models for binding, trafficking, and signalling
Douglas A. Lauffenburger, Jennifer J. Linderman
ISBN 0-19-506466-6; ISBN 0-19-510663-6 (pbk.)
1. Cellular signal transduction.   2. Cell receptors.
I. Linderman, Jennifer J.   II. Title.
QP517.C45L38   1993
574.87'6—dc20   92-37216

9 8 7 6 5 4 3 2 1

Printed in the United States of America
on acid-free paper

Contents

Preface

As discussed more fully in the Introduction chapter, this book was written to address the interface between the disciplines of chemical engineering and cell biology. Its earliest roots were in the course entitled "Cellular Bioengineering" taught by D.A.L. to a number of excellent and inquisitive graduate students at the University of Pennsylvania during the mid-to-late 1980s. This course sought to introduce students from engineering and physical sciences to contemporary issues in cell biology, primarily those involving quantitative, physicochemical aspects of receptor-mediated cell function. Before long, a text seemed desirable in order to summarize, explain, and critique the growing body of research literature in this area.

Our attempt to provide this sort of text was stimulated by a kind invitation from the Department of Chemical Engineering at the University of Wisconsin for D.A.L. to be the Olaf A. Hougen Visiting Professor for the 1989–90 academic year. As the unique and noble purpose of the Hougen Visiting Chair is specifically to encourage book-writing, the prospect of developing an original text in the emerging area of cellular bioengineering seemed feasible and attractive. Award of a Fellowship to D.A.L. from the John Simon Guggenheim Foundation further strengthened the support available for this project. But only when J.J.L. agreed to take part as a coauthor was it clear that the book would indeed eventually find its way to print. Manuscript notes of early versions of this text were used in courses taught by the two of us during the period 1990 to 1992 at the University of Illinois and University of Michigan, encouraging us to finish the project.

A number of individuals and institutions deserve acknowledgment for their many different kinds of help. First, we express our gratitude to our colleagues at the University of Pennsylvania, University of Illinois, and University of Michigan for their support in this effort. Second, we thank the Department of Chemical Engineering at the University of Wisconsin for its motivation and financial support and the J.S. Guggenheim Foundation for its additional financial assistance. Third, many friends must be recognized for their aid in critically reading manuscript drafts and suggesting improvements: Helen Buettner, Paul DiMilla, Leah Edelstein-Keshet, Byron Goldstein, Dan Hammer, Rick Horwitz, Geneva Omann, Alan Perelson, Cindy Stokes, Bob Tranquillo, Alan Wells, Steve Wiley, Dane Wittrup, and Sally Zigmond. Fourth, we thank Kim Forsten for special computational work, those who developed illustrations for figures, especially Ken Vandermuelen, Steve Wiley, Steven Charnick, Ann Saterbak, Patricia Mahama, and Christine Schmidt, along with those who provided original drawings from their work. We are also thankful for the excellent professional assistance of the staff at Oxford University Press, especially Bill Curtis, Kirk Jensen, and Anne Hegeman. Finally, and most importantly, we are profoundly grateful to our families for their forbearance of our seeming obsession with this project.

Douglas A. Lauffenburger
Jennifer J. Linderman
September 1992

RECEPTORS

1
Introduction

We intend for this book to build a bridge between cell biologists and engineers, over the ground that can be called quantitative cell biology or cellular bioengineering. Our aim is to communicate how insights can be gained into the relationship between receptor/ligand molecular properties and the cell functions they govern, by a judicious combination of cell biology experimentation and quantitative engineering models. At the heart of this combination is the use of molecular biology techniques to alter receptor and ligand properties so as both to test the validity of a model and exploit its predictive power.

Because we recognize that researchers from both disciplines will be treading on unfamiliar territory, we feel that we must take some care to present our reasons for building this bridge and to describe how we hope this book will do so.

1.1 PHILOSOPHY

A central goal of modern cell biologists, molecular biologists, biotechnologists, and bioengineers is to *understand cell behavior in terms of molecular properties.* All may be interested, as an example, in learning how cell proliferation is affected by properties that directly reflect molecular structure, such as the affinity of a key regulatory component for its site of action. At the same time, these investigators may also desire to know the effect of properties that are less related to molecular structure, such as the cellular concentrations of the key regulatory component and its site of action. In both cases, what is needed is an ability to know how a change in any such properties may influence a particular cell function.

Great advances have been made toward this goal by the application of tools from *molecular cell biology.* Specific cellular constituents can be isolated, identified, and altered, genetically or synthetically. Amounts of a particular component can be manipulated by expression levels or by microinjection. Molecular structures can be modified by site-directed mutagenesis or covalent chemistry. These tools provide an opportunity for investigating the effect of a molecular alteration, and for developing technologies based on it. A major portion of effort by the pharmaceutical industry, for example, is aimed at constructing drugs that mimic, replace, or interfere with natural compounds which regulate cell function. For reproducible success in such efforts, as well as those directed toward other applications, reliable prediction of the consequences of changes in molecular properties will be vital.

Moreover, this predictive capability must be as quantitative as possible. Information like that illustrated schematically in Figure 1–1 would be of great assistance to analysis and design of molecular-based processes for manipulation of cell function, whether in scientific or industrial (health care, bioprocessing) applications. Figure 1–1 shows that the dependence of a cell behavioral function on a particular molecular property may be relatively insensitive or highly sensitive, and further that the dependence may not be monotonic. In order to decide how best to alter relevant molecular properties so that we can bring about a desired cell behavior, we must be able to predict the corresponding relationships as accurately as possible.

Though there has been tremendous progress in the identification of molecular components involved in cell functions, this is only a beginning step in the process of understanding. It is not sufficient to study component molecules by themselves; we must figure out how they work together. Although this betrays a bias from our engineering training, we believe that many life scientists of this generation agree that quantitative integration of molecular parts will ultimately be required to understand the cellular whole.

A proven tool in the quantitative analysis of complicated systems, whether living or not, is *mathematical modeling*. Simply put, mathematical modeling is really a way to formulate hypotheses in a framework that allows their full implications to be recognized. Mathematical equations are merely the language by which the hypotheses are described and their implications communicated. When the systems are simple, the mathematical language can be discarded as superfluous. However,

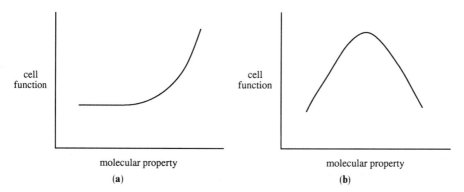

Figure 1–1 Schematic relationship between a cell function and a molecular property. As examples, the property may be affinity, valency, concentration, or a rate constant, and the cell function may be proliferation, contraction, secretion, motility, or adhesion. (a) The sensitivity of a cell function to a particular molecular property may vary from insensitive to highly sensitive. (b) The cell response may not simply be an increasing or decreasing function of a particular molecular property.

the history of many scientific disciplines indicates that discarding it for complicated systems leaves one a club short in the bag; that is, examination of hypotheses for such systems will be incomplete without use of the full range of tools. As Maddox (1992) writes ". . . the neglect of quantitative considerations [in molecular biology] may well be a recipe for overlooking problems inherently of great importance."

Our aim in this text is to describe the insights mathematical models can bring into the links between molecular properties and cell function. We intend this text to be a useful reference for both biologists and engineers and hope, in fact, that it helps to build a connection between the two disciplines. Models which are experimentally testable and are driven by a need to understand and predict experimental data at a mechanistic level are the goal. We will not dwell on mathematical subtleties or methods of analysis; the interested reader will find references on these matters but not an extensive discussion. Rather, we will explore first the assumptions and hypotheses that are involved in formulating model equations for problems of interest, and then will turn to the predictions of these model equations for the connection between molecular properties and cell function.

To narrow the scope of molecular components to be considered in this book, we limit our attention to cell surface *receptors*. Receptors, as shown in Figure 1–2, possess an extracellular domain for binding ligands (e.g., growth factors, adhesion molecules), a transmembrane domain, and an intracellular or cytoplasmic domain. We choose to focus on receptors for two reasons. First, receptor-driven cell

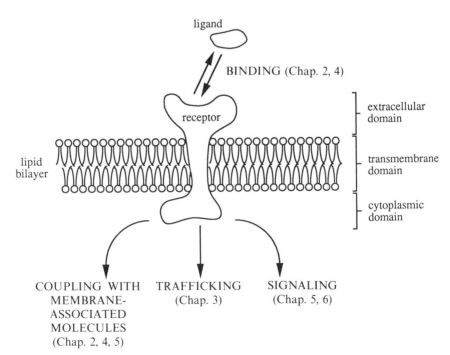

Figure 1–2 Schematic structure of cell receptors. Models for receptor binding, coupling with membrane-associated molecules, signaling, and trafficking are discussed in this text.

behavior is now known to be extremely important – for example, growth, secretion, contraction, motility, and adhesion and all receptor-mediated cell behaviors. Receptors are uniquely able to direct such cell behavior by virtue of their ability to sense the environment, through binding of ligands, and their ability to transmit this signal to the cell interior, through interaction of their intracellular domains with enzymes and proteins within the cell. Second, receptors and their ligands are ideal candidates for manipulation. For example, cell and molecular biologists can now alter the structure of a receptor and its ligand, thereby modifying association and dissociation rate constants and signaling capabilities, just as examples. This sort of manipulation provides a straightforward tool for the testing of hypotheses as to the role these molecular properties play in directing cell function.

As admitted above, in a simple system, intuition alone can guide us to a qualitative understanding of the effects of various parameters on observed system responses, and sometimes can even offer us semi-quantitative insight. However, receptor-mediated cell functions are highly complicated phenomena in two different ways (Figures 1–3 and 1–4). Receptor signaling induced by ligand binding is itself a process with multiple aspects, and these events are only a subset of the many concurrent dynamic events involving receptors.

Following Figure 1–3, the first step is the binding of ligands to receptors at the cell surface. Bound receptors can activate various intracellular enzymes and entire cascades of intracellular reactions. Some of these regions trigger short-term (of the order of milliseconds to minutes) responses. Long-term responses, such as

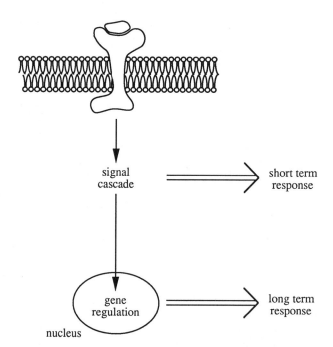

Figure 1–3 Levels of complexity in receptor signaling. Bound receptors may initiate a cascade of intracellular reactions that lead to short and/or long term responses.

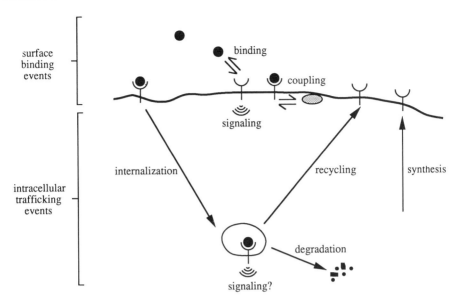

Figure 1-4 Levels of complexity in receptor state and location. Receptors may be unbound, bound, or coupled with other membrane-associated molecules. Receptors and their ligands may be internalized and routed through intracellular compartments. Both receptors at the cell surface and receptors inside the cell may have signaling capabilities, although these capabilities are likely a function of the receptor state and location.

those requiring protein synthesis, involve additional molecular interactions. Each step is characterized by parameters such as rate constants and concentrations which are in principle measurable and alterable. At the same time, as shown in Figure 1-4, the receptor population is undergoing events of coupling with other cell surface molecules, internalization, recycling, degradation, and synthesis – the whole pathway termed trafficking. When the complexities of both signaling and trafficking are considered together, it is difficult to intuit even qualitative trends, much less quantitative ones. Hence, there is a need for use of engineering modeling tools.

1.2   ORGANIZATION

The organization of this text is as follows. We first focus on determining the state (*e.g.*, bound, unbound, coupled with other molecules) and location (*e.g.*, cell surface, endosomes) of receptors. This is an important foundation, because we cannot hope to relate receptors to cell functions without a knowledge of their state and location. We discuss in Chapter 2 the details of receptor/ligand binding kinetics at the cell surface, as shown schematically in Figure 1-4. Because receptors and their ligands may also be internalized by the process termed endocytosis, we turn in Chapter 3 to this trafficking of receptors and ligands through the cell. Note that the location of a bound receptor may be important; for example, perhaps surface receptors, but not intracellular receptors, are able to produce the signals that elicit some cellular responses. In Chapter 4, we discuss the role of probability, diffusion, and valency in the determination of receptor state and location. With this background, we turn in Chapter 5 to the ability of receptors to transduce a signal to the cell interior by virtue of their ability to induce the generation of second messengers or to act as enzymes. In Chapter 6 we combine many of the ideas of the earlier chapters in order to model aspects of three central cellular responses – proliferation, adhesion, and motility – as they depend on receptor/ligand properties.

In each chapter, we will provide fundamental modeling frameworks to elucidate general trends and then proceed to examine specific problems. Here is a small sampling of particular examples. In Chapter 4, we discuss the dependence of histamine release by basophils on IgE–receptor crosslinking, and in Chapter 5 we investigate the relationship between β-adrenergic receptor occupancy and adenylate cyclase activation in erythrocytes. In Chapter 6, we study how alterations in trafficking of the EGF receptor can affect fibroblast proliferation and how the migration speed of smooth muscle cells depends on substratum-bound fibronectin density. In many cases, the reader will find that we generalize the behavior of related experimental systems. For example, in Chapter 3 we examine endocytic trafficking models that aim to explain the kinetics of receptor/ligand internalization and recycling in a variety of systems. We will try to make the point that qualitatively different results can be predicted for different systems based simply on quantitative differences in parameter values among those systems, despite a similar underlying model.

We intend this text to complement the more comprehensive discussions of modeling in biology that are found in Segel (1980), Edelstein-Keshet (1988), and Murray (1990). We will assume that the reader is familiar with receptor biology at the level of the texts of Alberts *et al.* (1989) or Darnell *et al.* (1990). We will also assume some familiarity with differential equations; a lucid and concise reference for this mathematical material is the Appendix in Segel (1980).

REFERENCES

Alberts, B., Bray, D., Lewis, J., Raff, M., Roberts, K. and Watson, J. D. (1989). *Molecular Biology of the Cell*. Second edition. New York: Garland Publishing, Inc. 1218 pp.

Darnell, J. E., Lodish, H. and Baltimore, D. (1990). *Molecular Cell Biology*. Second edition. New York: W.H. Freeman. 1105 pp.

Edelstein-Keshet, L. (1988). *Mathematical Models in Biology*. New York: Random House. 586 pp.

Maddox, J. (1992). Is molecular biology yet a science? *Nature*, **355**:201.

Murray, J. (1990). *Mathematical Biology*. New York: Springer-Verlag.

Segel, L. A. (1980). *Mathematical Models in Molecular and Cellular Biology*. Cambridge: Cambridge University Press. 757 pp.

Cell Surface Receptor/Ligand Binding Fundamentals

2.1 BACKGROUND

In order to model cell behavior regulated by receptor/ligand signaling, we must first quantify initial cell surface binding events as well as subsequent intracellular trafficking processes, as illustrated in Figure 1–4. This is necessary for following receptor/ligand complexes temporally and spatially throughout the cell. Once this foundation is established we can more easily analyze cell behavioral phenomena governed by the number, state, and location of these complexes. In this chapter our focus is on binding events on the cell surface, whereas in the next chapter we will turn to trafficking processes. Our attention throughout the book will be restricted to plasma membrane receptors, though some receptors (*e.g.*, those for steroid hormones) reside in the cytoplasm or on intracellular organelle membranes.

2.1.1 Experimental Methods

Binding of ligand to cell surface receptors has been amenable to direct experimental investigation for roughly the past two decades. Early work was largely limited to equilibrium binding properties, but recognition of the highly dynamic character of receptor phenomena has more recently stimulated interested in the kinetics of binding. Limbird (1986) and Hulme (1992) offer excellent background texts on details of experimental methods, and a helpful review is presented by Wiley (1985). We will provide an overview based primarily on these sources.

Five *major approaches* exist for distinguishing cell surface binding events from subsequent receptor trafficking processes: (1) performing experiments at lowered temperatures, typically 4–15 °C, because membrane trafficking processes – particularly internalization – appear to be drastically slowed at these temperatures as compared to the physiological temperature of 37 °C (Weigel and Oka, 1982; Tomoda *et al.*, 1989); (2) use of pharmacologic agents, such as phenylarsineoxide, that inhibit trafficking events (Low *et al.*, 1981); (3) use of isolated membranes possessing receptors (Kahn, 1976); (4) development of experimental protocols in which particular steps can be isolated by judicious choice of time scales or measurement procedure (*e.g.*, exploiting pH effects) (Wiley and Cunningham, 1982; Sklar *et al.*, 1984; Mayo *et al.*, 1989); and (5) estimation of all rate constants together using a complete model and comprehensive set of kinetic experiments involving a wider range of measurements, including not only bound ligand, but internalized ligand and receptors, degraded ligand and receptors, *etc.* (Gex-Fabry and DeLisi, 1984; Sklar and Omann, 1990; Waters *et al.*, 1990). A major shortcoming of the fifth approach is the need to validate each of the particular rate expressions used to model individual events. In the first four approaches, this validation will arise naturally from data obtained from experiments focused on specific events. However, use of lowered temperatures, pharmacological inhibitors, or isolated membranes raises the question of physiological relevance of the resulting data. Attempting to isolate particular steps by carefully chosen experimental protocols can fail when the time scales of confounding processes overlap or when the measurement procedures can give ambiguous results (*e.g.*, as with pH changes (Sklar and Omann, 1990)). Hence, no single approach is entirely satisfactory in general.

The most popular experimental method for monitoring the time course of receptor/ligand binding involves following the binding of *radioactively-labeled ligands*, principally using iodide (^{125}I) because of its relatively high specific radioactivity and long half-life (Limbird, 1986). For quantitative binding studies, one must determine the specific radioactivity of the fraction of ligand actually able to bind to receptors by using self-displacement methods in which concentrations of both labeled and unlabeled ligand are varied (Calvo *et al.*, 1983). Attachment of a label moiety can cause significant diminution of receptor/ligand binding by interfering with the active site (*e.g.*, see Kienhuis *et al.*, 1991). Heterogeneous ligand populations may also result if multiple susceptible amino acid residues exist.

To follow the binding of radiolabeled ligand to cell surface receptors, a typical experiment involves incubation of cells and ligand together for a specified time, rapid separation of unbound ligand from cells (*e.g.*, by centrifugation and washing or vacuum filtration), measurement of the radioactivity remaining with the cells using a gamma counter or scintillation counter, and then calculation of the amount of bound ligand per cell by consideration of total radioactivity counts, number of cells, and specific radioactivity of the ligand. The need for a step to separate unbound ligand from cells when radiolabeled ligands are used to follow the dynamics or receptor/ligand binding limits the application of this technique. Note that if the time scale for the binding event is of the order of the time scale for the separation step, the interpretation of the experimental data is compromised. Since these separation processes can require a few seconds to a few minutes, only those binding and trafficking processes that occur more slowly can be easily studied with radiolabeling methods.

Fluorescently-labeled ligands are also used to follow receptor/ligand binding; spectrofluorimetric methods, including fluorescence flow cytometry, can be used to measure the amount of labeled ligand associated with the cell surface (Muirhead *et al.*, 1985). Commonly used protein fluorophores include fluorescein and rhodamine. A particular advantage offered by these fluorescent labels is the ability to discriminate bound from unbound ligand without an intervening and time-consuming separation step, at least when the measurement of binding to whole cells (and not some particular cellular location) is the aim. With spectrofluorimetric approaches, bound and free ligand can be distinguished using spectral, physical, chemical, or optical differences (Sklar, 1987). Disadvantages to the use of fluorescence labels occur when high light intensities are needed to produce a strong fluorescence signal, since the fluorophore may photobleach; there may also be some cell damage. However, this is an extremely promising approach, and is likely to become more commonly applied in the near future.

A novel approach based on *avidin/biotin coupling* has been recently offered (King and Catino, 1990). Biotin binds with extremely high affinity to the protein avidin, so that biotinylated ligands can be detected upon incubation with enzyme-linked avidin conjugates followed by an appropriate enzymatic assay. Until methods for more rapid enzymatic assays are developed, however, this approach will be largely limited to equilibrium or steady-state binding studies.

Direct visualization of receptor/ligand complexes can be achieved by use of *electron-dense markers* such as the iron-containing protein ferritin, which can be quantitated by electron microscopy (McKanna et al., 1979; Gershon et al., 1981). This method requires not only a free/bound ligand separation step but also microscopy preparation steps, making it difficult to follow fast transients in receptor/ligand binding. In addition, quantitation is tedious. An alternative technique called *nanovid microscopy* has recently been demonstrated for direct visualization studies (Geerts et al., 1991). This technique involves using small (20–40 nm) colloidal gold particles which can be observed via video-enhanced microscopy to label ligands. An unsettled question at present is whether individual receptors, or only receptor aggregates, can be seen with this method.

Labeled ligands can be used in experiments with whole cells, cell membrane preparations, purified receptors in solution, purified receptors bound to a solid matrix, or purified receptors incorporated into membrane vesicles. Experiments with whole cells are the most difficult to interpret, because trafficking and other cellular processes that may affect binding can also be occurring. Cell membrane preparations are often a useful simplification, but these lack the cytoplasmic elements which may be involved in the regulation of binding. Isolation and purification of cell receptors has, in general, been a difficult undertaking since most receptors are present in extremely small quantities and are among a great number of other proteins with similar physical properties. Advances in cloning techniques are now allowing purified receptors to be studied more easily, since they can be produced in large quantities by microorganisms. When using these recombinant receptors, care must be taken to ensure that they are, in fact, functionally identical to the native receptor. Here, protein sequencing can provide the requisite information on primary structure, but it is more difficult to ensure that post-translational processes (*e.g.*, glycosylation, oligomerization) are accomplished properly. Regardless of how they are produced, isolated receptors also lack the accessory regulatory elements normally present in the cell. Thus, one can generally anticipate that ligand binding data generated from whole cells, cell membrane preparations, and isolated receptors may be significantly different. In the context of understanding receptor-mediated cell behavioral functions, clearly the ultimate goal must be a quantitative description of binding properties in whole cells.

When whole cells or cell membranes are used for receptor/ligand binding

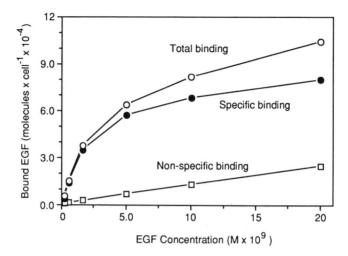

Figure 2–1 Binding of ligand to cells. Total binding is the sum of non-specific and specific, or receptor-mediated, binding. Notice that while receptor binding of ligand saturates at sufficiently high ligand concentrations, non-specific binding may continue to increase. These data are for epidermal growth factor binding to mouse (H. S. Wiley, University of Utah).

experiments, a ubiquitous problem is the occurrence of relatively weak, *non-specific binding* caused by association of ligand with membrane constituents or by trapping of ligand in the medium associated with the cells. Because it is the receptor-mediated ligand binding that is believed to be linked to the production of cellular responses, non-specific binding must be subtracted from total (specific or receptor-mediated plus non-specific) binding to obtain the true receptor binding level, as shown in Figure 2–1. This is usually accomplished by measuring the amount of labeled ligand binding in the presence of very high concentrations of unlabeled ligand concentrations that saturate receptor sites. The labeled ligand binding in such circumstances must be non-specific, since the receptors are occupied by the unlabeled ligand. Given this level of non-specific binding at a specific labeled ligand concentration, non-specific binding at other concentrations can be calculated by assuming that the non-specific binding is a linear function of the total ligand concentration. An alternative method is to determine the amount of non-specific binding by using a mathematical model that includes both receptor-specific and non-specific binding processes for labeled ligand in the absence of unlabeled ligand (Munson and Rodbard, 1980). The complication of non-specific binding is present with virtually all ligand labeling approaches.

2.1.2   Effects of Extrinsic Variables

It must be recognized that receptor/ligand binding phenomena are subject to the influence of a variety of *extrinsic variables, i.e.,* aspects of the experimental system that can be altered but are not intrinsic to the receptor and ligand molecules nor associated cellular components. Primary among these are temperature and medium composition, including pH and small ion concentrations.

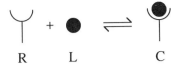

Figure 2–2 Schematic illustration for simple case of binding of ligand to receptor. Association and dissociation rates are characterized by rate constants k_f and k_r, respectively.

Consider first equilibrium binding for the simple reversible reaction scheme between a free receptor R and a free ligand L to form a receptor/ligand complex C, as illustrated in Figure 2–2 or written as

$$R + L \underset{k_r}{\overset{k_f}{\rightleftharpoons}} C$$

The relevant rate constants are the association rate constant k_f and the dissociation rate constant k_r. For typical units of the free receptor number R [#/cell], free ligand concentration L [moles/liter, or M], and receptor/ligand complex number C [#/cell], the units of k_f and k_r are $M^{-1} time^{-1}$ and $time^{-1}$, respectively. (Henceforth throughout the book, appropriate units for model variables and parameters will be listed in the Nomenclature sections at the end of each chapter.) We will use italic symbols (*e.g.*, *R*, *L*, and *C*) to denote numbers and concentrations of receptor and ligand species, while plain symbols (R, L, and C) will denote the entities themselves. Although receptors might be expressed in terms of concentration (moles/volume solution) or density (#/cell surface area), it is best for most purposes to express the receptor number as simply #/cell. Knowledge of the cell surface area or number of cells per volume solution can be used to convert R into density or concentration units when desired.

Thermodynamic principles dictate that the free receptor and free complex numbers and ligand concentration are related at chemical equilibrium through the *equilibrium dissociation constant*, $K_D = k_r/k_f$:

$$C = \frac{RL}{K_D} \tag{2–1}$$

A small value of K_D, corresponding to a large value of the equilibrium association constant, $K_A = k_f/k_r = 1/K_D$, indicates a high affinity of the receptor for the ligand. The values of K_D for various systems fall within a wide range, with 10^{-12} M near the high affinity end (the avidin/biotin bond with $K_D = 10^{-15}$ M (Green, 1975) is an extreme exception) to 10^{-6} at the low affinity end.

An experimental approach most frequently used to separate cell surface binding from intracellular trafficking events is to perform binding studies at lowered temperatures. It is therefore important to understand the effects that temperature may have on the binding parameters measured. The *Gibbs standard free energy change*, $\Delta G°$, for a chemical reaction such as receptor/ligand binding occurring at constant temperature and pressure is

$$\Delta G° = \Delta H° - T\Delta S° \tag{2–2}$$

where $\Delta H°$ is the enthalpy change between reactants and products at standard state, $\Delta S°$ is the entropy difference also at standard state, and T is the absolute temperature (*e.g.*, in degrees Rankine or Kelvin). K_D is related to the standard free energy change by

$$\Delta G° = R_g T \ln K_D \qquad (2\text{–}3)$$

(where R_g is the gas constant, 1.99 cal/mole-K), so small values of K_D correspond to large negative values of $\Delta G°$. Combining Eqns (2–2) and (2–3) allows expression of the effect of temperature on the equilibrium dissociation constant as

$$\ln K_D = \frac{\Delta H°(T)}{R_g T} - \frac{\Delta S°(T)}{R_g} \qquad (2\text{–}4)$$

where we have indicated that $\Delta H°$ and $\Delta S°$ may be functions of temperature.

Receptor/ligand binding processes are generally exothermic, *i.e.*, $\Delta H° < 0$, although many are primarily driven by positive entropy changes, $\Delta S° > 0$ (Klotz, 1985; Limbird, 1986). When the entropic term dominates, there may be only minor variation of equilibrium binding with temperature, but when the enthalpic term dominates there will be a strong temperature effect. Considerations regarding temperature variations are greatly complicated by the fact that $\Delta H°$ and $\Delta S°$ themselves generally vary significantly with temperature for protein/ligand inter-actions (Hinz, 1983). That is, the difference in the *constant-pressure heat capacity* between free and complexed species, $\Delta C_p°$,

$$\Delta C_p° = \left(\frac{\partial \Delta H°}{\partial T}\right)_P = T\left(\frac{\partial \Delta S°}{\partial T}\right)_P \qquad (2\text{–}5)$$

(where the derivatives are taken with pressure held constant) typically has a substantial value for proteins and can itself vary with temperature. This behavior is probably due to changes in three-dimensional protein conformation. Hence, van't Hoff plots of log K_D versus T^{-1} are rarely linear. An example for antibody/hapten binding (Szewczuk and Mukkur, 1977) is shown in Figure 2–3. There is substantial literature concerning equilibrium thermodynamic properties of protein binding interactions (Hinz, 1983), with advances in understanding due largely to modern microcalorimetric experimental techniques. Most of this work is devoted to enzyme/substrate or antibody/antigen binding, however, with little information available at present specifically for receptor/ligand binding thermodynamics.

Temperature variations of the individual kinetic rate constants for binding and dissociation, k_f and k_r, have not been widely examined. Rimon et al. (1980) measured an approximately 10-fold increase in k_r as temperature was increased $10\,°C$ to $37\,°C$ for dissociation of an "antagonist" ligand (*i.e.*, non-active ligand in terms of producing a cellular response, as opposed to an active, or "agonist", ligand) from the β-adrenergic receptor on turkey erythrocyte membranes. However, few quantitative studies have been specifically directed at the effects of temperature on rate constants for receptor/ligand systems. Following the above discussion regarding equilibrium binding behavior, it can be expected that temperature variations of binding rate constants will also prove to be quite complicated. Binding data obtained at low temperatures should therefore be held as quantitatively

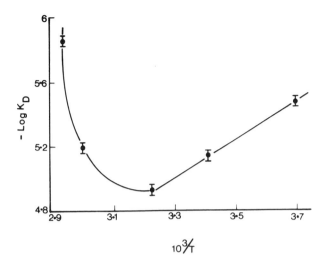

Figure 2–3 Van't Hoff plot for the effect of temperature on binding. The log of the equilibrium dissociation constant K_D is plotted against $1/T$, where T is absolute temperature. In this example showing the binding of ε-DNP-L-lysine to rabbit anti-DNP, redrawn from Szewczuk and Mukker (1977), the non-linear behavior suggests that the enthalpy change ΔH^0 is a function of temperature.

suspect in terms of understanding receptor/ligand binding at physiological temperature.

Receptor/ligand binding properties can also be strongly influenced by *pH and small ion concentrations*. It is often found that the equilibrium dissociation constant, K_D, increases as pH decreases below about 7. This is not universally true, however. In fact, Kaplan (1985) uses this as a point of distinction between "behavior" (signaling) and "uptake" receptors, with only the latter class showing pH dependence as a rule. Rarely is it determined clearly whether a decrease in the association rate constant, an increase in the dissociation rate constant, or both, is responsible for the dependence of K_D on pH. A dogma persists that the association rate constant is typically diffusion-limited, so that the pH effect would consequently be felt primarily due to changes in the dissociation rate constant. However, data indicate that the association rate constant can often be significantly smaller than diffusion-limited values (Kahn, 1976; Pecht and Lancet, 1977; Lauffenburger and DeLisi, 1983) (see Section 4.2), so this presumption is unwarranted in general. The concentrations of other small ions beside H^+, particularly physiologically important divalent cations such as Ca^{2+}, Mg^{2+}, and Mn^{2+}, have also been shown to influence receptor/ligand binding equilibria and rates (*e.g.*, Gailit and Ruoslahti, 1988; Grzesiak *et al.*, 1992). Little quantitative information on these effects is available at the present time, however.

2.1.3 Receptor Mobility in Cell Membranes

An additional influence on receptor interactions on the cell surface as well as during intracellular trafficking processes is *receptor mobility*, or diffusion, within cell membranes. Coupling of receptors with effector molecules, other receptors, cytoskeletal elements, and additional membrane-associated components requires protein movement or diffusion within the two-dimensional lipid bilayer. As will be seen later in this chapter, these sorts of coupling processes can significantly alter the equilibrium and kinetic binding properties of a receptor with its ligand. Therefore, even though we are not presently considering receptor trafficking but only surface binding, it will be helpful to have some information concerning receptor mobility. Two primary experimental methods for investigation of receptor mobility are *fluorescence recovery after photobleaching* (FRAP) (Axelrod et al., 1976; Edidin et al., 1976) and *post-electrophoresis relaxation* (PER) (Poo et al., 1979).

In the FRAP technique, the receptor or other membrane protein is labeled by a fluorescent antibody, lectin, or photoaffinity ligand, and the cell is irradiated by an attenuated laser beam focused on a small circular portion of the membrane (typically about 1–$10\ \mu m^2$). A brief pulse of sufficiently great beam intensity is used to irreversibly bleach the fluorophores within the beam focus area. The beam intensity is then lowered to its original value and the amount of fluorescence coming from the beam focus area is measured as a function of time (Figure 2–4a). As fluorescently-labeled (unbleached) receptors diffuse into the beam focus area, the fluorescence in that area increases. The rate of increase can be used to determine the receptor translational *diffusion coefficient*, D ((distance)^2/time). (Convective effects due to membrane flow may sometimes be present, but will be neglected here.) Rotational diffusion coefficients can also be defined for analysis of molecular orientation, but we will not have cause to use them in this book. Furthermore, the difference between a new, final asymptotic level of fluorescence and the initial, baseline level of fluorescence may be the fraction of receptors that is unable to diffuse freely on the cell surface; the fraction that is able to diffuse freely is termed the *mobile fraction*.

For determination of a protein's translational diffusion coefficient by PER, the chosen protein is labeled as in FRAP and then the cell is placed in an electric field (Figure 2–4b). Since proteins are typically charged, they will accumulate at one of the poles of the cell within the field, creating a protein concentration gradient. When the electric field is removed, the labeled proteins disperse down this gradient, and the time course of this redistribution can be followed by microscopic observation and used to quantify the receptor diffusion coefficient.

We note that these two methods actually measure two distinct quantities: FRAP measures the *self-diffusion coefficient*, D^{self}, whereas PER measures the *mutual diffusion coefficient*, D^{mutual}. The self-diffusion coefficient is defined by the increase in the *mean-square displacement* $\langle \bar{x}^2 \rangle$ of a single protein with time t:

$$\langle \bar{x}^2 \rangle = 4D^{self}t \qquad (2\text{–}6a)$$

in two dimensions. The mutual diffusion coefficient (neglecting diffusion-induced convection) is defined by Fick's Law for the flux, J (moles/area-time) of a protein population down its own concentration gradient, Δc:

$$J = -D^{mutual}\Delta c \qquad (2\text{–}6b)$$

In the limit of infinitely-dilute protein concentration these diffusion coefficients

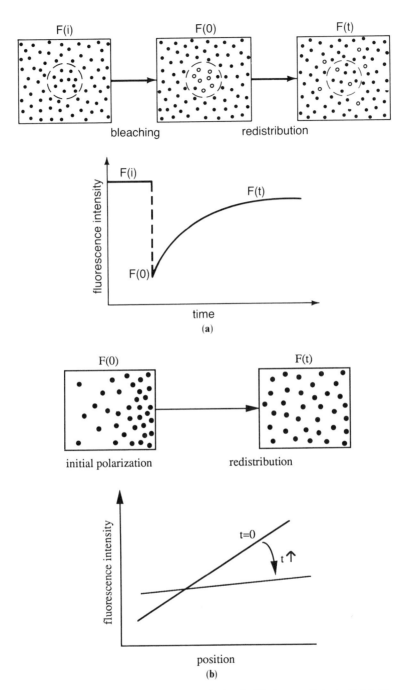

Figure 2–4 Measurement of the receptor translational diffusion coefficient, D. (a) Fluorescence recovery after photobleaching (FRAP). Fluorescence intensity drops when fluorophores within the beam focus area are bleached. Fluorescence intensity recovers as a result of diffusion of fluorescently-labeled receptors from other areas of the cell into the area of interest. Reprinted from Gennis (1989). (b) Post-electrophoresis recovery (PER). Fluorescence intensity redistributes as labeled receptors diffuse down their concentration gradient after a localizing electric field is removed.

are identical, $D^{self} = D^{mutual}$, but at high protein concentrations their values may diverge due to non-ideal, crowding effects. It has been shown theoretically that D^{self} should decrease and D^{mutual} should increase as protein concentration increases so that in general $D^{self} \leq D^{mutual}$, consistent with many experimental observations showing that the PER method can yield receptor mobilities 5–100-fold greater than the FRAP method (Schalettar et al., 1988). For the purposes of most of this book, we will assume that D^{self} and D^{mutual} are identical (that is, we will neglect possible crowding effects) and simply denote receptor diffusion coefficients as D.

Typical values of receptor translational diffusion coefficients in cell membranes found by both the FRAP and PER methods fall in the range 10^{-11} to 10^{-9} cm²/s, whereas those of mobile fractions typically fall between 0.1 and 1 (Gennis, 1989). From the mean-square displacement expression given above, we can estimate the time scale for receptor traverse across a cell surface. For a cell radius of 10 μm, $t \sim (10~\mu m)^2/4(10^{-9}$ to 10^{-11} cm²/s), yielding an approximate range of 4 min to 7 h.

Experimental measurements of D give values that are generally several orders of magnitude lower than *theoretical predictions* based on viscous interactions of proteins with membrane lipids (Wiegel, 1984), though a few membrane proteins – notably rhodopsin on photoreceptor cells (Wey and Cone, 1981) – possess diffusivities very close to the theoretical limit. A classical expression was derived by Saffman and Delbruck (1975) based on a hydrodynamic model for a cylindrical protein of radius s embedded in a membrane of thickness h of viscosity η_m, surrounded on both (extracellular and intracellular) sides by a solution of viscosity η_s:

$$D = \left[\frac{K_B T}{4\pi \eta_m h}\right](-\gamma + \ln \theta) \qquad (2-7)$$

where $K_B = 1.38 \times 10^{-16}$ g-cm²/s²-K is Boltzmann's constant, $\gamma = 0.5772$ is Euler's constant, $\theta = \eta_m h/\eta_s s$ is a ratio of membrane to solution viscosities (assumed $\gg 1$ for this derivation to hold). A value of approximately $D \sim 10^{-8}$ cm²/s is obtained for the set of parameter values $K_B T = 4 \times 10^{-14}$ g-cm²/s² (at 37 °C), $h = 10$ nm, $s = 2$ nm, $\eta_m = 2$ g/cm-s, and $\eta_s = 0.01$ g/cm-s. As mentioned, this is roughly one to three orders of magnitude greater than typical experimentally measured values for D.

The favored explanation for this discrepancy is that receptor diffusion is slowed by interaction with cytoskeletal components or other membrane-associated macromolecules (Ryan et al., 1988; Wade et al., 1989). This could also account for the existence of a fraction of receptors deemed to be immobile on typical experimental time scales. Experimental support for this explanation is provided by measurements of much larger values of D, close to the theoretical estimate, for LDL receptors on cell membrane regions known to be lacking in cytoskeletal connections (Barak and Webb, 1982) and for bacteriorhodopsin in reconstituted lipid vesicles (Peters and Cherry, 1982). An alternative, possibly complementary explanation is that the extracellular domains of membrane receptors may interact with other macromolecules present in the local extracellular environment (Wier and Edidin, 1988). Compilation of diffusion measurements for membrane proteins with different structural features may permit development of general rules governing mobility (Zhang et al., 1991). An excellent discussion of factors influencing glycoprotein diffusion in cell membranes is provided by Sheetz (1993).

Notice that Eqn. (2–7) predicts only a weak (logarithmic) dependence of D on the receptor radius, s. Experimentally, even fairly small receptor oligomers exhibit a diffusivity more than an order of magnitude lower than their subunit monomers (Menon et al., 1986; Baird et al., 1988). Again, a likely basis for this discrepancy is an interaction of receptors with membrane-associated components that depends on the number of receptors in an aggregate.

The effect of temperature on receptor diffusivity within the cell membrane should also be mentioned. Eqn. (2–7) predicts only a 10% reduction in D when T is decreased from 37 °C (310 K) to 4 °C (277 K). In contrast, experimental measurements have shown a roughly fourfold reduction in D over the same temperature interval (Hillman and Schlessinger, 1982), possibly due to changes in membrane structure. Nonetheless, it is important to recognize that receptor diffusivity is not reduced to zero at these low temperatures. Fixation of membrane components (e.g., with glutaraldehyde) offers an alternative method for reducing receptor diffusivity when desired.

2.2 CELL SURFACE RECEPTOR BINDING MODELS

2.2.1 Simple Monovalent Binding

As a base model of receptor/ligand binding, consider the case in which a monovalent ligand L binds reversibly to a monovalent receptor R to form a receptor/ligand complex C, with no further processes modifying this interaction (see also Figure 2–2):

$$R + L \underset{k_r}{\overset{k_f}{\rightleftharpoons}} C$$

Because monovalent receptors and ligands by definition have only one binding site, no additional receptors or ligands of the same type can bind to the complex. Let ligand be present in the extracellular medium at initial concentration L_0 [moles/volume medium, or M], receptors be present on the cell surface at constant number R_T [#/cell], and cells be present at concentration n [#/volume medium]. We will assume here that the medium is well-mixed, so that the ligand is available at uniform concentration from a macroscopic point of view.

Using principles of mass action kinetics, the equation describing the time rate of change of the receptor/ligand complex density C as a function of the free receptor number R and the ligand concentration L is:

$$\frac{dC}{dt} = k_f RL - k_r C \tag{2–8}$$

The association rate constant k_f (M^{-1} time^{-1}) characterizes the velocity of the second-order interaction between the receptor and ligand, while the dissociation rate constant k_r (time^{-1}) characterizes the velocity of the first-order breakdown of the receptor/ligand complex.

Eqn. (2–8) cannot yet be solved for $C(t)$ because both R and L change with time. We restrict our attention in this chapter to situations in which the cohort of surface receptors is unchanged, so that total surface receptor number R_T is constant. If we also assume for now that the total amount of ligand is unchanged, the following *conservation laws* apply at all times:

$$R_T = R + C \tag{2–9a}$$

$$L_0 = L + \left(\frac{n}{N_{Av}}\right)C \tag{2–9b}$$

where N_{Av} is Avogadro's number (6.02×10^{23} #/mole), needed for converting number of molecules into moles. Use of these two conservation laws together with Eqn. (2–8) allows us to describe the system with the single *ordinary differential equation*:

$$\frac{dC}{dt} = k_f[R_T - C]\left[L_0 - \left(\frac{n}{N_{Av}}\right)C\right] - k_r C \tag{2–10}$$

Because this equation follows only cell-associated ligand that is bound to cell surface receptors, this formulation assumes that non-specific binding of ligand to the cell surface has already been corrected for experimentally.

2.2.1.a Constant Ligand Concentration

Eqn. (2–10) is slightly more complicated in appearance than the more common starting point, one which neglects ligand depletion (or accumulation, for dissociation experiments) in the medium and instead assumes that the ligand concentration remains constant at its initial value, L_0. That assumption is equivalent to the limiting case $(n/N_{Av})C \ll L_0$, for which Eqn. (2–10) simplifies to

$$\frac{dC}{dt} = k_f[R_T - C]L_0 - k_r C \tag{2–11}$$

The solution to Eqn. (2–11) will be examined before returning to Eqn. (2–10). With imposition of an *initial condition* $C(t = 0) = C_0$, standard techniques for solving ordinary differential equations (*e.g.*, Ritger and Rose, 1968) yield the *transient solution*, $C(t)$:

$$C(t) = C_0 \exp\{-(k_f L_0 + k_r)t\} + \left(\frac{k_f L_0 R_T}{k_f L_0 + k_r}\right)[1 - \exp\{-(k_f L_0 + k_r)t\}] \tag{2–12}$$

The number of receptor/ligand complexes at *equilibrium*, C_{eq}, is in this case identical to the steady-state value at which $dC/dt = 0$ and is asymptotically approached for $t \gg (k_f L_0 + k_r)^{-1}$:

$$C_{eq} = \frac{R_T L_0}{K_D + L_0} \tag{2–13}$$

where K_D, the *equilibrium dissociation constant*, is equal to k_r/k_f.

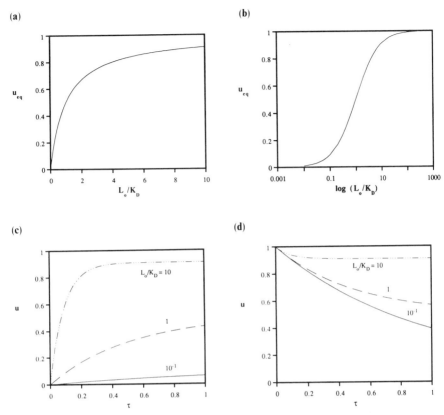

Figure 2–5 Transient and equilibrium binding of ligand to cell surface receptors for the case of ligand concentration L approximately constant and equal to L_0. The fraction of total surface receptors bound, u, is shown. (a), (b) The equilibrium value, u_{eq}, is plotted as a function of the ratio L_0/K_D and the logarithm of the ratio L_0/K_D. (c), (d) u is plotted as a function of scaled time τ for several values of the ratio L_0/K_D. The initial value and time course of u in (c) correspond to an association experiment. The initial value and time course of u in (d) correspond to a dissociation experiment with all receptors bound at time 0.

To illustrate the results given by Eqns. (2–12) and (2–13), it is convenient to define a dimensionless or scaled number of complexes, u,

$$u = C/R_T \qquad (2\text{–}14)$$

and a dimensionless or scaled time, τ,

$$\tau = k_r t \qquad (2\text{–}15)$$

u is clearly just the fraction of receptors occupied, and τ can be thought of as a complex "turnover" number during time t. In Figures 2–5a and 2–5b, the dimensionless equilibrium number of complexes, $u_{eq} = C_{eq}/R_T$, is plotted as a function of the dimensionless ratio L_0/K_D and the logarithm of the dimensionless ratio L_0/K_D. Although use of these *dimensionless groups* may seem unwieldly at first, they allow us to show the behavior of many similar systems very simply. For example, the curve in Figure 2–5a represents the equilibrium behavior of innumerable different receptor/ligand binding systems, each with particular values of the parameters R_T, L_0, and K_D, and is certainly a more concise representation of the behavior than pages of plots for different values of each of these parameters. One can easily convert from a point $(L_0/K_D, u)$ on the plot to the more physical (L_0, C) point if the values of K_D and R_T are known. It is convenient to note that $u_{eq} = C_{eq}/R_T = 0.5$ when $L_0/K_D = 1$; in other words, half the receptors are bound by ligand at equilibrium when the ligand concentration is equal to the value of K_D.

Figure 2–5c depicts the transient solution $u(\tau)$ for a sample association "experiment" when the initial condition is $u_0 = C_0/R_T = 0$. The *half-time*, $\tau_{1/2}$, for this transient, defined as the value of τ needed to yield half the change from $u_0 = 0$ to u_{eq}, is given by $\tau_{1/2} = 0.69/[1 + (L_0/K_D)]$. In this case, binding deviates from its equilibrium value by less than 5% after roughly three half-times, or $t \sim 2/(k_r[1 + (L_0/K_D)])$. Figure 2–5d shows a sample dissociation "experiment" starting with the initial condition $[C_0/R_T] = 1$; i.e., all receptors are initially bound.

It is helpful to notice that the dimensionless variables u and τ, along with the dimensionless parameter groups L_0/K_D and C_0/R_T, can be discovered in both the original kinetic model equation (Eqn. (2–11)) and its solution (Eqns. (2–12) and (2–13)). Rewriting those equations using these scaled quantities shows why the behavior of all particular systems can be captured together in this manner:

$$\frac{du}{d\tau} = (1 - u)\frac{L_0}{K_D} - u \tag{2–11'}$$

$$u(t) = u_0 \exp\left\{-\left(1 + \frac{L_0}{K_D}\right)\tau\right\} + \frac{(L_0 K_D)}{1 + (L_0/K_D)}\left[1 - \exp\left\{-\left(1 + \frac{L_0}{K_D}\right)\tau\right\}\right] \tag{2–12'}$$

$$u_{eq} = \frac{(L_0/K_D)}{1 + (L_0/K_D)} \tag{2–13'}$$

All dependence of transient and equilibrium complex numbers on any specified model parameter values are clearly represented by the results for this dimensionless formulation, not only saving much time and effort but also providing some insight into the roles played by the parameters in governing receptor behavior.

2.2.1.b Ligand Depletion Effects

Eqns. (2–11)–(2–13) (or (2–11')–(2–13')) and Figure 2–5 describe simple receptor/ligand binding when the ligand concentration in the medium is essentially unchanged throughout an experiment. When the inequality condition $(n/N_{Av})C \ll L_0$ is not satisfied, *ligand depletion* from the medium by cell binding can be significant. In such a situation, Eqn. (2–10) governs the dynamics of receptor/ligand complex formation, and the cell density n should be influential. Rewriting this equation in dimensionless form, using the scaled time and bound receptor number variables u and τ defined in Eqns. (2–14) and (2–15), we obtain:

$$\frac{du}{d\tau} = (1 - u)\left(1 - \left[\frac{nR_T}{N_{Av}L_0}\right]u\right)\frac{L_0}{K_D} - u \tag{2–10'}$$

We indeed see that one additional quantity is now of importance, the *scaled cell density* $\eta = (nR_T/N_{Av}L_0)$. The quantity η has a simple physical meaning; it is the ratio of total cell receptors to total ligand molecules present in the system.

In contrast to Eqn. (2–11′) for constant ligand concentration, Eqn. (2–10′) allowing ligand depletion is non-linear in the receptor/ligand complex number u. However, it possesses a special form, of the Riccati equation class, which permits analytical solution in implicit manner (O'Neil, 1991). We will not present this expression here, but instead illustrate the solution behavior in Figure 2–6a. The transient dimensionless receptor/ligand complex number, $u(\tau)$, is plotted for a range of the scaled cell density η and a few values of the scaled ligand concentration (L_0/K_D). The outcome for fractional receptor occupancy at equilibrium, u_{eq}, simply recognizes the depletion of ligand:

$$u_{eq} = \frac{L}{K_D + L} = \frac{L_0[1 - \eta u_{eq}]}{K_D + L_0[1 - \eta u_{eq}]} \tag{2–16}$$

This is an implicit equation since u_{eq} appears on both sides; it can be solved as a quadratic equation for u_{eq}, with results shown in Figure 2–6b. Both equilibrium and transient calculations demonstrate that significant ligand depletion effects arise when $\eta > 0.1$.

In experimental assays, η, the ratio of total receptors to total ligand molecules, is often large enough that ligand depletion effects are significant, especially when high cell densities are used to enhance a fluorescence or radioactivity signal. As an example, in experiments studying the binding of EGF to various fibroblast cell types, Schaudies *et al.* (1985) incubated 10^6 cells per well in 24-multiwell plates containing 0.2 ml medium of 10^{-10} to 10^{-9} M EGF in each well. R_T ranged from about 10^4 to 10^6 receptors/cell for the different cell types investigated. For this situation, then, η can be estimated to be in the approximate range 0.1–100. Figure 2–6 demonstrates that, especially for the higher end of this range corresponding

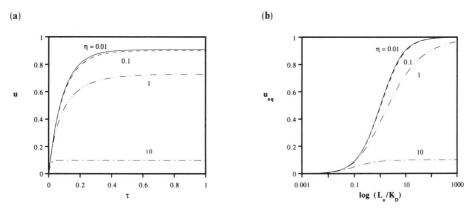

Figure 2–6 Transient and equilibrium binding of ligand to cell surface receptors for the case of ligand concentration L not constant. L_0 is the initial value of the ligand concentration. The fraction of total surface receptors bound, u, is shown. (a) u is plotted as a function of scaled time τ for several values of the scaled cell density η and $L_0/K_D = 10$. (b) The equilibrium value, u_{eq}, is plotted as a function of the ratio L_0/K_D for several values of η.

to the lower EGF concentrations and higher receptor numbers, depletion of EGF simply due to receptor binding substantially reduces equilibrium receptor binding. When ligand uptake occurs through receptor-mediated and fluid-phase endocytosis, the influence of depletion is even greater, and more complicated mathematical descriptions are required to assess the resulting effect (see Chapter 3).

2.2.1.c   Non-specific Binding

When non-specific binding is not subtracted from total binding to give specific receptor-mediated binding, a description of receptor/ligand binding requires that effects of *non-specific binding* be considered. The exact form of the mathematical description depends on the assumptions made regarding the non-specific binding process.

We can, for example, assume that non-specific binding is reversible and non-saturable at ligand concentrations normally used in experimental assays. This is the assumption made implicitly when experimental investigators correct for non-specific binding by measuring total binding of labeled ligand in the presence of saturating levels of unlabeled ligand and then assuming that non-specific binding is a linear function of ligand concentration. To briefly pursue the kinetic implications of this assumption, we can write an equation describing the rate of change in the concentration of non-specifically-bound ligand:

$$\frac{dB}{dt} = k_{fn}L - k_{rn}B \tag{2-17}$$

where B is the amount of non-specifically-bound ligand. The terms $k_{fn}L$ and $k_{rn}B$ describe the non-specific cell surface association and dissociation events. The corresponding rate constants characterizing non-specific ligand interaction with the cell surface are k_{fn} [(#/cell)(moles/volume)$^{-1}$ time^{-1}] and k_{rn} [time^{-1}]. Note that there are no identified "sites" for the non-specific binding of ligand; this binding is not saturable, at least in the range of ligand concentrations typically examined. To describe the system fully one would need to solve Eqns. (2–8) and (2–17) together with the receptor conservation relation, Eqn. (2–9a) and the ligand conservation relation:

$$L_0 = L + \left(\frac{n}{N_{Av}}\right)(C + B) \tag{2-18}$$

Note that we need not assume that ligand depletion effects are negligible, as depletion of ligand is explicitly accounted for in Eqn. (2–18).

One reasonable way to simplify the above equations for specific and non-specific ligand binding is to assume that the rates of non-specific association and dissociation are rapid compared to the rates of specific association and dissociation. This implies that $dB/dt = 0$ on the time scale relevant for receptor-mediated binding, so that

$$B = K_N L \tag{2-19}$$

$K_N = k_{fn}/k_{rn}$ [(#/cell)/(moles/volume)] is a *non-specific association equilibrium constant*. Munson and Rodbard (1980) make use of this relationship in their computer analysis of equilibrium binding, but it can also be used to help analyze

transient binding. Eqn. (2–18) can now be replaced by the conservation relation
$L_0 = [1 + (n/N_{Av})K_N]L + (n/N_{Av})C$ or, solving for L,

$$L = \frac{L_0 - \left(\dfrac{n}{N_{Av}}\right)C}{1 + \left(\dfrac{n}{N_{Av}}\right)K_N} \tag{2–20}$$

This relationship shows how free ligand concentration L is further diminished by non-specific binding. Specific ligand binding, then, is governed by a modification of Eqn. (2–10):

$$\frac{dC}{dt} = k_f[R_T - C]\left\{1 + \left(\frac{n}{N_{Av}}\right)K_N\right\}^{-1}\left[L_0 - \left(\frac{n}{N_{Av}}\right)C\right] - k_rC \tag{2–21a}$$

again with the initial condition $C(0) = C_0$. One can also show that total binding is governed by a multiple of this equation:

$$\frac{d(C + B)}{dt} = \left\{1 + \left(\frac{n}{N_{Av}}\right)K_N\right\}^{-1}\frac{dC}{dt} \tag{2–21b}$$

with the initial condition $[C(0) + B(0)] = C_0 + K_N L_0$ due to the quasi-equilibrium assumption for non-specific binding.

We can define a *dimensionless non-specific binding capacity* $v = nK_N/N_{Av}$. With this definition, we see that Eqns. (2–21a) and (2–21b) are simply modifications of Eqn. (2–10). The form of the equations are the same, but altered "effective" rate constants replace the true rate constants. In other words, Eqn. (2–21a) is identical to Eqn. (2–10) with k_f replaced by $k_f(1 + v)^{-1}$; hence, the solutions for $C(t)$ and C_{eq} are given in Figures 2–6a and 2–6b when $C(0) = 0$. Eqn. (2–21b) is identical to Eqn. (2–10) with k_f replaced by $k_f[1 + v]^{-2}$, k_r by $k_r[1 + v]^{-1}$, and C_0 by $C_0 + K_N L_0$ – which will, of course, be in error for short times. With this caveat, an approximately correct solution for $(C_{eq} + B_{eq})$ is therefore given in Figure 2–6b. In each case, the appropriate values of the modified rate constants must be used with the corresponding figures.

Equilibrium solutions to Eqns. (2–21a,b) will be influenced by the values of η and v. In this case, $u_{eq} = C_{eq}/R_T$ will be given by Eqn. (2–16) but with (L_0/K_D) replaced by $(1 + v)^{-1}(L_0/K_D)$. That is, non-specific binding causes further ligand reduction in the medium by a factor $(1 + v)^{-1}$. Munson and Rodbard (1980) report estimates of v in the range 0.004–0.04 from a series of example studies. These values indicate that non-specific binding should not usually cause significant ligand depletion effects.

When ligand depletion from both specific and non-specific binding is negligible, the equilibrium binding values are $C_{eq} = R_T(1 + [K_D/L_0])^{-1}$ (Eqn. 2–13) and $B_{eq} = K_N L_0$ (Eqn. 2–19). The fraction of binding that is non-specific is then $B_{eq}/(B_{eq} + C_{eq}) = \{1 + R_T/[K_N(K_D + L_0)]\}^{-1}$ (see Figure 2–7). This expression offers a rigorous criterion for determining conditions under which non-specific binding will become significant compared to specific binding: when $(L_0 + K_D)$ is sufficiently large to be the same order of magnitude as R_T/K_N. This

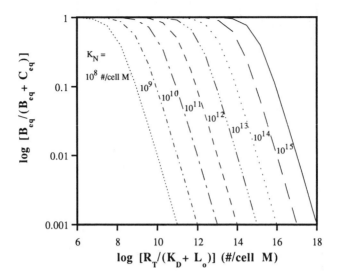

Figure 2–7 Non-specific binding. The fraction of total binding that is non-specific, $B_{eq}/(B_{eq} + C_{eq})$, is plotted as a function of the ratio $R_T/(K_D + L_0)$ for several values of the non-specific association equilibrium constant K_N.

result formalizes the approach of correcting for non-specific binding by a linear extension of labeled ligand binding in the presence of saturating levels of unlabeled ligand. Note also that given experimental measures of $B_{eq}/(B_{eq} + C_{eq})$ and $R_T/(K_D + L_0)$, one can estimate a value of K_N from Figure 2–7. This is essentially a graphical analog of the numerical procedure used by Munson and Rodbard (1980).

2.2.1.d Determination of Parameter Values

With an understanding of the mathematical terms describing simple homogenous receptor/ligand binding, our attention can now be focused on methods for determination of the model parameters. We turn first to the determination of the parameters relevant to *equilibrium binding*, R_T and K_D. Assuming that non-specific binding can be corrected for as discussed previously, measurements of the number of receptor/ligand complexes at equilibrium, C_{eq}, as a function of the ligand concentration L are necessary.

There are two major classes of parameter estimation methods, graphical and numerical, which can be used to obtain R_T and K_D values from the experimental data. Graphical methods, the most popular of which is the *Scatchard plot* (Scatchard, 1949), typically involve various linear transformations of the binding data. In this transformation, Eqn. (2–13) is rearranged into the form (note that L, rather than L_0, is used, to allow formally for cases in which ligand depletion is significant):

$$\frac{C_{eq}}{L} = -\frac{1}{K_D} C_{eq} + \left(\frac{R_T}{K_D}\right) \qquad (2\text{–}22)$$

A plot of (C_{eq}/L), sometimes termed "bound/free", versus C_{eq}, or "bound", should

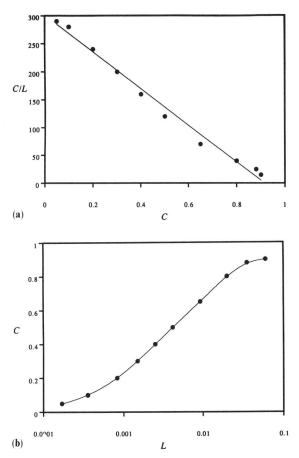

Figure 2–8 Equilibrium receptor binding data. (a) In a Scatchard plot of the data, the ratio of bound ligand to free ligand concentrations is plotted versus the bound ligand concentration. (b) The same data are plotted in the form of bound ligand concentration versus free ligand concentration. Note that at high free ligand concentration receptor saturation is approached. These data are for benzodiazepine binding to rat brain cells. Redrawn from Klotz (1985).

thus yield a straight line, with slope equal to $-1/K_D$, abscissa-intercept equal to R_T, and ordinate-intercept equal to (R_T/K_D). This will be true if there is actually equilibrium binding of single homogeneous, monovalent receptor and ligand populations. It is most important to note that the Scatchard analysis can be properly applied only to equilibrium binding data. Numerous other subtle pitfalls in interpreting Scatchard plots of binding data are discussed by Limbird (1986).

An example presented by Klotz (1985) in a review of common errors in receptor binding studies is shown in Figure 2–8, with the data plotted both as C_{eq} vs. log L and in the Scatchard format. The plotting of data in both of these formats is highly recommended, in order to avoid improper extrapolation in the Scatchard plot when receptor binding has not nearly approached its asymptotic

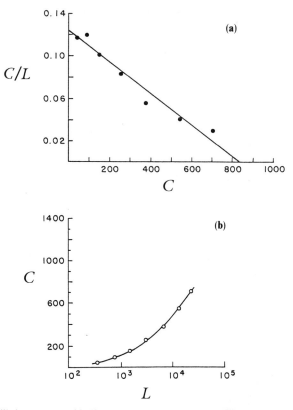

Figure 2–9 Equilibrium receptor binding data. (a) In a Scatchard plot of the data, the ratio of bound ligand to free ligand concentration is plotted versus the bound ligand concentration. (b) The same data are plotted in the form of bound ligand concentration versus free ligand concentration. Note that at high free ligand concentration receptor saturation has not been reached. These data are for diazepam binding to rat brain cells. Redrawn from Klotz (1985).

limit of R_T (Munson *et al.*, 1983). An example of such an error (Klotz, 1985) is shown for contrast in Figure 2–9. It is not always easy to take reliable binding data at ligand concentrations high enough to approach receptor saturation, though, because it is at these higher concentrations that non-specific binding becomes significant. Inaccuracies in estimation of non-specific binding will render such high concentration data less precise. Even when these issues are dealt with satisfactorily, there can still be some ambiguity in the determination of R_T and K_D because of the difficulty in exact determination of the value of L, since usually only L_0 is specified experimentally. Another subtle point is that experimental uncertainty about the data points shows up in a radial manner in Scatchard plots rather than in the more usual vertical or horizontal manner because C_{eq} appears in both the abscissa and ordinate variables (Zierler, 1989).

A better choice is to use numerical methods to determine the relevant binding parameters. *Numerical parameter-estimation algorithms* can be used to apply Eqn. (2–13) or Eqn. (2–16) directly to experimental data on specific binding (corrected for non-specific binding) to obtain best-fit values of R_T and K_D. Such algorithms do not require extrapolation as the linear transformation plots do. Munson and Rodbard (1980) offered one of the earliest applications of this approach to receptor/ligand binding, and today many computer-fitting algorithms are available. Wells (1992) gives an excellent summary of mathematical underpinnings and Press *et al.* (1989) provide sample routines. In addition, modern graphics software for personal computers often include automatic curve-fitting functions as part of their plotting capabilities.

A further advantage of numerical algorithms is that more complex models can be incorporated. For instance, Eqns. (2–13) and (2–19) (with $L = L_0$) for C_{eq} and B_{eq} can be applied directly to uncorrected, total equilibrium binding data on $[C_{eq} + B_{eq}]$ to obtain best-fit values for R_T, K_D, and K_N. The fact that underlying assumptions, such as linear non-specific binding, must be specified explicitly in the model formulation makes this approach attractive from the perspective of clarity.

Table 2–1 lists the values of R_T and K_D that have been found for some typical systems. Values of R_T are typically in the range of 10^4 #/cell to 10^6 #/cell, although much lower numbers have been reported for some receptors – particularly hematopoietic growth factors (Park *et al.*, 1990). We note that binding behavior is often more complicated than the simple case described here so far, so it is possible that at least a few of these values will be reevaluated with improved approaches in the near future. Very few investigators have reported rigorous quantitative measurements of K_N, with Munson and Rodbard (1980) being a notable exception.

Next, methods for estimation of the *kinetic parameters*, k_f and k_r are discussed; a nice exposition is given by Hollenberg and Goren (1985). A good starting point is the transient solution expressed by Eqn. (2–12), which assumes negligible ligand depletion or accumulation and that the data have been corrected for non-specific binding. This equation can be rearranged to the form

$$\ln[C_{eq} - C(t)] = \ln\left\{\frac{R_T}{1 + (K_D/L_0)} - C_0\right\} - \left[1 + \left(\frac{L_0}{K_D}\right)\right]k_r t \qquad (2\text{–}23)$$

Equilibrium experiments and their corresponding analysis should be performed first, as described above, so that R_T and K_D are fixed. Also, this permits C_{eq} to be already known as a function of L_0. To determine the remaining parameters, k_f and k_r, using graphical methods, we additionally need transient data describing $C(t)$ for a given added ligand concentration L_0. We then plot $\ln[C_{eq} - C(t)]$ versus t. According to Eqn. (2–23), a straight line with slope equal to $-[1 + (L_0/K_D)]k_r$ and ordinate-intercept equal to $\ln[R_T(1 + [K_D/L_0])^{-1} - C_0]$ should be obtained (see Figure 2–10a). Since L_0 is specified, and K_D has been already determined, the value of k_r can be evaluated from the slope. The remaining parameter k_f can then be calculated from $k_f = k_r/K_D$. If this procedure is repeated for different ligand concentrations, L_0, the same values for k_r should be found.

As an alternative graphical approach to determining k_f and k_r, the observed "net" rate constant for approach to equilibrium, $k_{obs} = [1 + (L_0/K_D)]k_r$, obtained from the plot in Figure 2–10a, can be plotted against L_0 as shown in Figure 2–10b. This should yield a straight line with slope equal to $k_r/K_D(=k_f)$ and ordinate-intercept equal to k_r.

Table 2–1 Sample receptor/ligand binding parameters

Receptor	Ligand	Cell type	R_T (#/cell)	k_f (M⁻¹ min⁻¹)	k_r (min⁻¹)	K_D (M)	$t_{95\%}(L_0=K_D)$ (min)	Reference
Transferrin	Transferrin	HepG2	5×10^4	3×10^6	0.1	3.3×10^{-8}	15	Ciechanover et al. (1983)
Fc_γ	2.4G2 Fab	Mouse macrophage	7.1×10^5	3×10^6	0.0023	7.7×10^{-10}	650	Mellman and Unkeless (1980)
Chemotactic peptide	FNLLP	Rabbit neutrophil	5×10^4	2×10^7	0.4	2×10^{-8}	3.7	Zigmond et al. (1982)
Interferon	Human interferon-α_2a	A549	900	2.2×10^8	0.072	3.3×10^{-10}	20	Bajzer et al. (1989)
TNF	TNF	A549	6.6×10^3	9.6×10^8	0.14	1.5×10^{-10}	11	Bajzer et al. (1989)
β-adrenergic	Hydroxybenzylpindolol	Turkey erythrocyte	—	8×10^8	0.08	1×10^{-10}	19	Rimon et al. (1980)
α_1-adrenergic	Prazosin	BC3H1	1.4×10^4	2.4×10^8	0.018	7.5×10^{-11}	83	Hughes et al. (1982)
Insulin	Insulin	Rat fat-cells	1×10^5	9.6×10^6	0.2	2.1×10^{-8}	7.5	Lipkin et al. (1986b)
EGF	EGF	Fetal rat lung	2.5×10^4	1.8×10^8	0.12	6.7×10^{-10}	12.5	Water et al. (1990)
Fibronectin	Fibronectin	Fibroblasts	5×10^5	7×10^5	0.6	8.6×10^{-7}	2.5	Akiyama and Yamada (1985)
Fc_ε	IgE	Human basophils	—	3.1×10^6	0.0015	4.8×10^{-10}	1000	Pruzansky and Patterson (1986)
IL-2 (heavy chain)	IL-2	T lymphocytes	2×10^3	2.3×10^7	0.015	6.5×10^{-10}	100	Smith (1988)
IL-2 (light chain)			1.1×10^4	8.4×10^8	24.	2.9×10^{-8}	0.06	
IL-2 (heterodimer)			2×10^3	1.9×10^9	0.014	7.4×10^{-12}	110	

Shown are the measured number of receptors per cell R_T, the association rate constant k_f, the dissociation rate constant k_r, and the equilibrium dissociation constant $K_D = k_r/k_f$. The time required to reach 95% of equilibrium receptor binding when no bound receptors are initially present, $t_{95\%}$ is calculated from $t_{95\%} = -\ln(0.05)/(k_f(1 + L_0/K_D))$ for the case of $L_0 = K_D$. HepG2 = human hepatoma cell line; 2.4G2 Fab = Fab portion of 2.4G2 antibody against receptor; FNLLP = N-formylnorleucylleucylphenylalanine; A549 = human lung alveolar carcinoma; TNF = tumor necrosis factor; hydroxybenzylpindolol is an antagonist to the receptor; EGF = epidermal growth factor; IgE = immunoglobulin E; IL-2 = interleukin 2; prazosin is an antagonist to the receptor; BC3H1 = smooth muscle-like cell line; RBL = rat basophilic leukemia cell line.

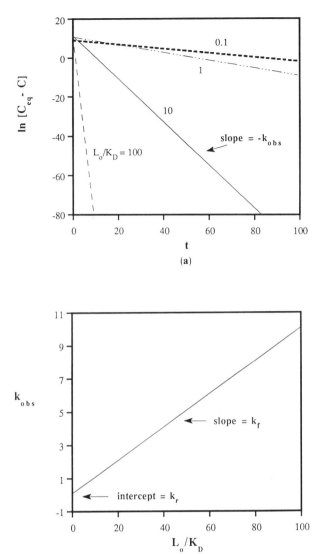

Figure 2–10 Graphical approaches to the determination of the association rate constant k_f and the dissociation rate constant k_r from experimental data. (a) The logarithm of the difference between the equilibrium and transient number of bound receptors, $\ln[C_{eq} - C(t)]$, is plotted versus time t. If K_D and L_0 are known, then k_r can be determined from the slope. k_f can then be determined from k_r/K_D. (b) k_{obs}, obtained from the slope of line plotted in (a), is graphed versus L_0. The rate constant k_f can be found from the slope and the rate constant k_r can be found from the ordinate intercept.

Procedures based on Eqn. (2–23) are applicable to both *association experiments* in which $L_0 > 0$ and $C_0 = 0$, and *dissociation experiments* in which $L_0 = 0$ and $C_0 > 0$. The rate constants k_f and k_r can each be found separately from short-time association and dissociation experiments without going to equilibrium.

In such circumstances, Eqn. (2–12) simplifies to

$$C(t) = k_f L_0 R_T t \tag{2–24}$$

for association experiments, and to

$$C(t) = C_0 \exp\{-k_r t\} \tag{2–25}$$

for dissociation experiments. Hence, k_f can be determined from plots of $C(t)$ versus t for short-time experiments with $L_0 > 0$ and $C_0 = 0$. k_r can be determined from plots of $\ln[C(t)/C_0]$ vs. t for short-time experiments with $L_0 = 0$ and $C_0 > 0$. This approach is especially useful when the presence of further events following receptor/ligand binding (see sections 2.2.2–2.2.4 and Chapter 3) complicate the relationship between a steady-state complex number and a true equilibrium complex number. Examples can be found in Mayo *et al.* (1989), Lipkin *et al.* (1986b), and Pollet *et al.* (1977), among others.

In a manner similar to that mentioned previously for equilibrium binding data, numerical curve-fitting algorithms can be employed for determination of kinetic binding parameters. The transient solution of Eqn. (2–11) expressed by Eqn. (2–12) can be fitted to association or dissociation data, $C(t)$ versus t, permitting estimation of k_f and k_r simultaneously. Press *et al.* (1989) and Caceci and Cacheris (1984) provide helpful computer routines for this, and most graphical software programs for personal computers offer automatic curve-fitting functions applicable to this situation. When more complicated models are considered, as in later sections of this chapter, an analytical expression for transient binding kinetics will likely not be available. In those cases, solutions for model predictions must be obtained by numerical integration of the ordinary differential equations (see Press *et al.*, 1989) for use with curve-fitting algorithms.

When L_0 is not constant during the course of an experiment, because the cell density is too great for *ligand depletion effects* to be negligible (recall the criterion $\eta \ll 1$, where η is the ratio of total receptors present to total ligand molecules), but non-specific binding is corrected for, then a transient solution to Eqn. (2–10) must be used to analyze association or dissociation kinetic data. This solution can be provided by either the implicit form for this Riccati equation (O'Neil, 1991) or by computational results using a numerical integration method for ordinary differential equations. If R_T and K_D are known from previous equilibrium experiments, and given a value for η (which also requires R_T), then the only unknown parameter is k_r. Application of a numerical parameter estimation algorithm fitting the $C(t)$ data to the transient solution for Eqn. (2–10) will allow k_r to be determined. As before, k_f can then be calculated from the values of K_D and k_r.

When *non-specific binding* is not corrected for before parameter estimation, then the solution for a more complete set of model equations, such as those given in Eqns. (2–21a) and (2–21b), is required. Eqns. (2–13) and (2–19) (with $L = L_0$) for C_{eq} and B_{eq} can be applied directly to uncorrected, total equilibrium binding data on $[C_{eq} + B_{eq}]$ to obtain best-fit values for R_T, K_D, and K_N. This is the approach followed by Munson and Rodbard (1980).

Typical values for k_f and k_r have been found to be in the range of 10^6–10^9 M^{-1} min^{-1}, and about 10^{-3}–1 min^{-1}, respectively. The temperature

Table 2–2   Example cellular responses to ligand binding

Ligand	Cellular response	Approximate time frame for response
Transferrin	Internalization of ligand	Minutes
Growth factor	Division	Hours–days
Growth factor	Second messenger generation	Seconds–minutes
Growth factor	Internalization of ligand	Minutes
ECM protein	Adhesion	Minutes–hours
Chemoattractant	Migration/chemotaxis	Minutes
IgE	Sensitization of basophils	Days
ATP	Contraction of smooth muscle cell	Seconds
Neurotransmitter	Secretion by nerve cell	Subsecond
Antigen + APC	IL-2 secretion by T lymphocyte	Hours

ECM = extracellular matrix; IgE = immunoglobulin E, APC = antigen presenting cell; IL-2 = interleukin 2.

dependence of the rate constants k_f and k_r has generally not been studied. Some values from the literature are listed in Table 2–1.

Table 2–1 shows the time required to reach 95% of the equilibrium binding level, starting from an initial state of no ligand present and assuming simple homogenous receptor/ligand binding. It is interesting to compare this time frame with that needed to achieve a cellular response. Representative cellular response times for some responses triggered by receptor/ligand binding are listed in Table 2–2. It is rarely certain whether cellular responses are related to, for example, initial changes in receptor binding or to the level of receptor binding achieved at some later time. If the first is the case, the assumption of equilibrium binding in modeling of the later response would be unreasonable; it is likely that the dynamics of receptor/ligand binding contribute significantly to the dynamics of the response. If the latter is the case, and the time allotted is sufficient for near-equilibrium, the assumption of equilibrium is reasonable and would simplify later modeling by eliminating consideration of the transients in the binding step.

Having presented this discussion of techniques to analyze binding data for this situation of single, homogeneous populations of receptors and ligands, it must be stated that very few real systems have been found to possess such simplicity. At physiological temperatures, trafficking events consequent to binding commonly occur at similar rates, complicating any analyses. Even at low temperatures, for which trafficking process rates are usually minimal, more complex behavior is frequently observed. We now move on to describe and then consider a number of different underlying causes of complex binding behavior, before turning to receptor trafficking models in the next chapter. The preceding discussion may nonetheless be of some use, because simple binding behavior can in some cases be found by careful restrictions of experimental time scales, by using pharmacologic agents to block complicating processes, by experimental protocols which distinguish between receptor states and/or locations, or by studying isolated membranes. Also, this discussion has served to establish foundational information and insight that will ease the way into the more complicated cases.

2.2.2 True and Apparent Cooperativity

Even when experimental conditions restrict observations to cell surface events, receptor/ligand binding is often more complicated than the previous analysis can successfully handle. A common finding is that a Scatchard plot of equilibrium data is not linear, as it ought to be from Eqn. (2–22). In such cases the simple model of Eqn. (2–1) is inadequate for accurate quantitative description of binding for use in models of receptor-mediated cell behavior.

Curvature observed in Scatchard plots is typically termed *cooperativity*. Because the magnitude of the slope of the Scatchard curve is the effective receptor/ligand binding affinity (the reciprocal of the equilibrium dissociation constant, K_D), cooperative binding refers to that in which K_D – and, correspondingly, one or both of the rate constants, k_f and k_r – appear to vary with the extent of receptor occupancy by ligand (see Figure 2–11). *Negative cooperativity* is the descriptive term used when K_D increases with occupancy (*i.e.*, L increasing),

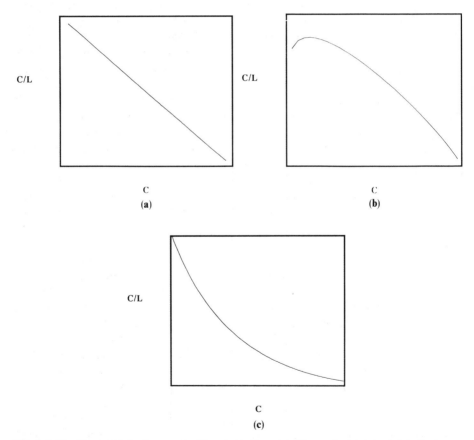

Figure 2–11 Cooperativity in receptor/ligand binding. (a) Linear Scatchard plot indicates no cooperativity effects. (b) Positive cooperativity is denoted by a concave downward curve in the Scatchard plot. (c) Negative cooperativity is denoted by a concave upward curve in the Scatchard plot.

thereby decreasing the slope, or affinity. This is illustrated by the concave-upward curve in Figure 2–11c. The term *positive cooperativity* is used when the concave-downward curve exhibited in Figure 2–11b appears. Note that the slope may be positive at low occupancy levels in this case, implying a negative value of K_D. Here the inappropriateness of the Scatchard analysis for more complicated situations reveals itself clearly. A popular method for pseudo-quantitation of the degree of cooperativity is the Hill plot (*e.g.*, Limbird, 1986), although this plot lacks fundamental meaning.

We note that many artifacts of experimental procedure can lead to curvature in Scatchard plots, as presented in detail by Limbird (1986). The focus of the present discussion, however, is on non-artificial deviation from simple binding behavior. Two categories of explanation exist for such deviations: true cooperativity and apparent cooperativity. *True cooperativity* in receptor/ligand binding, which may be either positive or negative, characterizes the situation in which the intrinsic receptor/ligand association and/or dissociation rate constants vary as a function of the number of occupied receptors, C, as in the classic – but non-receptor – system of oxygen binding to hemoglobin. As examples of phenomena underlying true cooperativity one could include feedback-modulation of receptor conformation consequent to ligand binding (or so-called "affinity conversion") and steric effects of multivalent receptors or ligands, although the dividing line can be somewhat murky.

Apparent cooperativity represents situations which result in curvature of the Scatchard plot but are not due to variation in the intrinsic binding and/or dissociation rate constants. In these cases, the rates of additional processes involved in overall receptor dynamics are affected by the extent of receptor occupancy, and it is the improper interpretation of binding in terms of a simple receptor/ligand interaction that leads to what seem to be changes in the binding parameters. Among examples of clearly apparent-cooperative effects are multiple receptor (or ligand) subpopulations, receptor and/or ligand binding multivalency, and cell surface interactions (including receptor aggregation, clustering, and ternary-complex formation). These phenomena will be analyzed in later sections of this chapter and in Chapter 4.

First, there may be some benefit in taking a brief look at efforts to analyze cooperative binding behavior with a phenomenological approach. Much work of this type was done during the late 1970s, when a substantial amount of quantitative data on receptor/ligand binding became available but before molecular mechanisms started to emerge. An excellent summary can be found in DeLean and Rodbard (1979), who proposed and analyzed several general phenomenological mathematical models for cooperative binding effects. These models were largely motivated by the non-linear Scatchard plots that had been obtained for a great number of experimental systems (*e.g.*, insulin, nerve growth factor, epidermal growth factor, and β-adrenergic receptors) and appeared to indicate negatively-cooperative binding.

In addition, the models of DeLean & Rodbard were motivated by kinetic data on receptor/ligand binding. Three types of experiments sometimes showed interesting behavior conflicting with the simple receptor/ligand binding model of Eqn. (2–8). The first type of experiment, an *accelerated dissociation experiment*, was originally popularized by DeMeyts and co-workers (De Meyts *et al.*, 1973; De Meyts, 1976) in their work on insulin binding to lymphocytes at 15 °C. The kinetics of radiolabeled ligand dissociation from receptors are compared for two protocol variations: in the standard ("dilution") protocol, the dissociation medium contains no ligand at all, whereas in a second ("unlabeled ligand") protocol, the dissociation medium contains only the unlabeled form of the ligand. If a simple process following Eqn. (2–8) is occurring, the kinetic behavior found for these two protocols should be identical. On the other hand, if negative/positive cooperative effects are present, the rate of dissociation might be increased/decreased in the presence of unlabeled ligand as compared to no ligand. It was found, for example, that the rate of insulin dissociation from lymphocytes was increased dramatically in the presence of unlabeled insulin; this is termed accelerated dissociation (Figure 2–12). Given this result in combination with the concave-upward Scatchard plots found for this system, an obvious interpretation is for negative binding cooperativity, with k_r an increasing function of C.

In the second type of kinetic experiment, dissociation rates are measured after associative incubation with different ligand concentrations, in order to determine the dissociation rate constant as a function of initial receptor occupancy C_0. For the insulin/lymphocyte case, however, k_r was found by Pollet *et al.* (1977) to be

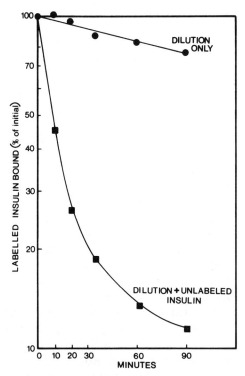

Figure 2–12   Accelerated receptor/ligand dissociation. The presence of unlabeled ligand, as compared to no ligand, enhances the rate of dissociation of labeled ligand from cell surface receptors. Redrawn from DeMeyts *et al.* (1973).

independent of C_0, casting doubt on a straightforward negative cooperativity model. In a third type of experiment the same group performed with this system, k_f was measured after incubation of cells with different levels of unlabeled ligand, with the result that k_f was also found to be independent of C_0.

Motivated at least in part by such inconclusive results, DeLean and Rodbard (1979) investigated the qualitative kinetic and equilibrium behavior of two *phenomenological models*; that is, models aimed at capturing features of a phenomenon without attempting to hypothesize an underlying mechanism. These models use the fundamental kinetic mass action equation for receptor/ligand binding (Eqn. 2–8), but incorporate the formal possibilities that the association or dissociation rate constants may vary with complex number. Expressed mathematically, these models allow $k_f = k_{fo}(1 + q_f C)$ or $k_r = k_{ro}(1 + q_r C)$ in Eqn. (2–8). The parameters q_f and q_r represent enhancement (when positive) or diminution (when negative) coefficients for association and dissociation rates, respectively.

The work of DeLean and Rodbard identified underlying mathematical descriptions of the rate constants k_f and k_r with the form of experimental data that would be produced. First, if only k_r varies with C, then Scatchard plots will be concave up or down for negative ($q_r > 0$) and positive ($q_r < 0$) cooperativity, respectively, but there can be no maxima or minima in these curves. If, on the other hand, only k_f varies with C, then there can be a maximum in the curve for positive ($q_f > 0$) cooperativity, but there will be little non-linearity for negative ($q_f < 0$) cooperativity. Second, transient dissociation experiments should indeed demonstrate accelerated dissociation in the presence of unlabeled ligand for $q_r > 0$ or < 0, but should exhibit no difference from simple dilution experiments (*i.e.*, in the absence of unlabeled ligand) for $q_f > 0$ or < 0. Third, kinetic association curves are likely to give very few clues for any of these cases, with little noticeable departure from standard behavior expected in plots of $\ln[C_{eq} - C(t)]$ versus t (Eqn. 2–23). If one wishes further to pursue qualitative behavior of phenomenological cooperative binding models in the abstract, contributions with just this aim exist (Rescigno *et al.*, 1982; Jose and Larralde, 1982). A very wide range of possible outcomes for equilibrium and kinetic binding plots can be obtained from model variations, though no insight into the biological mechanisms responsible for changes in k_f and k_r can be gained from these exercises.

In addition to characterizing the general properties of these two models, DeLean and Rodbard (1979) applied them specifically to the insulin controversy, and found that neither adequately accounted for all the reported behavior in that system. From a retrospective view, a satisfactory model for this application must be based on some of the mechanisms that will be discussed in the next subsections (see De Meyts *et al.*, 1989).

This has also been the case for a second widely-studied application, discussed in detail by Limbird (1986), the β-adrenergic receptor. In this situation, mechanistic models were applied to equilibrium and kinetic binding data in an attempt to discriminate among them. An important lesson is that model discrimination requires a variety of experimental data, both equilibrium (including studies of competition between labeled and unlabeled ligand) and kinetic (association and dissociation, with different initial ligand concentrations), combined with comparison to model computations. Examination of qualitative features can carry model evaluation only so far before rigorous parameter estimation is required to help decide on the possible consistency (though not to say proven equivalence) of a particular model with experimental data. An excellent example of rigorous model discrimination has been published by Lipkin et al. (1986a,b) for the case of insulin binding to fat cells. This work includes mathematical analysis of non-specific binding (considered as trapping of ligand in medium pelleted with the cells during centrifugation) and insulin dimerization, and statistical model selection from among many alternative possibilities by means of the Akaike information criterion (Akaike, 1974). This criterion permits comparison of model agreement with data with regard for the number of model parameters involved.

Finally, a quantitative inquiry into what may be true cooperative binding behavior is that for EGF binding to human fibroblasts (Wiley et al., 1989). Equilibrium binding data at $4\,^\circ$C obtained for control cells yield a linear Scatchard plot with $K_D = 2.2 \times 10^{-9}$ M and $R_T = 6.5 \times 10^4$ receptors/cell. (It should be noted that Roy et al. (1989) and Mayo et al. (1989) have observed non-linear, concave-upward Scatchard plots at $4\,^\circ$C, though with the non-linearity showing up at levels of bound ligand greater than those examined by Wiley et al. (1989).) When the fibroblasts were pretreated with unlabeled EGF for 5 min at $37\,^\circ$C before cooling to $4\,^\circ$C to perform binding experiments, resulting Scatchard plots were again linear but indicated a decrease in affinity. K_D increased by a factor of up to five as the pretreatment EGF concentrations were increased. This was not a straightforward blocking effect; rather, the affinity reduction was nearly complete when fewer than 5% of the cell receptors were occupied during the preincubation (see Figure 2–13). The half-time for this affinity reduction was about 2 min, with binding data at intermediate time points showing a heterogeneous receptor population with respect to affinity, and was reversible upon removal of EGF with a half-time of 20–30 min.

Using kinetic binding experiments and mathematical analysis similar to that presented on pp. 26–33. Wiley et al. found that both the association and dissociation rate constants were altered by preincubation with EGF: k_f decreased almost 3-fold from $1.5 \times 10^6\,\mathrm{m}^{-1}\,\mathrm{s}^{-1}$ to $5.6 \times 10^5\,\mathrm{m}^{-1}\,\mathrm{s}^{-1}$ whereas k_r decreased slightly from 9.2×10^{-3} to $6.5 \times 10^{-3}\,\mathrm{s}^{-1}$. The precise values of these parameters for the pretreated cells are somewhat in question, since the fact that they vary with time was not accounted for in the mathematical analysis. Nevertheless, the investigators concluded that the major cause of the affinity reduction was the decrease in k_f, and thus derived from a change in unoccupied receptors' ability to bind ligand. Although others (Lin et al., 1986) have observed decreased EGF-binding affinity with chemical treatments that activate protein kinase C (leading to phosphorylation of the EGF receptor at the threonine residue at site 654 just inside the cell membrane), Wiley et al. found identical behavior when their cells were depleted of this enzyme. They speculate that phosphorylation of the threonine 669 residue, independent of protein kinase C, may be involved in their observed affinity modulation. Despite all uncertainty, this at present seems to be an unusually well-documented case of true cooperativity in receptor/ligand binding.

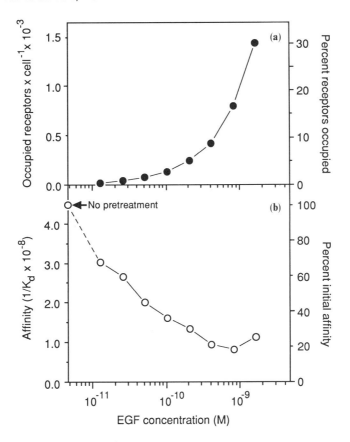

Figure 2–13 Data suggesting true cooperative behavior in the system of EGF binding to normal human fibroblasts. Cells were pretreated with ^{131}I-labeled EGF at the concentrations indicated. Experiments were then performed with ^{125}I-labeled EGF. (a) The number of receptors initially occupied with ^{131}I-labeled EGF is plotted as a function of ^{131}I-labeled EGF concentration. (b) The apparent equilibrium association constant $(1/K_D)$ for ^{125}I-labeled EGF binding is plotted as a function of the concentration of ^{131}-labeled EGF initially used. The equilibrium association constant was determined by Scatchard analysis. Reprinted from Wiley *et al.* (1989).

2.2.3 Multiple Receptor States

A possible mechanistic explanation of complicated equilibrium and kinetic ligand binding behavior, including apparent cooperativity, is that subpopulations of receptors exist with respect to association and dissociation rate constants. We will explore the ways in which *receptor subpopulations* may influence binding data, in order to aid in the interpretation of such binding data and the accurate representation of binding with mathematical models.

The existence of receptor subpopulations may be the result of two situations: (a) there may be *molecularly distinct receptor types*; or, (b) there may be *conversion* of a single molecular type between different forms with different binding properties. Conversion may occur through allosteric or covalent modification or by interaction by a third molecular component to form what is termed a ternary complex.

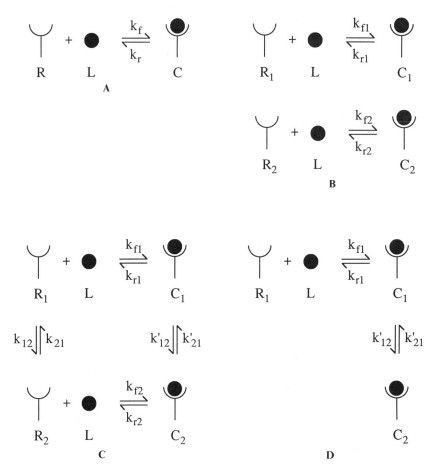

Figure 2–14 Schematic representation of several possible relationships between multiple receptor states. Ligand binding and dissociation from receptors are shown by arrows and associated rate constants k_{fi} and k_{ri}, respectively (all models). Interconversions of receptor and complex states are shown with rate constants k_{12}, k_{21}, k'_{12}, and k'_{21} (models C and D only).

A very clear categorization of some classes of models for multiple receptor states (although restricted to explicit inclusion of just the receptor and ligand species, with additional components involved only implicitly) is given by Lipkin *et al.* (1986b). The most important of these models are drawn schematically in Figure 2–14, including the simple base model for sake of formal comparison.

Model A exhibits the simple behavior described in detail in the earlier sections of this chapter. Model B represents the situation of true receptor subpopulations with molecularly distinct receptor subclasses which, in principle, can be isolated and identified biochemically. Such isolation is more difficult in the situation described by model C, for the two species of receptors interconvert, perhaps by covalent modification such as phosphorylation or methylation of particular receptor amino acid residues. Notice that model D is a limiting case of model C: only one receptor form can bind or release ligand.

Having delineated these various classes of models describing multiple receptor states, we can summarize their equilibrium and kinetic behavior and discuss their application. We will assume for the remainder of this presentation that ligand depletion, non-specific binding, and diffusion limitations (see Chapter 4) are absent or already taken into account. We will also assume for all of these models that there is negligible receptor synthesis, degradation, or trafficking during the time for which the model is to apply to an experimental system.

Before beginning discussion of the mathematical treatment of these various classes of models, it should be noted that when binding experiments are performed for labeled ligand in the presence of some concentration of unlabeled ligand, an additional set of binding relationships may be needed to keep track of the unlabeled ligand if one is to test the effect of receptor occupancy by either labeled or unlabeled ligand on binding and dissociation kinetics. For example, this is necessary to describe the observation of accelerated dissociation which was detailed in section 2.2.2. In these equations, the rate constants are usually assumed to be identical for both labeled and unlabeled forms of the ligand, although this may be incorrect in practice when the labeling moiety affects binding.

2.2.3.a Simple One-step Binding

To illustrate the model equations for a simple one-step binding case, consider model A when both labeled and unlabeled ligand are present, at concentrations L and L^u, respectively. Two kinetic equations based on Eqn. (2–8) can be written to describe the rates of formation of labeled and unlabeled complexes. With the substitution of $R = R_T - C_1^u - C_1$, along with subsequent rearrangement, these equations are

$$\frac{dC_1}{dt} = k_{f1}L[R_T - C_1^u] - (k_{f1}L + k_{r1})C_1 \qquad (2\text{–}26a)$$

$$\frac{dC_1^u}{dt} = k_{f1}L^u[R_T - C_1] - (k_{f1}L^u + k_{r1})C_1^u \qquad (2\text{–}26b)$$

where C_1 and C_1^u are the numbers of labeled and unlabeled receptor/ligand complexes, respectively, and the rate constants are unaffected by the values of C_1 or C_1^u. Because the apparent rate constant for labeled ligand dissociation is $(k_{f1}L + k_{r1})$, this model predicts that the dissociation kinetics of labeled ligand are identical when $L^u > 0$ or $L^u = 0$, as long as $L = 0$ during the dissociation experiment. Thus the phenomenon of accelerated dissociation will not be observed, and there is no need to account simultaneously for both labeled and unlabeled ligand binding. Recall from the models of DeLean and Rodbard (1979) described previously that accelerated dissociation can be observed if k_{r1}, but not k_{f1}, is a function of the number of receptor/ligand complexes. This result is consistent with Eqns. (2–26a) and (2–26b).

2.2.3.b Non-Interconverting Receptor Classes

For Model B, by writing the equivalent of Eqn. (2–12) for both complex species and summing, the total complex number C as a function of time can be expressed as

$$C(t) = C_1(t) + C_2(t) = \sum_{i=1}^{2} C_i(t)$$

$$= \sum_{i=1}^{2} C_{i0} \exp\left\{-\left(1 + \frac{L_0}{K_D}\right)k_{ri}t\right\} + R_{Ti}\left(1 + \frac{K_{Di}}{L_0}\right)^{-1}\left[1 - \exp\left(1 + \frac{L_0}{K_{Di}}\right)k_{ri}t\right\}\right]$$

$$\text{(2–27)}$$

where L_0 is the ligand concentration (assumed constant at its initial value), C_{10} and C_{20} are the initial numbers of occupied receptors for the two classes, $K_{Di} = k_{ri}/k_{fi}$, and R_{Ti} is the total surface receptor number for class i. Transient association or dissociation experiments will exhibit double-exponential behavior because of the appearance of the two exponential terms, $\exp[-(1 + (L_0/K_{D1})k_{r1}t)]$ and $\exp[-(1 + (L_0/K_{D2})k_{r2}t)]$, in Eqn. (2–27). At equilibrium,

$$C_{eq} = C_{1eq} + C_{2eq} = \sum_{i=1}^{2} C_{ieq} = \sum_{i=1}^{2} R_{Ti}\left(1 + \frac{K_{Di}}{L_0}\right)^{-1} \qquad \text{(2–28)}$$

and Scatchard plots will demonstrate concave-upward curvature. Equilibrium behavior shows apparent negative cooperativity because at higher ligand concentrations the lower affinity receptor will increasingly come into play.

For this simple two-class model, no accelerated dissociation effect should be seen in the presence of unlabeled ligand. The basis for this result is analogous to the argument made using Eqns. (2–26a) and (2–26b) for the one-class model.

Because the proper graphical methods for estimation of the equilibrium parameters R_{Ti} and K_{Di} for model B are not straightforward and incorrect analyses have appeared in the literature (as discussed by Zierler, 1989), *numerical non-linear parameter estimation* algorithms are recommended (see Press et al., 1989). Great care must be taken in interpretation of these parameter estimation results, however, A critical question is whether the two receptor-class model, model B, is truly superior to the simple one-class model, model A, or whether a better agreement with experimental data by the former is merely due to the increased number of available parameters. To decide this, there exist some statistical criteria. Munson and Rodbard (1980) use a sum of squares principles (Dowdy and Wearden, 1991), whereas Lipkin et al. (1986b) apply the Akaike information criterion (Akaike, 1974).

In addition, one must be suspicious of parameter estimates which yield values of K_{D1} and K_{D2} of similar magnitude. DeLean et al. (1982) performed *Monte Carlo computer simulations* of equilibrium binding for a two-class model, generating statistically random "data" based on various sets of R_{Ti} and K_{Di} combinations, and then analyzed these data with a parameter estimation algorithm. They found that to achieve a resolution of the two class parameters within 90% confidence limits, the ratio of the K_{Di} values must be greater than 100; for resolution within only 50% confidence limits, the ratio must be greater than 30. These results also depend on the ratio of total receptor numbers. For situations with the lesser receptor class comprising greater than about 20% of the total number, resolution is less difficult but still requires ratios of 6 to 10 in K_D values to satisfy 50% and 90% confidence limits. Without ratios of these orders of magnitude, expected uncertainty in the binding data is likely to render resolution problematic. Transient association or dissociation data, on the other hand, may be analyzed reasonably well with graphical techniques, well-known in the pharmacokinetics literature. Plots of $\ln[C_{eq} - C(t)]$ versus t should generally resolve into two exponentials provided that the values of k_{r1} and k_{r2} differ by a significant factor.

Several examples of receptor systems exist which are believed to behave according to model B. Two examples of two-class receptor systems include the α- and β-adrenergic receptors on most mammalian cells. Lefkowitz and coworkers (Kent et al., 1980; De Lean et al., 1982) have used numerical parameter estimation techniques to demonstrate agreement of equilibrium binding data with a two-receptor model. Of course, confirmation of the two-class model ultimately required isolation and identification of α_1- and α_2-, and β_1- and β_2-receptors for these two systems (Lefkowitz et al., 1983).

Another example is the T-lymphocyte growth factor interleukin-2 (IL-2) (Smith, 1988). In fact, IL-2 is more complicated yet. Two distinct individual receptor molecules have been isolated and identified, termed heavy chain (a 75 kDa molecular weight glycoprotein) and light chain (a 55 kDa molecular weight glycoprotein). Each type of receptor can bind IL-2, with parameter values as follows. For the heavy-chain form, $k_f = 3.8 \times 10^5 \, \text{M}^{-1} \, \text{s}^{-1}$, $k_r = 2.5 \times 10^{-4} \, \text{s}^{-1}$ (giving $K_D = 7.0 \times 10^{-10}$ M), while for the light-chain form $k_f = 1.4 \times 10^7 \, \text{M}^{-1} \, \text{s}^{-1}$, $k_r = 4.0 \times 10^{-1} \, \text{s}^{-1}$ (yielding $K_D = 3.0 \times 10^{-8}$ M). Notice that the 75 kDa heavy chain has small association and dissociation rate constants compared to the 55 kDa light chain. A further complication arises in that the two chains can interact reversibly to form an *heterodimeric receptor*. This receptor has a large association rate constant characteristic of the light chain, $k_f = 3.1 \times 10^7 \, \text{M}^{-1} \, \text{s}^{-1}$, but a small dissociation rate constant characteristic of the heavy chain, $k_r = 2.3 \times 10^{-4} \, \text{s}^{-1}$, yielding an extremely high affinity corresponding to $K_D = 7.0 \times 10^{-12}$ M. The physiological significance of these various classes is not known, although it is speculated that they may relate to proliferation control via regulation of receptor biosynthesis. Because of the apparently reversible interaction of the two chains, the behavior of this system will be more complicated than that of the basic two-class model or of the other models discussed in this section.

2.2.3.c Interconverting Receptor States

Model 3 represents a single molecular class of receptors which can interconvert between two forms possessing different binding properties. This interconversion may be due to allosteric or covalent modification, with the latter being strongly indicated by the activity of many known phosphorylating and methylating enzymes (along with their dephosphorylating and demethylating counterparts) on a number of receptors. Perhaps surprisingly, the equilibrium description of this fairly complicated model has the same form, but with a modified equilibrium constant, as the simple one-class receptor model, Eqn. (2–13). Using the four equilibrium relations, $K_{D1} = R_1 L / C_1$, $K_{D2} = R_2 L / C_2$, $K_{21} = R_1 / R_2$, and $K'_{21} = C_1 / C_2$, with the receptor conservation relation $R_T = R_1 + R_2 + C_1 + C_2$, we obtain for the equilibrium binding expression

$$C_{1eq} + C_{2eq} = \frac{R_T L}{K_{Dapp} + L} \tag{2-29a}$$

where the apparent equilibrium dissociation constant is:

$$K_{Dapp} = K_{D1} \frac{(1 + K_{21}^{-1})}{(1 + K_{21}'^{-1})} \tag{2-29b}$$

and $K_{D1} = k_{r1}/k_{f1}$, $K_{21} = k_{21}/k_{12}$, and $K_{21}' = k_{21}'/k_{12}'$.

This is an extremely important result: even though multiple states possessing different binding properties may be available to a receptor, when these states can freely interconvert (without stoichiometric limitation by accessory elements) only a single, average affinity will be found under equilibrium conditions. This situation will be encountered shortly for some real experimental systems.

When the system is at true thermodynamic equilibrium the *principle of detailed balance* applies. In this situation, the forward and reverse velocities of each elementary step must be equal, so that $k_{f1}k_{12}'k_{r2}k_{21} = k_{12}k_{f2}k_{21}'k_{r1}$ (Wyman, 1975). Under these circumstances, one can show that the following relation is identically valid for model C:

$$K_{Dapp} = K_{D2} \frac{(1 + K_{21})}{(1 + K_{21}')} \tag{2-29c}$$

where $K_{D2} = k_{r2}/k_{f2}$. Detailed balance only applies to elementary reaction steps at equilibrium. For example, if several intermediate steps involving enzymatic catalysis are required to accomplish the conversion from R_2 to R_1, then this conversion is not an elementary step. Hence, detailed balance is unlikely to be a valid constraint in biochemical reactions involving protein phosphorylation, methylation, and so on.

As emphasized above, the Scatchard plot of $(C_{1eq} + C_{2eq})/L_0$ vs. $(C_{1eq} + C_{2eq})$ will be linear for systems obeying model C, just as it was for the simple one-class model, model A. In this case, the apparent binding affinity will be an average value governed by Eqns. (2-29b) or (2-29c).

Equilibrium data obeying model D will similarly give a linear Scatchard plot, with a single, average apparent binding affinity, because the interconversions between receptor states are also completely reversible. Here, K_{Dapp} will be given by Eqn. (2-29b) with $1/K_{21} = 0$.

The linear Scatchard plots predicted for these models are a consequence of neglecting any accessory components participating in the receptor affinity conversion mechanism. When such additional components are limiting in terms of either stoichiometry or rate, they must be explicitly accounted for in the model equations. Most present models that include mechanistic details of accessory components fall into the category of ternary complex models, in which the stoichiometric interaction of a free and/or occupied receptor with a G-protein, cytoskeletal component, coated pit protein, or similar molecule leads to a change in binding properties. Such models will be discussed in Section 2.2.4.

The transient equations describing models C and D are linear ordinary differential equations when the ligand concentration L can be assumed constant, and thus can be solved by matrix methods (*e.g.*, Strang, 1986). The solution to model C applied to association or dissociation experiments will consist of a sum of four exponentials representing linear combinations of the four reversible reactions, asymptotically approaching the equilibrium solution. The solution to model D will be composed of two exponential terms, also approaching the equilibrium solution asymptotically. Both models predict accelerated dissociation in the presence of unlabeled ligand, because the total ligand concentration (labeled plus unlabeled) will influence the proportion of labeled receptor complexes in the different forms through the interconversion reactions.

Table 2–3 summarizes aspects of the behavior of these apparent cooperativity models, together with those of true cooperative receptor/ligand binding. Equilibrium properties (characterized by Scatchard analysis) and one type of kinetic property (possible accelerated dissociation in the presence of unlabeled ligand) are listed. We note that these properties are listed primarily because of their use as benchmarks by experimentalists. Definitive development of a model for a particular system will require a variety of experiments, non-linear parameter estimation, and biochemical identification of the various receptor forms or states. Therefore, we do not offer this table as a checklist for comprehensive model identification, but merely as a convenient compilation of the results presented here for some idealized models.

2.2.3.d Applications

Current literature offers several applications of the affinity conversion models described by models C and D. The first is that of the *cAMP receptor* on the eukaryotic microorganism *Dictyostelium*. Chemotactic response to cAMP mediated by this receptor leads to cell aggregation for fruiting body formation under conditions of nutrient deprivation. Devreotes and Sherring (1985) demonstrated that quantitative data on binding of cAMP to its receptor can be successfully described by model C, with the two receptor forms corresponding to a 40 kDa unmodified state and a 43 kDa modified state that have been identified electrophoretically (Klein *et al.*, 1985). It was found that the rates of association and dissociation of cAMP to the two receptor forms are fast relative to the time scale of receptor conversion (Van Haastert and De Wit, 1984). In this case one can assume that the ligand binding reactions are at equilibrium when examined on the time scale appropriate for receptor conversion events. With this assumption to simplify the mathematical analysis, values were obtained for the interconversion rate constants k_{12}, k_{21}, k'_{12}, and k'_{21} from kinetic data such as that shown in Figure 2–15a. Using electrophoresis methods, the relative proportions of receptor and complex in the two different molecular weight forms were measured as a function of time following incubation with cAMP.

Mathematical analysis of this simplified version of model C is as follows. We begin with kinetic conservation equations for the occupied receptor numbers C_1 and C_2:

$$dC_1/dt = k_{f1}LR_1 - k_{r1}C_1 - k'_{12}C_1 + k'_{21}C_2 \qquad (2\text{–}30a)$$

$$dC_2/dt = k_{f2}LR_2 - k_{r2}C_2 + k'_{12}C_1 - k'_{21}C_2 \qquad (2\text{–}30b)$$

Table 2–3 Summary of model predictions

Mechanism	Appearance of Scatchard plot (equilibrium data)	Observation of accelerated dissociation? (kinetic data)	Text reference
Simple one state binding (model A)	Linear	No	2.2.1, 2.2.3.a
True positive cooperativity	Concave down	Yes if $k_r = f(C)$ No if $k_f = f(C)$	2.2.2
True negative cooperativity	Concave up	Yes if $k_r = f(C)$ No if $k_f = f(C)$	2.2.2
Two non-interconverting receptor states (model B)	Concave up	No	2.2.3.b
Two fully interconverting receptor states (model C)	Linear	Yes	2.2.3.c
Two interconverting complex states (model D)	Linear	Yes	2.2.3.c
Ternary complex	Linear or concave up, depending on value of χ	Yes	2.2.4.b
Diffusion-limited binding	Linear	Yes	4.2.2
Multivalent ligand, monovalent receptor	Concave up	Yes	4.3.1, 4.3.2
Multivalent ligand, multivalent receptor	Concave up	Yes	4.3.3

Predictions for the form of the Scatchard plot and the presence or absence of accelerated dissociation are listed. Although the equilibrium and kinetic properties of a system can be partially elucidated through Scatchard analysis and experiments to detect accelerated dissociation, these techniques must be accompanied by others before a characterization of the experimental system can be made with certainty. C is the number of receptor/ligand complexes per cell. χ is the ratio of total accessory molecules to total receptors.

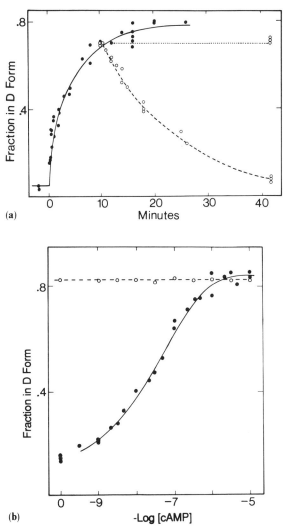

Figure 2–15   Conversion of receptors from state "1" (R_1 or C_1) to state "2" (R_2 or C_2), as described by model C of Figure 2–14. Data on binding of cAMP to its receptor on *Dictyostelium* is shown. (a) Cells were incubated with cAMP and at various times afterwards the fraction of receptors in each form was determined. Solid curve shows the conversion of receptors from state "1" to state "2" during incubation with cAMP; dashed curve shows the conversion of receptors back to the state "1" following removal of the ligand. (b) The steady state fraction of receptors in state "2" is plotted as a function of ligand concentration. Reprinted from Devreotes and Sherring (1985).

The assumption of equilibrium binding of ligand to both receptor forms, relative to the time scale for $R_1 \leftrightarrow R_2$ and $C_1 \leftrightarrow C_2$ receptor conversion, means that we can set $k_{f1}R_1L = k_{r1}C_1$ and $k_{f2}R_2L = k_{r2}C_2$. (A more mathematically rigorous perturbation approach, which leads to more complicated though not conflicting results, is provided by Segel and Slemrod (1989), and an intuitively satisfying discussion of the underlying concepts by Segel (1988).) With this simplification,

$$\frac{dC_1}{dt} = -\frac{dC_2}{dt} = -k'_{12}C_1 + k'_{21}C_2 \qquad (2\text{--}31)$$

When cells are incubated in a constant ligand concentration L and the assumption of equilibrium binding holds, the total number of bound receptors is constant, $[C_1(t) + C_2(t)] = [C_{1eq} + C_{2eq}]$, where C_{ieq} refers to the equilibrium value of complex species i ($=1$ or 2). Thus

$$C_1(t) = [C_{1eq} + C_{2eq}] - C_2(t) \qquad (2\text{--}32)$$

Eqn. (2–31), after substitution of Eqn. (2–32) for C_1, can be expressed as the single kinetic equation

$$\frac{dC_2}{dt} = k'_{12}(C_{1eq} + C_{2eq}) - (k'_{12} + k'_{21})C_2 \qquad (2\text{--}33)$$

For initial condition $C_2(0) = C_{20}$ (a non-zero value can result from preincubation with ligand during $t < 0$), the transient solution for the number of bound receptors of the C_2 form following addition of ligand at concentration L for $t > 0$ is:

$$\frac{C_2(t) - C_{20}}{C_{2eq} - C_{20}} = 1 - \exp\{-(k'_{12} + k'_{21})t\} \qquad (2\text{--}34)$$

where $C_{2eq} = [k'_{12}/(k'_{12} + k'_{21})](C_{1eq} + C_{2eq})$, and $(C_{1eq} + C_{2eq})$ is given by Eqn. (2–29a). The solution for $C_1(t)$ immediately follows from the conservation relation for bound receptors, Eqn. (2–32).

An analogous procedure can be followed for $L = 0$, leading to a similar transient solution for the number of free receptors in the R_2 form following removal of ligand for $t > 0$:

$$\frac{R_2(t) - R_{20}}{R_{2eq} - R_{20}} = 1 - \exp\{-(k_{12} + k_{21})t\} \qquad (2\text{--}35)$$

where $R_{2eq} = [k_{12}/(k_{12} + k_{21})]R_T$. Here the initial condition $R_2(0) = R_{20}$ is similarly a result of preincubation with ligand. Again, the solution for $R_1(t)$ follows from Eqn. (2–32).

This analysis forms the basis for the model *parameter determination* procedure followed by Devreotes and Sherring (1985). The apparent rate constant $(k'_{12} + k'_{21}) = \ln 2/t_{1/2}$ was calculated from the half-time, $t_{1/2} = 2.5$ min, for formation of C_2 upon addition of cAMP. The apparent rate constant $(k_{12} + k_{21}) = \ln 2/t_{1/2}$ was calculated from the half-time, $t_{1/2} = 6.0$ min, for formation of R_2 upon addition of cAMP. At a steady state in the absence of cAMP, 10% of the receptors were found in the "2" state; thus $R_2/(R_1 + R_2) = 0.10$ and the ratio $k_{21}/k_{12} = R_1/R_2 = 9.0$. At a steady state in the presence of saturating levels of cAMP, 80% of the receptor were found in the "2" state; thus $C_2/(C_1 + C_2) = 0.80$ and the ratio $k'_{21}/k'_{12} = C_1/C_2 = 0.25$. This set of information yields values for the individual rate constants of $k_{12} = 0.012$ min^{-1}, $k_{21} = 0.104$ min^{-1}, $k'_{12} = 0.222$ min^{-1}, and $k'_{21} = 0.055$ min^{-1}. These represent interconversion rates which are about an order of magnitude slower than the ligand association/dissociation rates, which occur on the time scale of seconds, supporting the assumption of equilibrium receptor/ligand binding on the time scale of the interconversion events.

Equilibrium ligand binding data at low and high cAMP concentrations give linear Scatchard plots, as expected for a model C system, with less than a twofold difference in binding affinity and approximately identical receptor numbers. This observation is in rough agreement with Eqn. (2–29b), which predicts that the apparent equilibrium dissociation constant, K_{Dapp}, as well as the total receptor number, R_T, must be independent of ligand concentration.

Eqn. (2–29b) further predicts that K_{Dapp} should be equal to $0.2 \times K_{D1}$ and Eqn. (2–29c) that it should be equal to $8 \times K_{D2}$. Literature values of K_{D1} vary in the range 10^{-8}–10^{-7} M (Martiel and Goldbeter, 1987), whereas Figure 2–15b gives an average value for $K_{Dapp} \sim 2 \times 10^{-9}$ M. This yields $K_{D1} = 10^{-8}$ M, consistent with the lower end of the measured range. Literature values for K_{D2} are in the range 3×10^{-9} to 9×10^{-9} M (Martiel and Goldbeter, 1987). The value predicted from Eqn. (2–29c) is 2×10^{-10} M, which is off by an order of magnitude. However, it is not clear whether the measured values in the literature are for the isolated receptor form or are taken from the overall system at high ligand concentrations. Also, since it is now known that the interconversion mechanism requires receptor phosphorylation (Devreotes, 1989), the validity of Eqn. (2–29c) based on detailed balance is in question.

An additional prediction from model C is the steady-state receptor distribution dependence on ligand concentration, L, according to the expression:

$$\frac{(R_{2eq} + C_{2eq})}{R_T} = \frac{\left[\left(\dfrac{K'_{21}}{K_{21}}\right) + \left(\dfrac{L}{K_{D1}}\right)\right]}{\left[K'_{21}\left(1 + \dfrac{1}{K_{21}}\right) + (1 + K'_{21})\left(\dfrac{L}{K_{D1}}\right)\right]} \tag{2–36}$$

Given the experimentally determined values of the ratios $K_{21} = k_{21}/k_{12} = 8.7$ and $K'_{21} = k'_{21}/k'_{12} = 0.25$, with $K_{D1} = 10^{-8}$ M, increasing ligand concentrations lead to increasing proportions of the "2" form, as shown in Figure 2–15b. This experimental plot corresponds quite well to the predictions of Eqn. (2–36) with the estimated parameter values substituted to get

$$[(R_{2eq} + C_{2eq})/R_T] = [0.028 + L/10^{-8} \text{ M}]/[0.28 + 1.25L/10^{-8} \text{ M}].$$

Overall, model C does a satisfactory job of accounting for much of the experimental binding data on the *Dictyostelium* cAMP receptor system. Its major shortcomings appear to stem from the additional involvement of non-reversible covalent modification reactions which will affect predictions based on microscopic reversibility arguments. The results of this analysis, however, will be found applicable to the cAMP receptor signal transduction problem discussed in Chapter 5.

An example model D system, in which one of two interconverting complex states cannot dissociate ligand, is that of *insulin receptor* binding on fat-cells at 15 °C, as analyzed by Lipkin et al. (1986a,b). Their equilibrium binding data presented a linear Scatchard plot, with $K_{Dapp} = 8.8 \times 10^{-9}$ M. Kinetic dissociation experiments exhibited two exponential terms with half-times of 12 min and 103 min. Association kinetics also appeared to show biphasic behavior. A substantial amount of kinetic data was obtained for both transient association and dissociation experiments in the presence of a number of different insulin concentrations. Transient solutions for the receptor and complex conservation equations appropriate to model B can be obtained fairly easily with matrix methods (*e.g.*, Strang, 1986). Instead of using analytical expressions, however, Lipkin *et al.* chose to generate solutions computationally using a standard fourth-order Runge–Kutta numerical integration method (Press *et al.*, 1989). With these solutions, the model rate constants were evaluated using a numerical parameter estimation algorithm, obtaining $k_{f1} = 1.6 \times 10^5 \, M^{-1} \, s^{-1}$, $k_{r1} = 3.4 \times 10^{-3} \, s^{-1}$, $k'_{12} = 3.2 \times 10^{-4} \, s^{-1}$, and $k'_{21} = 2.0 \times 10^{-4} \, s^{-1}$ (Figure 2–16). Using Eqn. (2–29b), these values predict $K_{Dapp} = 1.8 \times 10^{-8}$ M, approximately twice the independently measured value but likely within experimental uncertainty.

Ross *et al.* (1977) applied the form of model D to binding of iodohydroxybenzylpindolol (IHYP) to the β-adrenergic receptor on S49 lymphocytes, a system which exhibited equilibrium and kinetic behavior quite similar to that described here. They derived a simpler analytical expression for the transient association and dissociation experiment solution by assuming that the ligand binding step was relatively fast and hence in quasi-equilibrium. Because the system behavior was so similar to the insulin results just mentioned, this work is not discussed in any detail here. Our motivation for mentioning this contribution is the significance these investigators presented for the role of an accessory component in receptor affinity conversion. Ross *et al.* demonstrated that increased levels of purine nucleotides such as GTP strongly shifted the distribution of the receptor from higher to lower affinity form. Further, in the absence of these modulators, equilibrium Scatchard plots possessed a more noticeable amount of concave-upward curvature, suggesting a lesser degree of receptor interconversion. It is now known that, for this receptor and many others, G-proteins (guanine nucleotide-dependent regulatory proteins) couple with the receptor to effect an affinity conversion as well as signal transduction (Taylor, 1990; Birnbaumer *et al.*, 1990). The models discussed thus far have neglected explicit consideration of accessory components such as G-proteins and others. In the next section, we introduce models that account for these sorts of accessory species, whereas models analyzing their role in signal transduction are discussed in Chapter 5.

2.2.4 Ternary Complex Formation and Receptor Aggregation

The function of a receptor in signal transduction, ligand internalization, and other events may require its interaction with membrane-associated proteins such as G-proteins, coated pit adaptors, cytoskeletal elements, or other receptors (see Figure 2–17). In preparation for later analysis of these receptor-mediated events, in this section we will discuss these types of interactions and their effects on receptor/ligand binding properties. These models will be termed *ternary complex* models, in reference to the involvement of receptor, ligand, and at least one other "accessory" species.

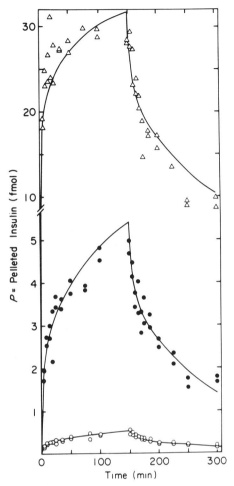

Figure 2–16 Binding of labeled insulin to fat-cells and fit of the data with model D of Figure 2–14. The binding of labeled insulin at three concentrations (triangles $= 1.04 \times 10^{-8}$ M; solid circles $= 1.03 \times 10^{-9}$ M; open circles $= 0.96 \times 10^{-10}$ M) was followed as a function of time. Then 10^{-6} M unlabeled insulin was added to the incubation medium at 150 min and dissociation of labeled insulin was followed as a function of time. Solid curves represent the best fit of model D to the data. Reprinted from Lipkin *et al.* (1986).

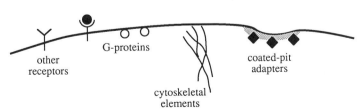

Figure 2–17 Ternary complex possibilities. Receptors may interact with G-proteins, coated-pit binding proteins, cytoskeletal elements, or other receptors to generate ternary complexes.

2.2.4.a Types of Interactions

G-proteins, or guanine nucleotide-binding proteins, serve to couple receptors to cell membrane-associated enzymes for purposes of signal transduction (Gilman, 1987; Birnbaumer *et al.*, 1990). Coupling of a quiescent G-protein with a bound receptor leads to release, following GTP binding to the ternary complex, of a form of the G-protein which is capable of regulating particular intracellular enzymes. In this manner, signaling information inherent in ligand binding to the receptor is transmitted to the cell interior via the G-protein. Details of this interaction will be presented in Chapter 5 when models for signal transduction events are discussed. Our present focus is on the effects of G-protein coupling on receptor/ligand binding properties, which can be quite significant. Association and dissociation rate constants for receptor/ligand binding can be dramatically different depending on the state of receptor/G-protein coupling.

Many receptors, including the *β*-adrenergic receptor, *Dictyostelium* cAMP receptor, and neutrophil leukocyte chemotactic peptide receptor, have been demonstrated to activate G-proteins (Birnbaumer *et al.*, 1990). The two former examples utilize the adenylate cyclase pathways, whereas the latter utilizes the phospholipase C pathway; involvement of both pathways is also possible in these and other systems. It is thought that the receptor/ligand/G-protein ternary complex possesses higher affinity for the ligand than does the receptor/ligand binary complex. Evidence for this comes from studies demonstrating that GTP reduces receptor binding affinity by converting the high affinity state to the low affinity state, and that the proportion of receptors in the high affinity state can be quantitatively correlated with signal generation and cell behavioral function (*e.g.*, Ross *et al.*, 1977; Koo *et al.*, 1982).

Similar behavior may also occur when the receptor couples with a *cytoskeleton-associated* accessory molecule. The neutrophil chemotactic peptide receptor appears to be an example of this second case, as it can couple with unidentified cytoskeletal components as well as G-proteins in separate interactions (Sklar *et al.*, 1989). Ligand binding properties of the receptor can be altered by this sort of coupling. A cytoskeleton-associated receptor state appears to be more slowly dissociating for the neutrophil chemotactic peptide receptor (Sklar and Omann, 1990) and the EGF receptor (Roy *et al.*, 1989).

A third type of membrane-associated molecule involved in receptor ternary complexes are the *adaptors* (Pearse and Robinson, 1990), located within coated pits. Coated pits are specialized membrane regions characterized by the cytoplasmic presence of the protein clathrin, by which receptor/ligand complexes and free receptors may be internalized. Structural signals in the cytoplasmic tail domains of cell receptors appear to be responsible for reversible coupling interactions with the adaptors (Trowbridge, 1991). The affinity of this interaction may be influenced by covalent modification of the receptor, especially by phosphorylation of tyrosine, serine, and/or threonine residues, allowing the internalization rate of a receptor to depend upon ligand binding and subsequent receptor modification events. For example, the EGF receptor shows this type of behavior (Wiley *et al.*, 1989). Involvement of receptor/ligand/coated pit protein ternary complexes in trafficking phenomena will be discussed in Chapter 3, but the apparent effects of such interactions on receptor/ligand binding are discussed here. Effects of adaptor coupling to receptors on ligand binding properties have not yet been documented, however.

Finally, we include in this treatment the *aggregation* of a receptor with another receptor of the same type, either free or bound. This coupling interaction can be thought of as leading to a ternary complex. For multivalent ligands this is a well-studied phenomenon which will be examined in more detail in Chapter 4. In addition, some receptors may aggregate even when bound to monovalent ligands; the EGF receptor system may fall into this category (Schlessinger, 1986). In such instances, ligand binding to a receptor might cause a conformational change (perhaps due to covalent modification through a signal transduction cascade) allowing the occupied receptor to complex with a second receptor to form a dimer. Again, this complex may alter the association and/or dissociation properties of the first receptor, through either structural or diffusion effects. Diffusion effects (see Chapter 4) come into play because the distribution of binding sites on a cell have an effect on the overall observed binding and dissociation rate constants (Goldstein, 1989). These may also appear in the coated-pit adaptor interactions described above.

The point of considering these receptor coupling interactions is twofold. First, if they affect ligand binding then they must be understood in order to analyze experimental data properly. Second, if they are related to receptor signaling, they must be understood in order to predict cell responses.

2.2.4.b Basic Ternary Complex Model

Interactions of a receptor/ligand complex with one other molecule or *accessory component*, including the G-proteins and coated pit proteins described above, can be described by models similar to those we have already developed. The schematic representation of a simple ternary complex model is shown in Figure 2–18. In

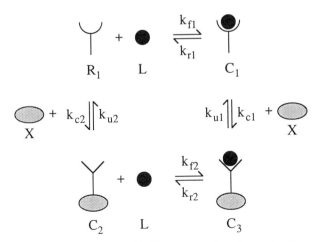

Figure 2–18 Simple ternary complex model. The receptor/ligand complex C_1 may bind with an accessory component X to form a ternary complex C_3. Alternatively, the free receptor/effector binary complex C_2 may bind with a ligand molecule L to form the ternary complex.

describing this system, we will use the notation C_1 to represent the number of simple receptor/ligand complexes, X the accessory components, C_2 the free receptor/accessory binary complexes, and C_3 the ternary complexes of receptor/ligand/accessory, with units of #/cell. In this general formulation, we allow for the binding of both free receptors and receptor/ligand complexes with the accessory component. k_{c1} [(#/cell)$^{-1}$ time^{-1}], k_{c2} [(#/cell)$^{-1}$ time^{-1}], k_{u1} (time^{-1}), and k_{u2} (time^{-1}) are the coupling and uncoupling rate constants for the occupied and free receptor, respectively. We note for future reference that if receptor density, rather than receptor number, is followed, then one would write k_{c1} and k_{c2} with units of (#/area)$^{-1}$ time^{-1}. The conversion between the two sets of units simply involves the cell surface area. Because the coupling rate constant may very well depend on how rapidly receptors and accessory components find each other, and hence on the distances between molecules (*i.e.*, their density), we will use the units of (#/area)$^{-1}$ time^{-1} in the discussions in Chapters 4 and 6.

It is assumed that the total number of receptors and effector molecules on the cell surface do not change with time and that there are no trafficking events, so that conservation relations $R_T = R + C_1 + C_2 + C_3$ and $X_T = X + C_2 + C_3$ apply.

Before beginning quantitative examination of the predictions of this simple model, it is helpful to note the similarity between it and the affinity conversion models, particularly model C, presented in section 2.2.3. This similarity can be noted by comparing the schematic illustrations in Figures 2–14c and 2–18. The primary difference between the ternary complex model and the receptor inter-conversion model is that the former requires the addition of a molecule X, the accessory component, to move from the receptor states at the top of the figure to those at the bottom.

Key to the relationship between a ternary complex model and an affinity conversion model is the *stoichiometric ratio* of total accessory molecules, X_T, to total receptors, R_T. We can define this formally for analysis purposes as the parameter $\chi = X_T/R_T$, which will be used shortly. When the accessory component is in great excess, *i.e.* when the fraction of accessory molecules complexed with receptors is very small, the two models are essentially identical. To see this, note that in the ternary complex model the forward rate of conversion of C_1 into C_3 is given according to bimolecular reaction kinetics as $k_{c1}XC_1$. In model C, the analogous conversion of C_1 into C_2 is given according to first order reaction kinetics as $k_{12}C_1$. Thus, when the number of free accessory molecules X is essentially constant, the bimolecular reaction kinetics reduce to pseudo-first order kinetics and there are no mathematical differences between the two models. In fact, most of those simpler models introduced in section 2.2.3 have been used for the very examples we will present here, with some degree of success. We will point out what important new results arise from including the accessory component as we proceed with our analysis.

The ternary complex model introduced above was first posed by Jacobs and Cuatrecasas (1976), who termed it the *mobile receptor hypothesis* in reference to the presumed ability of receptors to move in the membrane and reversibly associate with accessory molecules. We first examine the equilibrium behavior predicted by the model. An immediate result can be found when detailed balance can be invoked. Defining equilibrium dissociation constants $K_{D1} = k_{r1}/k_{f1}$ and $K_{D2} = k_{r2}/k_{f2}$ along with equilibrium coupling constants $K_{C1} = k_{c1}/k_{u1}$ and $K_{C2} = k_{c2}/k_{u2}$, detailed balance requires that $K_{C1}K_{D2} = K_{C2}K_{D1}$. It is usually assumed that an occupied receptor will couple more strongly with the accessory component than will a free receptor, for it is presumably occupied receptors which elicit a cellular response. This implies that $K_{C1} > K_{C2}$. Indeed, a frequent simplifying assumption is that free receptors cannot appreciably couple with accessory components, so that $K_{C2} = 0$. In any event, given $K_{C1} > K_{C2}$ it must be the case that $K_{D1} > K_{D2}$. The extreme case is $K_{D2} = 0$, meaning that the ternary complex cannot release ligand without first uncoupling. The relationship between K_{D1} and K_{D2} implies that the ternary complex must be of higher affinity for ligand binding than the simple receptor/ligand binary complex.

If the accessory component is in stoichiometric excess, the model is identical to model C (section 2.2.3) and linear Scatchard plots will be obtained. In contrast, when the accessory component is stoichiometrically limiting, apparent negative cooperativity should be observed for *equilibrium binding*. This is because as more ligand binds, the accessory components that provide for high affinity interactions become depleted.

To confirm this result, Jacobs and Cuatrecasas (1976) derived analytical expressions for the various bound ligand forms C_1, C_2, and C_3 and constructed theoretical Scatchard plots. Defining dimensionless free receptor number $w = R/R_T$, dimensionless ligand concentration $\lambda = L/K_D$, ratio of total accessory components per cell to total receptors per cell $\chi = X_T/R_T$, and dimensionless coupling equilibrium constants $\kappa_{C1} = K_{C1}R_T$ and $\kappa_{C2} = K_{C2}R_T$, the total amount of bound ligand is

$$\frac{C_{1eq} + C_{3eq}}{R_T} = \lambda w + \chi \left\{ \frac{1 - \dfrac{1}{1 + (\kappa_{C2}w)^{-1}}}{1 + \dfrac{1}{\lambda \kappa_{C1}w} - \dfrac{1}{1 + (\kappa_{C2}w)^{-1}}} \right\} \tag{2--37a}$$

where the dimensionless free receptor number, w, is found from solving the implicit quadratic equation

$$1 = w(1 + \lambda) + \chi \left\{ 1 - \frac{\left[1 - \dfrac{1}{1 + (\kappa_{C2}w)^{-1}} \right] (\lambda \kappa_{C1}w)^{-1}}{1 + (\lambda \kappa_{C1}w)^{-1} - \dfrac{1}{1 + (\kappa_{C2}w)^{-1}}} \right\} \tag{2--37b}$$

An example plot is shown in Figure 2--19. The central parameter effect illustrated in this graph is that of the ratio $\chi = X_T/R_T$, the number of total accessory components per cell divided by the number of total receptors per cell. Strongly non-linear behavior shows up for values of χ equal to or less than unity, suggestive of negative cooperativity or multiple receptor classes (*i.e.*, model B, section 2.2.3). When χ is greater than unity, however, the accessory component is present in stoichiometric excess over the receptor, and the model simplifies to model C (section 2.2.3) and a linear Scatchard plot is obtained as argued earlier.

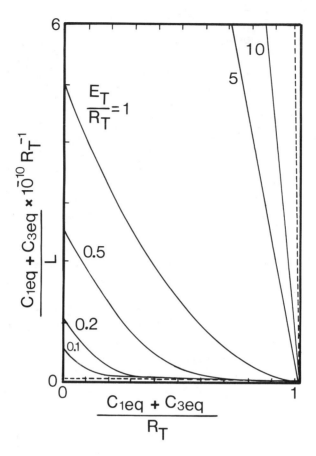

Figure 2–19 Theoretical Scatchard plot predicted by the simple ternary complex model. Because bound ligand is present in two types of complexes, the total amount of bound ligand is given by $C_1 + C_3$. Curves are shown for different values of the ratio of the number of accessory components to the number of receptors, χ (given in the figure as E_T/R_T). Apparent negative cooperativity is predicted for low values of χ. Redrawn from Jacobs and Cuatrecasas (1976).

To elucidate the *kinetic behavior* of this ternary complex model, one can write population balance equations for the three types of receptor complexes:

$$\frac{dC_1}{dt} = k_{f1}(R_T - C_1 - C_2 - C_3)L - k_{r1}C_1 - k_{c1}(X_T - C_2 - C_3)C_1 + k_{u1}C_3 \quad (2\text{--}38a)$$

$$\frac{dC_2}{dt} = k_{c2}(R_T - C_1 - C_2 - C_3)(X_T - C_2 - C_3) - k_{u2}C_2 - k_{f2}C_2L + k_{r2}C_3 \quad (2\text{--}38b)$$

$$\frac{dC_3}{dt} = k_{c1}(X_T - C_2 - C_3)C_1 - k_{u1}C_3 + k_{f2}C_2L - k_{r2}C_3 \quad (2\text{--}38c)$$

Note that we have used the conservation relations $R_T = R + C_1 + C_2 + C_3$ and $X_T = X + C_2 + C_3$ to eliminate R and X from the equations.

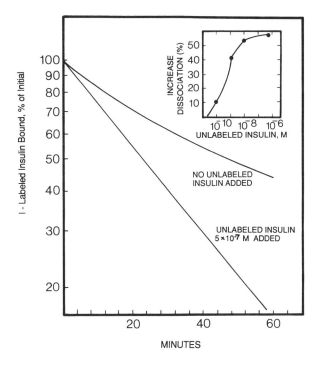

Figure 2–20 Theoretical time course of ligand dissociation as predicted with the simple ternary complex model. The rate constants used in the simulation are: $k_{f1} = 5 \times 10^5 \, M^{-1} s^{-1}$, $k_{r1} = 4 \times 10^{-3} s^{-1}$, $k_{f2} = 5 \times 10^6 \, M^{-1} s^{-1}$, $k_{r2} = 1.6 \times 10^{-6} s^{-1}$, $k_{c1} = 2 \times 10^{-1} R_T^{-1} \, M^{-1} s^{-1}$, $k_{u1} = 5 \times 10^{-4} s^{-1}$, $k_{c2} = 8 \times 10^{-5} R_T^{-1} \, M^{-1} s^{-1}$, $k_{u2} = 5 \times 10^{-3} s^{-1}$, $R_T = 2 \times 10^{-15} \, M$, and $X_T = 10^{-16} \, M$. Note that calculations were done using different conventions on units from those in this text. All receptors are initially occupied by labeled ligand. Accelerated dissociation in the presence of unlabeled ligand is predicted. Redrawn from Jacobs and Cuatrecasas (1976).

Because the kinetic behavior of this ternary complex model cannot be elucidated analytically, Jacobs and Cuatrecasas (1976) solved the differential equations numerically for parameter values appropriate to their system, insulin binding to fat-cells. Figure 2–20 shows some example results, using $\chi = 0.2$. Accelerated dissociation in the presence of unlabeled ligand is predicted by the model, providing an alternative explanation for the experimental observations by De Meyts (1976) of apparent negative cooperativity for insulin binding (discussed in Section 2.2.2). Note that accelerated dissociation behavior should also result for values of χ greater than one, since it is also predicted for model C of the previous section. In other words, accelerated dissociation arises from the presence of the second receptor/ligand complex form, regardless of whether or not the accessory component is in stoichiometric excess over the receptor.

2.2.4.c Applications

A simplified ternary complex model was applied by Mayo *et al.* (1989) to binding of *epidermal growth factor* (EGF) to its receptor on fibroblasts. This growth factor stimulates proliferation of a wide range of tissue cells. At 4 °C, equilibrium binding data showed concave upward Scatchard plots, whereas kinetic data for both association and dissociation experiments demonstrated two time scales (fast and slow) in approach to steady state (see Figures 2–21a–c). These investigators used the ternary complex model outlined above with further assumptions that $K_{C2} = 0$ and $K_{D2} = 0$; that is, that free receptors do not "precouple" appreciably to the accessory component, and that ternary complexes do not release ligand to a significant degree. These simplifications reduce the ternary complex model to its barest form.

In this case, the equilibrium relationships specified in Eqns. (2–37a) and

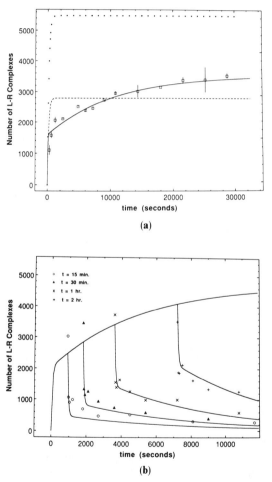

(a)

(b)

Figure 2–21   EGF binding data and ternary complex model fits. (a) The number of receptor/ligand complexes is shown as a function of time for incubation of fibroblasts with 5×10^{-10} M EGF at $4\,°$C. The data (open squares) are fitted well by a ternary complex model (solid line) but not by a simple one-state receptor binding model, (model A of Figure 2–14, dotted lines). (b) The binding of EGF at $4\,°$C is interrupted by the removal of EGF from the incubation medium at the times indicated. The ensuing decrease of receptor/ligand complexes was measured and the ternary complex model was fitted to the data (solid lines).

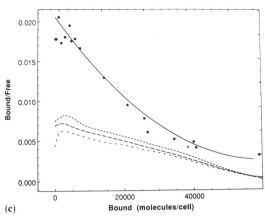

Figure 2–21 (*continued*)   (c) Plots of ternary complex model predictions in "pseudo-Scatchard" form for transient binding data corresponding to 1 min, 3 min, and 10 min incubation of cells with 2×10^{-11} M EGF at 4 °C are shown to possess concave downward curvature, while model and data for equilibrium binding exhibit concave upward curvature (270 min). Experimental transient data are in qualitative agreement with these predictions, demonstrating concave downward curvature in this type of plot (see original reference). Reprinted from Mayo *et al.* (1989).

(2–37b) reduce to

$$\frac{C_{1eq} + C_{3eq}}{R_T} = \lambda w + \chi\left(\frac{\lambda w}{\lambda w + \kappa_{C1}^{-1}}\right) \tag{2–39a}$$

with

$$1 = (1 + \lambda)w + \chi\left(1 - \frac{1}{1 + \lambda\kappa_{C1}w}\right) \tag{2–39b}$$

Thus, equilibrium binding is affected by κ_{C1}, the dimensionless coupling equilibrium constant, and χ, the ratio of accessory components to receptors. As with the more complete ternary complex model discussed earlier, these equations predict a concave-upward Scatchard plot of $[(C_1 + C_3)/R_T]/\lambda$ versus $[(C_1 + C_3)/R_T]$.

Once again, the transient solutions governing association and dissociation experiments required numerical computations, and two time scales are predicted for each of these situations. The ligand binding rate constants, k_{f1} and k_{r1}, were determined from short time-scale kinetic data using a range of ligand concentrations and applying the analysis method following Eqn. (2–23), obtaining $k_{f1} = 1.2 \times 10^6$ M^{-1} s^{-1} and $k_{r1} = 1.7 \times 10^{-2}$ s^{-1}. R_T was calculated as 6×10^4 receptors/cell from the long-time asymptote of kinetic association experiments at high ligand concentrations. Next, $K_{C1} = 6 \times 10^{-5}$ (#/cell)$^{-1}$ and $X_T = 2.4 \times 10^4$ #/cell were determined from a non-linear parameter estimation algorithm applied to the equilibrium data. Notice that $\chi = 0.4$ for this system, so that the accessory component is clearly stoichiometrically limiting. Finally, the long-time, slow time-scale dissociation data were used to determine $k_{u1} = 8 \times 10^{-5}$ s^{-1}, with $k_{c1} = 5 \times 10^{-9}$ (accessory components/cell)$^{-1}$ s^{-1} consequently calculated from K_{C1}. Evidence points to a cytoskeleton-associated component as the coupling molecule in this situation, responsible for alteration of receptor binding properties to yield the apparent high-affinity form (Roy *et al.*, 1989).

Another example of a ternary complex system is the *chemotactic peptide receptor* on neutrophils. This receptor enables these phagocytic cells of the inflammatory system to sense concentration gradients of peptides released by bacteria invading host tissue. Some investigators (Mackin *et al.*, 1982; Marasco *et al.*, 1985) have observed Scatchard plots with concave upward curvature, along with accelerated dissociation in the presence of unlabeled ligand, for peptide chemoattractant binding to whole cells and purified membrane preparations at 4 °C. Sklar *et al.* (1984, 1989) have for this system posed a series of models which has roots in the ternary complex concept, in which interactions of the receptor both with G-proteins and with cytoskeletal components are allowed. These models are directed toward kinetic association and dissociation data obtained from neutrophils using fluorescent-labeled ligands and spectrofluorometry (Figure 2–22). Fluorescence methods are superior to radioactivity methods when data on a fast (seconds) time-scale are desired. The primary features of these data are biphasic dissociation kinetics along with a shift in predominance from the more rapidly to the more slowly dissociating fraction for increasing incubation times. Further, this shift between dissociation rates is accompanied by a loss of receptor signaling activity (assayed by membrane depolarization, calcium generation, and other cellular events). Therefore, ligand and receptor may form an actively signaling complex with fast ligand dissociation rate constant which rapidly converts into an inactive complex with slow ligand dissociation rate constant. This latter complex likely corresponds to a cytoskeleton-associated state whereas the former complex is likely a free receptor able to couple with a G-protein.

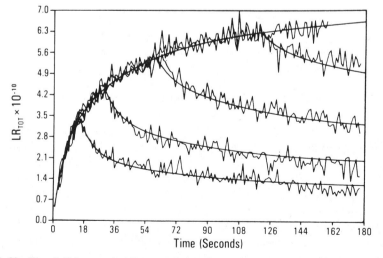

Figure 2–22 Fit of Sklar *et al.* (1989) model to spectrofluorometer data on association and dissociation kinetics for a neutrophil chemotactic peptide, allowing determination of the model parameters. Reprinted from Sklar and Omann (1990).

In more recent work, this group has focused on the receptor/G-protein interaction in permeabilized neutrophils, permitting alteration of intracellular components (Fay *et al.*, 1991). Guanine nucleotide (primarily GTP) levels were controlled in order to examine their effects on receptor/ligand binding. Kinetic equations similar to Eqns. (2–38) were written for a variety of model schemes related to that illustrated in Figure 2–18, neglecting receptor interactions with the cytoskeleton. A crucial point to note is that the total number of accessory components, X_T, was assumed to be sufficiently large compared to the total receptor number, R_T, that it remains constant. Hence, this model formally belongs with interconverting receptor models in which an accessory component does not appear in the model explicitly. We treat it here because it is, in principle, a ternary complex model transformed to an interconverting receptor model by a simplifying assumption.

When these model equations were applied to transient ligand binding experiments in the presence of the guanine nucleotide GTP, the data could be fitted by a simple one-step ligand-binding model (as in Figure 2–2) with $K_{D1} = 2.6 \times 10^{-9}$ M. In contrast, when guanine nucleotides were absent, a more complicated ternary complex model was required to account for the data. This seems to be another example of a system which exhibits an apparent single affinity when receptor states can interconvert freely (in the presence of GTP here) but shows multiple affinities when the states cannot interconvert (in the absence of GTP). We note that this can only occur when the accessory component is not stoichiometrically limiting (see Figure 2–19).

Pursuing the experiments in the absence of GTP, the Sklar group compared various submodels related to the general scheme of Figure 2–18 (recall, however, that X remains constant here at X_T). The best fit of computer-generated model results with the data was obtained for a submodel in which $K_{C2}^{-1} = 0$, and values for the various model parameters were determined. An important implication of this submodel is that a substantial fraction of the chemotactic peptide receptors, in this case approximately 50%, are apparently precoupled to G-proteins even without ligand binding. At equilibrium in the presence of ligand, however, 20% of the binding sites retain the original low-affinity value of K_{D1} but 80% are found to possess a high affinity with $K_{D2} = 4.0 \times 10^{-11}$ M. That is, ligand binding shifts the ratio of high to low affinity receptor states from roughly 1:1 ($C_2 : R_1$ when $L = 0$) to about 4:1 ($C_3 : C_1$ when $L > 0$). Interestingly, both receptor states exhibit identical association rate constants for the chemotactic peptide ligand ($k_{f1} = k_{f2} = 3 \times 10^7$ M^{-1} s^{-1}) but vastly different dissociation rate constants ($k_{r1} = 1.7 \times 10^{-1}$ s^{-1} for the C_1 form, $k_{r2} = 1.1 \times 10^{-3}$ s^{-1} for the C_3 form).

The question of precoupling of free receptors to accessory molecules in the absence of ligand is also addressed by a mathematical model for the α_2-*adrenergic receptor* on platelets (Neubig *et al.*, 1988), with the accessory molecule again being a G-protein. Here also, significant precoupling was found, with about 30% of the receptors residing in the C_2 state of Figure 2–18.

Finally, a more complicated system that can be mentioned in this context is the *IL-2* receptor on T lymphocytes. As mentioned earlier, there are actually three forms of this receptor: a low-affinity single chain (p55), an intermediate-affinity single chain (p75), and a complex of these two to form a high-affinity heterodimeric receptor (Smith, 1988). This case can essentially be described by Figure 2–18, where R is p55, X is p75, and C_2 is the heterodimer; however, the scheme must be modified to allow ligand binding to the "accessory" molecule X. A key issue in this system is again that of whether there is significant precoupling of the p55 and p75 chains or whether ligand binding to one of these separately first is necessary for the high-affinity heterodimer form. Goldstein et al. (1992) constructed a mathematical model for the IL-2 system to analyze data on equilibrium binding as a function of ligand concentration. These investigators showed that Scatchard plots for this scheme are, not surprisingly, non-linear and concave upward, because increased ligand binding saturates the high-affinity form. An important prediction is that the initial (low ligand concentration) slope is directly proportional to the number of low-affinity chains when precoupling is negligible while it would be independent of that number when precoupling is significant. Experimental data from Robb et al. (1984, 1986, 1987) for cells possessing different numbers of p55 chains showed no differences in the Scatchard plot initial slope, consistent with significant precoupling of p55 and p75 chains in high-affinity heterodimeric receptor form in the absence of IL-2.

2.2.4.d Receptor Aggregation

The EGF system can also be used to illustrate the possible effects of *receptor aggregation* on ligand binding behavior. It is believed that, despite the fact that both EGF and its receptor are apparently monovalent, the EGF receptor can dimerize (Schlessinger, 1988). (It should be noted, however, that some experimental studies caution against this as a substantial phenomenon (Carraway and Cerione, 1991; Koland and Cerione, 1988; Northwood and Davis, 1988).) If indeed real, this receptor dimerization seems to be enhanced by EGF, with the affinity of EGF for dimerized receptors possibly higher than for receptor monomers (Yarden and Schlessinger, 1987a,b). Combining these observations would lead to the prediction that equilibrium binding data ought to show apparent positive cooperativity, because increasing concentrations of ligand lead to increasing proportions of the higher affinity receptor state. However, positive cooperativity has never been found experimentally for EGF equilibrium binding; only neutral or negative cooperative behavior has been reported.

To understand this situation, Wofsy et al. (1992) have developed a mathematical model similar to the ternary complex model but with receptor/receptor coupling instead of receptor/accessory component coupling. The model schematic is given in Figure 2–23. Dimers are hypothesized to be capable of forming between two free receptors, two receptor/ligand complexes, and one free receptor with one receptor/ligand complex, in order to allow for the widest generality, by second-order mass action kinetics. Dimers can uncouple into their separate moieties with first-order kinetics. Ligand is permitted to bind to free receptors in both monomer and dimer forms, and to dissociate from complexes whether separate or dimerized. The multiplicative factor 2 for certain reactions accounts for statistical factors in the reaction stoichiometry. For example, there are two ways in which the receptor–receptor dimer denoted D_0 can bind ligand, because there are two free binding sites on the dimer.

Six equilibrium parameters can be defined: $K_A = k_f/k_r$, $K_{A1} = k_{f1}/k_{r1}$,

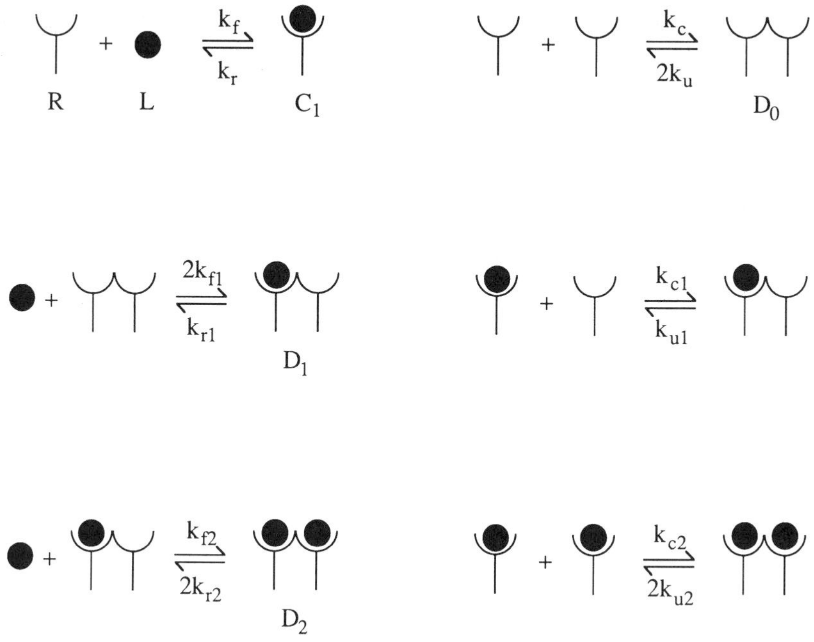

Figure 2–23 Receptor dimerization model. Ligand binding and dissociation steps are described by the rate constants k_f, k_{f1}, k_{f2}, k_r, k_{r1}, and k_{r2}. Free receptor/free receptor coupling, complex/free receptor coupling, and complex/complex coupling are described by the rate constants k_c, k_{c1}, k_{c2}, k_u, k_{u1}, and k_{u2}.

$K_{A2} = k_{f2}/k_{r2}$, $K_C = k_c/k_u$, $K_{C1} = k_{c1}/k_{u1}$, and $K_{C2} = k_{c2}/k_{u2}$, but of these six only four are independent. This can be shown by invoking detailed balance, as we require $K_{C1} = K_C(K_{A1}/K_A)$ and $K_{C2} = K_C(K_{A1}K_{A2}/K_A^2)$. Using these relationships, Wofsy et al. (1992) prove that in order for Scatchard plots not to show positive cooperativity the following condition must hold:

$$\frac{K_{A1}(K_{A1} - K_{A2})}{(K_{A1} - K_A)^2} > \frac{\sqrt{1 + 4K_c R_T} - 1}{2K_c R_T \sqrt{1 + 4K_c R_T}} \tag{2-40}$$

Since the right-hand side of this inequality is positive, the left-hand side must also be positive if there is no apparent positive cooperativity. Therefore, a necessary but not sufficient condition for the experimentally-observed equilibrium binding behavior is $K_{A1} > K_{A2}$. That is, binding of ligand to receptor dimers must exhibit true negative cooperativity, so that the affinity for binding to a single receptor in a dimer when both binding sites are free is greater than when only one site is free.

In conclusion, there are a number of unresolved issues concerning ternary complexes that may well be addressed by future experiments and mathematical models. The identification, measurement of rate constants, and elucidation of various modes of interaction of many of the molecular species are still under way in each of the areas addressed here, and models that attempt to describe these systems are still evolving. Yet even at present, it is heartening to see increasing attention to mechanistic mathematical models for complicated cell surface binding phenomena to go along with the growing amount of molecular information being generated experimentally.

NOMENCLATURE

Symbol	Definition	Typical Units
B	non-specifically bound ligand number	#/cell
B_{eq}	non-specifically bound ligand number	#/cell
C	receptor/ligand complex number	#/cell
C_{eq}	equilibrium receptor/ligand complex number	#/cell
C_i	receptor/ligand complex number (state i)	#/cell
C_{ieq}	equilibrium receptor/ligand complex number (state i)	#/cell
C_0	initial receptor/ligand complex number	#/cell
C_1	receptor/ligand complex number (state 1)	#/cell
C_{1eq}	equilibrium receptor/ligand complex number (state 1)	#/cell
C_{10}	initial receptor/ligand complex number (state 1)	#/cell
C_1^u	unlabeled receptor/ligand complex number	#/cell
C_2	receptor/ligand (state 2) or receptor/effector complex number	#/cell
C_{2eq}	equilibrium receptor/ligand complex number (state 2)	#/cell
C_{20}	initial receptor/ligand complex number (state 2)	#/cell
C_3	ternary complex number	#/cell
C_{3eq}	equilibrium ternary complex number	#/cell
D	translational diffusion coefficient	cm^2/s
D^{self}	self translational diffusion coefficient	cm^2/s
D^{mutual}	mutual translational diffusion coefficient	cm^2/s
h	membrane thickness	nm
k_c	rate constant for coupling	$(\#/\text{cell})^{-1}\,\text{min}^{-1}$
k_{c1}	rate constant for coupling	$(\#/\text{cell})^{-1}\,\text{min}^{-1}$
k_{c2}	rate constant for coupling	$(\#/\text{cell})^{-1}\,\text{min}^{-1}$
k_f	rate constant for association of receptor/ligand complexes	M^{-1} min^{-1}
k_{fi}	rate constant for association of receptor/ligand complexes	M^{-1} min^{-1}
k_{fn}	rate constant for association of non-specifically bound ligand	$(\#/\text{cell})\,\text{M}^{-1}\,\text{min}^{-1}$
k_{f1}	rate constant for association of receptor/ligand complexes	M^{-1} min^{-1}
k_{f2}	rate constant for association of receptor/ligand complexes	M^{-1} min^{-1}
k_{f0}	base value of k_f	M^{-1} min^{-1}
k_{obs}	rate constant observed for approach to binding equilibrium	min^{-1}
k_r	rate constant for dissociation of receptor/ligand complexes	min^{-1}
k_{ri}	rate constant for dissociation of receptor/ligand complexes	min^{-1}
k_{rn}	rate constant for dissociation of non-specifically bound ligand	min^{-1}
k_{r1}	rate constant for dissociation of receptor/ligand complexes	min^{-1}
k_{r2}	rate constant for dissociation of receptor/ligand complexes	min^{-1}
k_{r0}	base value of k_r	$(\#/\text{cell})^{-1}\,\text{min}^{-1}$
k_u	rate constant for uncoupling	min^{-1}
k_{u1}	rate constant for uncoupling	min^{-1}
k_{u2}	rate constant for uncoupling	min^{-1}
k_{12}	rate constant for receptor state conversion	min^{-1}
k_{21}	rate constant for receptor state conversion	min^{-1}
k'_{12}	rate constant for receptor state conversion	min^{-1}
k'_{21}	rate constant for receptor state conversion	min^{-1}
K_A	equilibrium association constant	M^{-1}
K_{A1}	equilibrium association constant	M^{-1}
K_{A2}	equilibrium association constant	M^{-1}

Symbol	Definition	Typical Units
K_B	Boltzmann's constant ($=1.38 \times 10^{-16}$ g-cm^2/s^2-degree Kelvin)	
K_C	equilibrium coupling constant	$(\#/\text{cell})^{-1}$
K_{C1}	equilibrium coupling constant	$(\#/\text{cell})^{-1}$
K_{C2}	equilibrium coupling constant	$(\#/\text{cell})^{-1}$
K_D	equilibrium dissociation constant	M
K_{Dapp}	apparent equilibrium dissociation constant	M
K_{D1}	equilibrium dissociation constant	M
K_{D2}	equilibrium dissociation constant	M
K_{Di}	equilibrium dissociation constant (state i)	M
K_N	non-specific association equilibrium constant	$(\#/\text{cell})$-M^{-1}
K_{21}	equilibrium constant for receptor state conversion	
K'_{21}	equilibrium constant for receptor state conversion	
L	ligand concentration	M
L_0	initial concentration of ligand	M
L^u	unlabeled ligand concentration	M
n	cell concentration	#/volume
N_{Av}	Avogadro's number ($=6.02 \times 10^{23}$ #/mole)	
q_f	enhancement/diminution factor for association	$\#/\text{cell})^{-1}$
q_r	enhancement/diminution factor for dissociation	$(\#/\text{cell})^{-1}$
R	free receptor number	#/cell
R_g	gas constant	cal/mole-degree Kelvin
R_T	total surface receptor number	#/cell
R_{Ti}	total surface receptor number (state i)	#/cell
R_1	free receptor number (state 1)	#/cell
R_2	free receptor number (state 2)	#/cell
R_{20}	initial free receptor number (state 2)	#/cell
s	receptor radius	nm
t	time	min
$t_{1/2}$	half-time	min
T	absolute temperature	degrees Kelvin (K)
u	scaled receptor/ligand complex number	
u_{eq}	equilibrium scaled receptor/ligand complex number	
u_0	initial scaled receptor/ligand complex number	
w	scaled free receptor number	
X	accessory component number	#/cell
X_T	total accessory component number	#/cell
γ	Euler's constant ($=0.5772$)	
ΔC_p°	difference in constant pressure heat capacity between two states	cal/mole-K
ΔG°	Gibbs standard free energy change	cal/mole
ΔH°	enthalpy change between reactants and products at standard state	cal/mole
ΔS°	entropy change between reactants and products at standard state	cal/mole-K
η	scaled cell density	
η_m	membrane viscosity	g/cm-s
η_s	solution viscosity	g/cm-s
θ	scaled ratio of membrane to solution viscosities	
κ_{C1}	scaled equilibrium coupling constant (state 1)	
κ_{C2}	scaled equilibrium coupling constant (state 2)	
λ	scaled ligand concentration	
ν	non-specific binding capacity	
τ	scaled time	
$\tau_{1/2}$	scaled half-time	
χ	stoichiometric ratio of accessory molecules to receptors	

REFERENCES

Akaike, H. (1974). A new look at the statistical model identification. *IEEE Trans. Automatic Control*, **19**:716–723.

Akiyama, S. K. and Yamada, K. M. (1985). The interaction of plasma fibronectin with fibroblastic cells in suspension. *J. Biol. Chem.*, **260**:4492–4500.

Axelrod, D., Koppel, D. E., Schlessinger, J., Elson, E. and Webb, W. W. (1976). Mobility measurement by analysis of fluorescence photobleaching recovery kinetics, *Biophys. J.*, **16**:1055–1069.

Baird, B., Erickson, J., Goldstein, B., Kane, P., Menon, A. K., Robertson, D. and Holowka, D. (1988). Progress toward understanding the molecular details and consequences of IgE-receptor crosslinking. *In* A. S. Perelson (Ed.), *Theoretical Immunology, Part One*, pp. 41–59. Redwood City, California: Addison-Wesley Publishing Company.

Bajzer, Z., Meyers, A. C. and Vuk-Pavlovic, S. (1989). Binding, internalization, and intracellular processing of proteins interacting with recycling receptors: a kinetic analysis. *J. Biol. Chem.*, **264**:13623–13631.

Barak, L. S. and Webb, W. W. (1982). Diffusion of low density lipoprotein–receptor complex on human fibroblasts. *J. Cell Biol.*, **95**:846–852.

Birnbaumer, L., Abramowitz, J. and Brown, A. M. (1990). Receptor–effector coupling by G proteins. *Biochim. Biophys. Acta*, **1031**:163–224.

Caceci, M. S. and Cacheris, W. P. (1984). Fitting curves to data. *Byte*, **9**:340–362.

Calvo, J. C., Radicella, J. P. and Charreau, E. H. (1983). Measurement of specific radioactivities in labelled hormones by self-displacement analysis. *Biochem. J.*, **212**:259–264.

Carraway, K. L., III and Cerione, R. A. (1991). Comparison of epidermal growth factor (EGF) receptor–receptor interactions in intact A431 cells and isolated plasma membranes: large scale receptor micro-aggregation is not detected during EGF-stimulated early events. *J. Biol. Chem.*, **266**:8899–8906.

Ciechanover, A., Schwartz, A. L., Dautry-Varsat, A. and Lodish, H. F. (1983). Kinetics of internalization and recycling of transferrin and the transferrin receptor in a human hepatoma cell line: effect of lysosomotropic agents. *J. Biol. Chem.*, **258**:9681–8689.

De Lean, A., Hancock, A. A. and Lefkowitz, R. J. (1982). Validation and statistical analysis of a computer modeling method for quantitative analysis of radioligand binding data for mixtures of pharmacological receptor subtypes. *Mol. Pharmacol.*, **21**:5–16.

De Lean, A. and Rodbard, D. (1979). Kinetics of cooperative binding. *In* R. D. O'Brien (Ed.), *The Receptors: A Comprehensive Treatise*, pp. 143–192. New York: Plenum Press.

De Meyts, P. (1976). Cooperative properties of hormone receptors in cell membranes. *J. Supramol. Struct.*, **4**:241(201)–258(218).

De Meyts, P., Gu, J., Smal, J., Kathuria, S., Gonzales, N., Rotella, C. M. and Shymko, R. M. (1989). The Insulin receptor gene and its product: structure and function. *In* J. Nerup, T. Mandrup-Poulsen and B. Hokfelt (Eds.), *Genes and Gene Products in the Development of Diabetes Mellitus: Basic and Clinical Aspects*, pp. 185–203. Amsterdam: Elsevier Science Publishers B.V.

De Meyts, P., Roth, J., Neville, D. M., Jr., Gavin, J. R., III and Lesniak, M. A. (1973). Insulin interactions with its receptors: experimental evidence for negative cooperativity. *Biochem. Biophys. Res. Commun.*, **55**:154–161.

Devreotes, P. (1989). *Dictyostelium discoideum*: a model system for cell–cell interactions in development. *Science*, **245**:1054–1058.

Devreotes, P. N. and Sherring, J. A. (1985). Kinetics and concentration dependence of reversible cAMP-induced modification of the surface cAMP receptor in dictyostelium. *J. Biol. Chem.*, **260**:6378–6384.

Dowdy, S. and Wearden, S. (1991). *Statistics for Research* (second ed.). New York: John Wiley.

Edidin, M., Zagyansky, Y. and Lardner, T. J. (1976). Measurement of membrane protein lateral diffusion in single cells. *Science*, **191**:466–468.

Fay, S. P., Posner, R. G., Swann, W. N. and Sklar, L. A. (1991). Real-time analysis of ligand, receptor, and G-protein by quantitative fluorescence flow cytometry. *Biochem.*, **30**:5066–5075.

Gailit, J. and Ruoslahti, E. (1988). Regulation of the fibronectin receptor affinity by divalent cations. *J. Biol. Chem.*, **263**:12927–12932.

Geerts, H., de Brabander, M. and Nuydens, R. (1991). Nanovid microscopy. *Nature*, **351**:765–766.

Gennis, R. B. (1989). *Biomembranes: Molecular Structure and Function.* New York: Springer-Verlag.

Gershon, N. D., Smith, R. M. and Jarett, L. (1981). Computer assisted analysis of ferritin–insulin receptor sites on adipocytes and the effects of cytochalasin B on groups of insulin receptor sites. *J. Membr. Biol.*, **58**:155–160.

Gex-Fabry, M. and DeLisi, C. (1984). Model for kinetic and steady state analysis of receptor mediated endocytosis. *Math. Biosci.*, **72**:245–261.

Gilman, A. G. (1987). G proteins: transducers of receptor-generated signals. *Annu. Rev. Biochem.*, **56**:615–649.

Goldstein, B. (1989). Diffusion limited effects of receptor clustering. *Comments Theor. Biol.*, **1**:109–127.

Goldstein, B., Jones, D., Kevrekidis, I. G. and Perelson, A. S. (1992). Evidence for p55–p75 heterodimers in the absence of IL-2 from Scatchard plot analysis. *Int. Immunol.*, **4**:23–32.

Green, N. M. (1975). Avidin. *Adv. Protein Chem.*, **29**:85–133.

Grzesiak, J. J., Davis, G. E., Kirchhofer, D. and Pierschbacher, M. D. (1992). Regulation of $\alpha_2\beta_1$-mediated fibroblast migration on type I collagen by shifts in the concentrations of extracellular Mg^{+2} and Ca^{+2}. *J. Cell Biol.*, **117**:1109–1117.

Hillman, G. M. and Schlessinger, J. (1982). Lateral diffusion of epidermal growth factor complexed to its surface receptors does not account for the thermal sensitivity of patch formation and endocytosis. *Biochemistry*, **21**:1667–1672.

Hinz, H. (1983). Thermodynamics of protein-ligand interactions: calorimetric approaches. *Annu. Rev. Biophys. Bioeng.*, **12**:285–317.

Hollenberg, M. D. and Goren, H. J. (1985). Ligand–receptor interactions at the cell surface. *In* G. Poste and S. T. Crooke (Eds.), *Mechanisms of Receptor Regulation*, pp. 323–373. New York: Plenum Press.

Hughes, R. J., Boyle, J. R., Brown, R. D., Taylor, R. and Insel, P. A. (1982). Characterization of coexisting alpha$_1$- and beta$_2$-adrenergic receptors on a cloned muscle cell line, BC3H-1. *Mol. Pharmacol.*, **22**:258–266.

Hulme, E. C. (Ed.) (1992). *Receptor–Ligand Interactions: A Practical Approach.* Oxford: IRL Press. 458 pp.

Jacobs, S. and Cuatrecasas, P. (1976). The mobile receptor hypothesis and "cooperativity" of hormone binding. *Biochim. Biophys. Acta*, **433**:482–495.

Jose, M. V. and Larralde, C. (1982). Alternative interpretation of unusual Scatchard plots: contribution of interactions and heterogeneity. *Math. Biosci.*, **58**:159–170.

Kahn, C. R. (1976). Membrane receptors for hormones and neurotransmitters. *J. Cell Biol.*, **70**:261–286.

Kaplan, J. (1985). Patterns in receptor behavior and function. *In* G. Poste and S. T. Crooke (Eds.), *Mechanisms of Receptor Regulation*, pp. 13–36. New York: Plenum Press.

Kent, R. S., De Lean, A. and Lefkowitz, R. J. (1980). A quantitative analysis of beta-adrenergic receptor interactions: resolution of high and low affinity states of the receptor by computer modeling of ligand binding data. *Mol. Pharmacol.*, **17**:14–23.

Kienhuis, C. B. M., Heuvel, J. J. T. M., Ross, H. A., Swinkels, L. M. J. W., Foekens, J. A. and Benraad, T. J. (1991). Six methods for direct radioiodination of mouse epidermal growth factor compared: effect of nonequivalence in binding behavior between labeled and unlabeled ligand. *Clin. Chem.*, **37**:1749–1755.

King, I. C. and Catino, J. J. (1990). Nonradioactive ligand binding assay for epidermal growth factor receptor. *Anal. Biochem.*, **188**:97–100.

Klein, P., Theibert, A., Fontana, D. and Devreotes, P. N. (1985). Identification and cyclic AMP-induced modification of the cyclic AMP receptor in *Dictyostelium discoideum*. *J. Biol. Chem.*, **260**:1757–1764.

Klotz, I. M. (1985). Ligand–receptor interactions: facts and fantasies. *Q. Rev. Biophys.*, **18**:227–259.

Koland, J. G. and Cerione, R. A. (1988). Growth factor control of epidermal growth factor receptor kinase activity via an intramolecular mechanism. *J. Biol. Chem.*, **263**:2230–2237.

Koo, C., Lefkowitz, R. J. and Snyderman, R. (1982). The oligopeptide chemotactic factor receptor on human polymorphonuclear leukocyte membranes exists in two affinity states. *Biochem. Biophys. Res. Commun.*, **106**:442–449.

Lauffenburger, D. and DeLisi, C. (1983). Cell surface receptors: physical chemistry and cellular regulation. *Int. Rev. Cytol.*, **84**:269–302.

Lefkowitz, R. J., Stadel, J. M. and Caron, M. G. (1983). Adenylate cyclase-coupled beta-adrenergic receptors: structure and mechanisms of activation and desensitization. *Annu. Rev. Biochem.*, **52**:159–186.

Limbird, L. E. (1986). *Cell Surface Receptors: A Short Course on Theory and Methods.* Boston: Martinus Nijhoff Publishing.

Lin, C. R., Chen, W. S., Lazar, C. S., Carpenter, C. D., Gill, G. N., Evans, R. M. and Rosenfeld, M. G. (1986). Protein kinase C phosphorylation at Thr 654 of the unoccupied EGF receptor and EGF binding regulate functional receptor loss by independent mechanisms. *Cell*, **44**:839–848.

Lipkin, E. W., Teller, D. C. and de Haen, C. (1986a). Equilibrium binding of insulin to rat white fat cells at 15 °C. *J. Biol. Chem.*, **261**:1694–1701.

Lipkin, E. W., Teller, D. C. and de Haen, C. (1986b). Kinetics of insulin binding to rat white fat cells at 15 °C. *J. Biol. Chem.*, **261**:1702–1711.

Low, D. A., Baker, J. B., Koonce, W. C. and Cunningham, D. D. (1981). Released protease-nexin regulates cellular binding, internalization, and degradation of serine proteases. *Proc. Natl. Acad. Sci. USA*, **78**:2340–2344.

Mackin, W. M., Huang, C. and Becker, E. L. (1982). The formylpeptide chemotactic receptor on rabbit peritoneal neutrophils. *J. Immunol.*, **129**:1608–1611.

Marasco, W. A., Feltner, D. E. and Ward, P. A. (1985). Formyl peptide chemotaxis receptors on the rat neutrophil: experimental evidence for negative cooperativity. *J. Cell Biochem.*, **27**:359–375.

Martiel, J. and Goldbeter, A. (1987). A model based on receptor desensitization for cyclic AMP signaling in *Dictyostelium* cells. *Biophys. J.*, **52**:807–828.

Mayo, K. H., Nunez, M., Burke, C., Starbuck, C., Lauffenburger, D. and Savage, C. R., Jr. (1989). Epidermal growth factor receptor binding is not a simple one-step process. *J. Biol. Chem.*, **264**:17838–17844.

McKanna, J. A., Haighler, H. T. and Cohen, S. (1979). Hormone receptor topology and dynamics: morphological analysis using ferritin-labeled epidermal growth factor. *Proc. Natl. Acad. Sci. USA*, **76**:5689–5693.

Mellman, I. S. and Unkeless, J. C. (1980). Purification of a functional mouse Fc receptor through the use of a monoclonal antibody. *J. Exp. Med.*, **152**:1048–1069.

Menon, A. K., Holowka, D., Webb, W. W. and Baird, B. (1986). Clustering, mobility, and triggering activity of small oligomers of immunoglobulins E on rat basophilic leukemia cells. *J. Cell Biol.*, **102**:534–540.

Muirhead, K. A., Horan, P. K. and Poste, G. (1985). Flow cytometry: present and future. *Bio/Technology*, **3**:337–356.

Munson, P. J. and Rodbard, D. (1980). Ligand: a versatile computerized approach for characterization of ligand-binding systems. *Anal. Biochem.*, **107**:220–239.

Munson, P. J., Rodbard, D. and Klotz, I. M. (1983). Number of receptor sites from Scatchard and Klotz graphs: a constructive critique. *Science*, **220**:979–981.

Neubig, R. R., Gantzog, R. D. and Thomsen, W. J. (1988). Mechanism of agonist and antagonist binding to α_2 adrenergic receptors: evidence for a precoupled receptor–guanidine nucleotide protein complex. *Biochemistry*, **27**:2374–2384.

Northwood, I. C. and Davis, R. J. (1988). Activation of the epidermal growth factor receptor tyrosine protein kinase in the absence of receptor oligomerization. *J. Biol. Chem.*, **263**:7450–7453.

O'Neil, P. V. (1991). *Advanced Engineering Mathematics* 3rd edition. Belmont, CA: Wadsworth Publishing.

Park, L. S., Gillis, S. and Urdal, D. L. (1990). Hematopoietic growth-factor receptors. *In* J. M. G. T. M. Dexter and N. G. Testa (Eds.), *Colony-Stimulating Factors: Molecular and Cellular Biology*, New York: Marcel Dekker.

Pearse, B. M. F. and Robinson, M. S. (1990). Clathrin, adaptors, and sorting. *Annu. Rev. Cell Biol.*, **6**:151–171.

Pecht, I. and Lancet, D. (1977). Kinetics of antibody–hapten interactions. *Mol. Biol. Biochem. Biophys.*, **24**:306–338.

Peters, R. and Cherry, R. J. (1982). Lateral and rotational diffusion of bacteriorhodopsin in lipid bilayers: experiments test of the Saffman–Delbruck equations. *Proc. Natl. Acad. Sci. USA*, **79**:4317–4321.

Pollet, R. J., Standaert, M. L. and Haase, B. A. (1977). Insulin binding to the human lymphocyte receptor. *J. Biol. Chem.*, **252**:5828–5834.

Poo, M., Lam, J. W. and Orida, N. (1979). Electrophoresis and diffusion in the plane of the cell membrane. *Biophys. J.*, **26**:1–22.

Press, W. H., Flannery, B. P., Teukolsky, S. A. and Vetterling, W. T. (1989). *Numerical Recipes: The Art of Scientific Computing.* New York: Cambridge University Press.

Pruzansky, J. J. and Patterson, R. (1986). Binding constants of IgE receptors on human blood basophils for IgE. *Immunology*, **58**:257–262.

Rescigno, A., Beck, J. S. and Goren, H. J. (1982). Determination of dependence of binding parameters on receptor occupancy. *Bull. Math. Biol.*, **44**:477–489.

Rimon, G., Hanski, E. and Levitzki, A. (1980). Temperature dependence of β receptor, adenosine receptor, and sodium fluoride stimulated adenylase cyclase from turkey erythrocytes. *Biochemistry*, **19**:4451–4460.

Ritger, P. D. and Rose, N. J. (1968). *Differential Equations with Applications.* New York: McGraw-Hill.

Robb, R. J., Greene, W. C. and Rusk, C. M. (1984). Low and high affinity cellular receptors for interleukin-2: implications for the level of Tac antigen. *J. Exp. Med.*, **160**:1126–1144.

Robb, R. J. and Rusk, C. M. (1986). High and low affinity receptors for interleukin-2: implications of pronase, phorbol ester, and cell membrane studies upon the basis for differential ligand affinities. *J. Immunol.*, **137**:142–149.

Robb, R. J., Rusk, C. M., Yodoi, J. and Greene, W. C. (1987). Interleukin-2 binding molecule distinct from the Tac protein: analysis of its role in formation of high-affinity receptors. *Proc. Natl. Acad. Sci. USA*, **84**:2002–2006.

Ross, E. M., Maguire, M. E., Sturgill, T. W., Biltonen, R. L. and Gilman, A. G. (1977). Relationship between the β-adrenergic receptor and adenylate cyclase: studies of ligand binding and enzyme activity in purified membranes of S49 lymphoma cells. *J. Biol. Chem.*, **252**:5761–5775.

Roy, L. M., Gittinger, C. K. and Landreth, G. E. (1989). Characterization of the epidermal growth factor receptor associated with cytoskeletons of A431 cells. *J. Cell. Physiol.*, **140**:295–304.

Ryan, T. A., Myers, J., Holowka, D., Baird, B. and Webb, W. W. (1988). Molecular crowding on the cell surface. *Science*, **239**:61–64.

Saffman, P. G. and Delbruck, M. (1975). Brownian motion in biological membranes. *Proc. Natl. Acad. Sci. USA*, **72**:3111–3113.

Scalettar, B. A., Abney, J. R. and Owicki, J. C. (1988). Theoretical comparison of the self diffusion and mutual diffusion of interacting membrane proteins. *Proc. Natl. Acad. Sci. USA*, **85**:6726–6730.

Scatchard, G. (1949). The attractions of proteins for small molecules and ions. *Ann. NY Acad. Sci.*, **51**:660–672.

Schaudies, R. P., Harper, R. A. and Savage, C. R., Jr. (1985). ^{125}I-EGF binding to responsive and nonresponsive cells in culture: loss of cell-associated radioactivity relates to growth induction. *J. Cell. Physiol.*, **124**:493–498.

Schlessinger, J. (1986). Allosteric regulation of the epidermal growth factor receptor kinase. *J. Cell Biol.*, **103**:2067–2072.

Schlessinger, J. (1988). Signal transduction by allosteric receptor oligomerization. *Trends Biochem. Sci.*, **13**:443–447.

Segel, L. A. (1988). On the validity of the steady state assumption of enzyme kinetics. *Bull. Math. Biol.*, **50**:579–593.

Segel, L. A. and Slemrod, M. (1989). The quasi-steady state assumption: a case study in perturbation. *SIAM Review*, **31**:446–477.

Sheetz, M. P. (1993). Glycoprotein mobility and dynamic domains in fluid plasma membranes. *Annu. Rev. Biophys. Biomolec. Struct.*, **22** (in press).

Sklar, L. A. (1987). Real-time spectroscopic analysis of ligand-receptor dynamics. *Annu. Rev. Biophys. Biophys. Chem.*, **16**:479–506.

Sklar, L. A., Finney, D. A., Oades, Z. G., Jesaitis, A. J., Painter, R. G. and Cochrane, C. G. (1984). The dynamics of ligand–receptor interactions: real-time analyses of association, dissociation, and internalization of an N-formyl peptide and its receptors on the human neutrophil. *J. Biol. Chem.*, **259**:5661–5669.

Sklar, L. A., Mueller, H., Omann, G. and Oades, Z. (1989). Three states for the formyl peptide receptor on intact cells. *J. Biol. Chem.*, **264**:8483–8486.

Sklar, L. A. and Omann, G. M. (1990). Kinetics and amplification in neutrophil activation and adaptation. *Semin. Cell Biol.*, **1**:115–123.

Smith, K. A. (1988). Interleukin-2: Inception, impact, and implications. *Science*, **240**:1169–1176.

Strang, G. (1986). *Introduction to Applied Mathematics*. Wellesley, Massachusetts, Wellesley–Cambridge Press.

Szewczuk, M. R. and Mukkur, T. K. S. (1977). Enthalpy–entropy compensation in dinitrophenyl–anti-dinitrophenyl antibody interaction(s). *Immunology*, **32**:111–119.

Taylor, C. W. (1990). The role of G-proteins in transmembrane signalling. *Biochem. J.*, **272**:1–13.

Tomoda, H. Kishimoto, Y. and Lee, Y. C. (1989). Temperature effect on endocytosis and exocytosis by rabbit alveolar macrophages. *J. Biol. Chem.*, **264**:15445–15450.

Trowbridge, I. S. (1991). Endocytosis and signals for internalization. *Curr. Opin. Cell Biol.*, **3**:634–641.

Van Haastert, P. J. M. and De Wit, R. J. W. (1984). Demonstration of receptor heterogeneity and affinity modulation by nonequilibrium binding experiments: the cell surface cAMP receptor of *Dictyostelium discoideum*. *J. Biol. Chem.*, **259**:13321–13328.

Wade, W. F., Freed, J. H. and Edidin, M. (1989). Translational diffusion of class II major histocompatibility complex molecules is constrained by their cytoplasmic domains. *J. Cell Biol.*, **109**:3325–3331.

Waters, C. M., Oberg, K. C., Carpenter, G. and Overholser, K. A. (1990). Rate constants for binding, dissociation, and internalization of EGF: Effect of receptor occupancy and ligand concentration. *Biochemistry*, **29**:3563–3569.

Weigel, P. H. and Oka, J. A. (1982). Endocytosis and degradation mediated by the asialoglycoprotein receptor in isolated rat hepatocytes. *J. Biol. Chem.*, **257**:1201–1207.

Wells, J. W. (1992). Analysis and interpretation of binding at equilibrium. *In* E. C. Hulme (Ed.), *Receptor–Ligand Interactions: A Practical Approach*, pp. 289–395. Oxford: IRL Press.

Wey, C. and Cone, R. A. (1981). Lateral diffusion of rhodopsin in photoreceptor cells measured by fluorescence photobleaching and recovery. *Biophys. J.*, **33**:225–232.

Wiegel, F. W. (1984). Diffusion of proteins in membranes. *In* A. S. Perelson, C. DeLisi and F. W. Wiegel (Eds.), *Cell Surface Dynamics: Concepts and Models*, pp. 135–150. New York: Marcel Dekker.

Wier, M. and Edidin, M. (1988). Constraint of the translational diffusion of a membrane glycoprotein by its external domains. *Science*, **242**:412–414.

Wiley, H. S. (1985). Receptors as models for the mechanisms of membrane protein turnover and dynamics. *Curr. Top. Membr. Transp.*, **24**:369–412.

Wiley, H. S. and Cunningham, D. D. (1982). The endocytotic rate constant: a cellular parameter for quantitating receptor-mediated endocytosis. *J. Biol. Chem.*, **257**:4222–4229.

Wiley, H. S., Walsh, B. J. and Lund, K. A. (1989). Global modulation of the epidermal growth factor receptor is triggered by occupancy of only a few receptors: evidence for a binary regulation system in normal human fibroblasts. *J. Biol. Chem.*, **264**:18912–18920.

Wofsy, C., Goldstein, B., Lund, K. and Wiley, H. S. (1992). Implications of epidermal growth factor (EGF) induced EGF receptor aggregation. *Biophys. J.*, **63**:98–110.

Wyman, J. (1975). The turning wheel: A study in steady states. *Proc. Natl. Acad. Sci. USA*, **72**:3983–3987.

Yarden, Y. and Schlessinger, J. (1987a). Epidermal growth factor induces rapid, reversible aggregation of the purified epidermal growth factor receptor. *Biochemistry*, **26**:1443–1451.

Yarden, Y. and Schlessinger, J. (1987b). Self-phosphorylation of epidermal growth factor receptor: Evidence for a model of intermolecular allosteric activation. *Biochemistry*, **26**:1434–1442.

Zhang, F., Crise, B., Su, B., Rose, J. K., Bothwell, A. and Jacobson, K. (1991). Lateral diffusion of membrane-spanning glycosylphosphatidylinositol-linked proteins:

towards establishing rules governing the lateral mobility of membrane proteins. *J. Cell Biol.*, **115**:75–84.

Zierler, K. (1989). Misuse of nonlinear Scatchard plots. *Trends Biochem. Sci.*, **14**:314–317.

Zigmond, S. H., Sullivan, S. J. and Lauffenburger, D. A. (1982). Kinetic analysis of chemotactic receptor modulation. *J. Cell Biol.*, **92**:34–43.

Receptor / Ligand Trafficking

3.1 BACKGROUND

In Chapter 2, we explored models of receptor/ligand binding on the cell surface. For each of those models, we assumed that the number and location of receptors are constant, *i.e.*, that no significant synthesis, degradation, internalization, or recycling of receptors occurs over the time frame for which the models apply. Under normal physiological conditions, however, these *dynamic trafficking events* take place concurrently with receptor/ligand binding. In particular, cell surface receptors and receptor/ligand complexes can be internalized in a process known as *receptor-mediated endocytosis* (RME). The basic steps of the endocytic cycle have been described in a number of reviews (Brown *et al.*, 1983; Wileman *et al.*, 1985b; Mellman *et al.*, 1986; van Deurs *et al.*, 1989; Schwartz, 1990) and are shown schematically in Figure 3–1. The selection and internalization of ligands via RME plays a role in nutrition (*e.g.*, LDL, transferrin), clearance or retrieval of molecules (*e.g.*, immune complexes, asialoglycoproteins, mannose-6-phosphate glycoproteins) and presumably the elicitation or attenuation of cellular responses (*e.g.*, EGF, chemotactic peptides).

Trafficking processes can alter the number of receptors present on the cell surface as well as the amount of ligand remaining in solution (see Figure 3–2), Receptor/ligand binding may consequently be affected in significant ways, complicating the models of Chapter 2. One immediate result of trafficking is that receptor/ligand binding will never be at mere chemical equilibrium. Rather, at its simplest, binding can be part of a dynamic steady state involving all the rest of the trafficking steps. This circumstance has been a major cause of ambiguity in the experimental literature on ligand binding kinetic and equilibrium properties. To suppress trafficking for purposes of focusing on the surface ligand binding events, the usual approach is to lower the temperature to near 4 °C. At these low temperatures, trafficking processes are slowed dramatically (Thilo and Vogel, 1980; Weigel and Oka, 1982; Tomoda *et al.*, 1989). This approach is not without drawbacks, and section 2.1.2 discussed the poorly-characterized influence of temperature on the association and dissociation rate constants for receptor/ligand binding. In addition, temperature shifts can cause shedding of cell surface

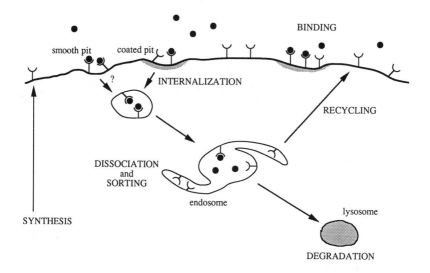

Figure 3–1 The endocytic cycle. Receptors mediate the selective transport of ligand from the extracellular environment to the interior of the cell. Receptor and ligand molecules may be sorted intracellularly within endosomes, governing recycling to the cell surface and delivery to lysosomes. Current evidence suggests that smooth-pit and coated-pit internalization pathways meet at the endosome compartment.

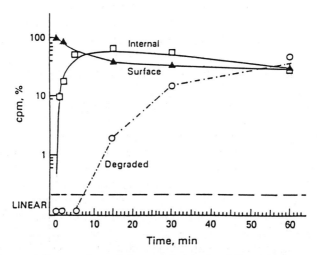

Figure 3–2 Representative experimental data on the kinetics of ligand internalization and degradation. Data from Bridges *et al.* (1982) show the amounts of surface bound, intracellular, and degraded radiolabeled ligand as a function of time for the asialoglycoprotein receptor system on rat hepatocytes with the ligand asialoorosomucoid. In this experiment, ligand was incubated with cells at 4 °C prior to the start of the experiment. The *x*-axis gives the time elapsed after raising the temperature to 37 °C.

74

receptors, compromising observed behavior (Kaplan and Keogh, 1982). Alternatively, pharmacological inhibitors of trafficking such as phenylarsineoxide can be used (Low et al., 1981), also with possible attendant side effects.

Furthermore, it is believed that receptor/ligand trafficking may play a major role in receptor-mediated cell responses. Internalization of receptors may contribute to generation of a cellular response (e.g., adhesion receptor recycling in cell migration (Bretcher, 1988)) or provide a mechanism for attenuation of a response (e.g., as in neutrophil chemotaxis (Zigmond, 1981) and EGF stimulation of fibroblast proliferation (Wells et al., 1990)). Numbers of cell surface receptors can also increase after exposure to stimuli, suggesting that the regulation of receptor number may be important in cellular responses (e.g., expression of adhesion receptors on endothelial cells for leukocytes during inflammation (Springer, 1990)). Thus the discussion of Chapter 2 must be extended to include the movement of receptors and ligands throughout the cell.

3.1.1 The Endocytic Cycle

The endocytic trafficking cycle has been recognized and investigated in detail only in recent years. Helpful background material is presented by Pastan and Willingham (1985), Pugsley (1989), Holtzman (1989), and Wiley (1992). For a more quantitative perspective, reviews by Thilo (1985) and Wiley (1985), and a monograph by Linderman and Lauffenburger (1989), may be useful.

Receptor/ligand complexes can accumulate, most likely by diffusion in the plasma membrane followed by trapping, in localized membrane regions on the cell surface. Such regions include both *coated pits* and *smooth pits*. The adjective coated refers to an associated lattice made of the protein clathrin on the cytoplasmic side of the membrane at the location of the pit structures (Pearse and Crowther, 1987; Keen, 1990); smooth pits are similar structures lacking associated clathrin (van Deurs et al., 1989). Coated pits generally cover between about 0.5% and 4% of the total plasma membrane surface area at any time, depending on the cell type (Orci et al., 1978; Steinman et al., 1983). For a typical coated pit diameter of about 100 nm, this translates to roughly 10^3 coated pits present on the cell surface at any given time. Coated pits invaginate and pinch off to form intracellular vesicles which carry molecular contents to further trafficking processes. Smooth pits are similarly involved in endocytic internalization (Hansen et al., 1991). Cells can internalize the equivalent of their entire surface membrane on time scales of roughly 15 min (hepatocytes) to 5 h (adipocytes), depending on the type (see Gennis, 1989, p. 339), and not all of this turnover occurs through coated pits. Indeed, McKinley and Wiley (1988) have determined that coated and smooth pathways probably contribute about equally to the volumetric rate of fluid-phase endocytosis by confluent human fibroblasts.

A central feature of these endocytic structures is their potential for *selectivity* in trapping specific membrane proteins (Goldstein et al., 1985). Receptors whose primary function is transport of nutritional molecules, such as low-density lipoprotein (LDL) and transferrin (Tf) receptors, preferentially congregate within coated pits whether bound or free. Ligand binding alters neither the distribution on the cell surface nor the rate of internalization of these types of receptors.

In contrast, receptors for hormones and growth factors, such as insulin and EGF receptors, are distributed uniformly on the cell surface in the absence of ligand but congregate within coated pits after ligand binding. Ability of receptors to congregate within coated pits appears to involve a reversible biochemical inter-action between receptor cytoplasmic tail domains and accessory molecules, termed *adaptors*, found associated with the clathrin lattice (Pearse, 1988; Glickman *et al.*, 1989; Pearse and Robinson, 1990). Trowbridge (1991) provides a compilation of the key structural signals governing coated-pit endocytic internalization for various receptor types; a key determinant favoring receptor localization in coated pits appears to be a "tight turn" in the three-dimensional conformation of the receptor cytoplasmic domain (for instance, in the LDL receptor (Chen *et al.*, 1990) and the Tf receptor (Collawn *et al.*, 1990)).

Endocytic vesicles formed at the cell surface may contain free receptors, if these were present on the membrane taken in, as well as receptor/ligand complexes. In addition, free ligand and other molecules can also be taken up in the bulk fluid contained within the vesicles; this is known as *non-specific uptake* or *fluid-phase endocytosis*.

Upon internalization, the clathrin lattice is removed from coated vesicles (those formed from coated pits) by ATP-dependent enzymes in the cytoplasm. The vesicular contents then accumulate in larger intracellular organelles called *endosomes*. Experimental evidence supports the view that multiple vesicles fuse to generate the larger endosomal vesicles, with the time window for incoming vesicles to join together being roughly 2–5 minutes (Salzman and Maxfield, 1988, 1989). The coated-pit and smooth-pit internalization pathways appear to merge at the endosomal compartment (Tran *et al.*, 1987; Hansen *et al.*, 1991).

From endosomes, receptors and ligands have at least two possible fates: recycling to the cell surface (often termed exocytosis in the case of ligand recycling) or degradation in lysosomes. It is also possible that some receptors and ligands are neither recycled nor degraded; rather, they may be sent to an undetermined intracellular compartment, in which they carry out further aspects of their signaling function, or routed to the Golgi. Particular routes of receptors and ligands depend on the outcome of separation or sorting events which occur within the endosome.

Endosomal sorting is a critical step in the endocytic pathway, for at this point the cell determines the destinations of the endocytosed molecules. If receptors are efficiently recycled, these receptors can participate in future rounds of endocytosis. On the other hand, by routing receptors to the degradative pathway or to another intracellular location instead of back to the cell surface, the cell can decrease, or "downregulate", the number of cell surface receptors and therefore also its ability to respond to future doses of the same ligand. Understanding of the sorting mechanism is, therefore, critical to an understanding of the kinetics and efficiency of receptor recycling and downregulation, of ligand degradation and recycling, and of the overall kinetics of the endocytic cycle.

The mechanism of the endosomal sorting process is not entirely clear. It is known that receptors and their ligands can in some cases be directed along different intracellular routes following endocytosis. For example, LDL and Tf receptors are preferentially recycled to the cell surface whereas their ligands primarily have intracellular destinations. EGF, on the other hand, appears to stay with its receptor for either degradation or recycling. For the systems in which receptor and ligand pathways separate in the endosome, complex dissociation may be aided by its acidic environment. Measurements made using pH-dependent fluorochromes have indicated that the endosome pH is in the range 5–6 (Tycko et al., 1983; Roederer et al., 1987). In many systems, the binding of a ligand to its receptor is pH-dependent: as the pH is lowered, the binding affinity decreases (Mellman et al., 1984; Dunn and Hubbard, 1984; Wileman et al., 1985a). Dissociation of a receptor–ligand complex is certainly a prerequisite when the receptor and ligand are routed along different intracellular pathways. We note that the outcome of the endosomal sorting may differ not only among receptor species but also among identical receptors bound by different ligands. Further description and a proposed model for the sorting process will be given later in this chapter.

The dynamics of the endocytic cycle are also influenced by the synthesis of new receptors and their delivery to the cell surface. It is now believed that the trafficking pathway for synthesized proteins merges with the endocytic trafficking pathway somewhere outside the Golgi region (Klausner, 1989). For example, Stoorvogel et al. (1988) have demonstrated that endocytosed transferrin and newly-synthesized α_1-antitrypsin, a secreted protein, are found together in the trans-Goldi reticulum in a hepatoma cell line. The rate of receptor synthesis may be influenced by the presence or absence of ligand (e.g., EGF receptor (Earp et al., 1986) and IL-2 receptor (Smith and Cantrell, 1985).

The endocytic pathway operates not only for simple molecular ligands but also for more complex receptor-binding species. Many viruses, including the influenza and Semliki Forest viruses, use the endocytic cycle to penetrate a cell (White et al., 1983). Upon binding to surface receptors, they are internalized via endocytic vesicles. In the acidic environment of the endosome, these viruses can fuse with the plasma membrane and escape into the cytoplasm. As another example, antigen-presenting cells of the immune system, such as macrophages and B-lymphocytes, internalize native antigen for intracellular degradation and subsequent presentation with major histocompatibility complex (MHC) molecules on the cell surface for purposes of stimulating helper T-lymphocytes (Unanue and Allen, 1987). Depending on the antigen, this internalization can occur either by non-specific uptake or receptor-mediated endocytosis.

3.1.2 Experimental Methods

Quantitative studies are more difficult for receptor/ligand trafficking than for surface receptor/ligand binding because the molecular species must be followed as they move through various cellular organelles. For the most part, only the ligand has been subject to labeling, by the same types of fluorescence- or radio-label approaches described in Chapter 2. However, the trafficking routes of the receptor need not always be identical to those of the ligand. Hence, investigation of trafficking greatly benefits from labeling the receptor as well. Labeling particular receptors in cell membranes is difficult because the other proteins present will also take up label to some degree. Covalent labels directed at a given receptor type (such as photoaffinity labels) can provide satisfactory specificity, but may alter the course of trafficking if the routing of receptors is a function of the type of ligand. Some investigators have reported successful tracking of receptor movement throughout cells using general radiolabeling of surface proteins and subsequent immunoreactive procedures to isolate and quantify the desired receptor (Mellman and Plutner, 1984; Krangel, 1987). Another approach, useful for following a cohort of receptors starting with biosynthesis, are synthetic labeling methods (Stoscheck and Carpenter, 1984). A pulse of isotopically-heavy amino acids are utilized during protein synthesis, yielding receptors that can be later identified either by radioactivity or as a band of above-normal density in a centrifugation assay.

Localization of labeled receptors and/or ligands within particular cellular compartments is also required for the investigation of trafficking. The amount of surface-bound ligand can be distinguished from the amount internalized by acid-stripping (Haigler *et al.*, 1980). That is, after incubation with ligand for a certain period of time, the cells are washed in a low-pH medium. Since receptor/ligand binding affinity is reduced at low pH, bound ligand on the cell surface will dissociate while that inside the cell will remain. The amount of undegraded cell-associated ligand can be determined by lysing cells and measuring how much ligand can be recovered as a precipitate in trichloroacetic acid. The difference between this quantity and the total cell-associated ligand label is taken to be the amount of cell-associated ligand that has been degraded.

Alternatively, localization of species in whole cells can be inferred from fluorescence studies, if the label's fluorescence intensity varies with pH (Sklar *et al.*, 1984; Roederer *et al.*, 1987). Because the pH of the receptor/ligand environment changes during its movement within the cell, alterations in fluorescence can be used to follow movement of label; this requires a confident understanding of pH changes through the trafficking route, however. Although it is well established that the pH of the endosome is in the range 5–6 and that of the lysosome about 4.5–5, pH is not necessarily a monotonic function of movement. Recycling vesicles moving material from endosomes to the cell surface appear to increase in pH, and the pH of endosomes may vary with time (Roederer *et al.*, 1987).

Two other approaches for following receptor/ligand trafficking dynamics are available. Ability to separate trafficking organelles on the basis of lipid membrane density via gradient sedimentation (Beaumelle *et al.*, 1990) or density and charge via free-flow electrophoresis (Marsh *et al.*, 1987) has been demonstrated. This enables quantification of the amount of a labeled species in a variety of organelles, including small incoming vesicles, "early" and "late" endosomes, and tubular vesicles. Others (Anderson *et al.*, 1978; Orci *et al.*, 1978) have made use of visual approaches, usually either radioactive labels with autoradiography or electron-dense labels (*e.g.*, ferritin, colloidal gold) with electron microscopy, to identify subcellular locations of receptors or ligands. The latter procedure usually involves incubation with native ligand for varying times prior to fixation for microscopy, then incubation with the electron-dense label conjugated to an antireceptor or antiligand antibody. Since colloidal gold is available in a range of specific sizes, simultaneous labeling and thus localization of different molecular species is possible (Geuze *et al.*, 1983, 1987). Further details on experimental approaches to trafficking kinetics can be found in Limbird (1986).

3.2 MATHEMATICAL MODELS

We can introduce two levels of mathematical models for receptor/ligand trafficking: (1) models describing the kinetics of receptor/ligand movement through whole cells; and (2) models describing specific mechanisms involved in the endocytic cycle. Each of these two levels addresses a distinct set of issues, though they are ultimately related. *Whole-cell kinetic models* seek to quantify the rate constants for the various trafficking processes. This information is crucial for predicting receptor and ligand states and locations relevant to cell behavioral functions, which is required for construction of models for those functions. Values for these parameters also help to discover differences and similarities among cell/receptor/ligand systems. On the other hand, *mechanistic models* aim to account for the trafficking parameter values in terms of biochemical and biophysical properties of the receptors, ligands, and other components of the endocytic cycle. Combining the understanding gained from both models will allow quantitative relationships between molecular properties and cell function to be developed.

3.2.1 Whole-Cell Kinetic Models

A detailed analysis of receptor trafficking is best understood in the context of a whole-cell kinetic analysis of receptor and ligand traffic during endocytosis (see Figure 3–3). Many models of this type have been offered for various specific experimental systems. We can list here some representative examples with no attempt to be exhaustive: EGF receptor on various cell types (Wiley and Cunningham, 1981; Gex-Fabry and DeLisi, 1984; Lauffenburger *et al.*, 1987; Waters *et al.*, 1990; Starbuck and Lauffenburger, 1992), asialoglycoprotein receptor on hepatocytes (Schwartz *et al.*, 1982; Bridges *et al.*, 1982), chemotactic peptide receptor on neutrophils (Zigmond *et al.*, 1982), transferrin receptor on hepatoma cells (Ciechanover *et al.*, 1983), and interferon-α and tumor necrosis factor receptors on carcinoma cells (Myers *et al.*, 1987; Bajzer *et al.*, 1989). Additionally, some efforts (Beck and Goren, 1983, 1985; Beck, 1988; Wiley, 1985) offer analyses of general model behavior, not developed for a particular set of experimental data.

In all of these models, rate constants are assigned to each of the steps in the endocytic trafficking cycle. Although there are differences among the various models, many of the essential features are quite similar. In order to provide a context for discussion of these models, we will first present a base model containing many, though not all, of the features found in the examples cited above. We next discuss ways to apply such models for analysis of kinetic trafficking data and parameter determination. A few examples will then be examined to study some useful extensions of this base model.

3.2.1.a *Base Model for Endocytosis*

As shown in Figure 3–3, we begin by assigning the initial steps of *receptor/ligand binding and dissociation* at the cell surface the rate constants k_f [(concentration)$^{-1}$ time^{-1}] and k_r (time^{-1}), assuming simple bimolecular, non-cooperative binding (section 2.2.1). If the ligand has more than one binding site for the receptor, additional rate constants must be assigned for the crosslinking reactions (see

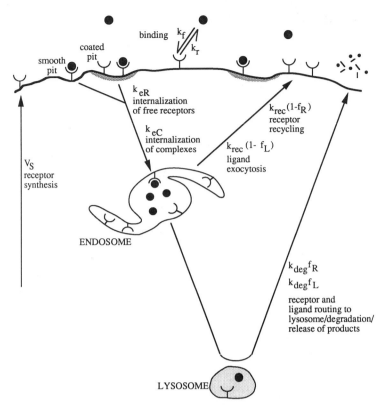

Figure 3–3 Base model for endocytosis. Rate constants are assigned to the steps in the cycle corresponding to binding and dissociation of ligand at the cell surface, internalization, recycling, and degradation of molecules. The outcome of the endosomal sorting process is reflected in the values of the sorting fractions f_R, the fraction of receptors degraded, and f_L, the fraction of ligand molecules degraded. New receptor synthesis occurs at rate V_S.

Chapter 4). Ternary complex formation between a bound receptor and a coated-pit adaptor will also not be considered explicitly in this formulation. Rate constants describing the *internalization* of receptor/ligand complexes and free receptors are k_{eC} (time^{-1}) and k_{eR} (time^{-1}), respectively; allowing these to be different reflects the possible selectivity of endocytosis for receptor states. Given the limited data on smooth-pit internalization in most cells, we will not distinguish between coated-pit and smooth-pit pathways for now. V_S [(#/cell)-time^{-1}] is the rate of new receptor synthesis and expression on the cell surface as free receptors.

Terms must also be included to characterize the *endosomal sorting* step. k_{rec} (time^{-1}) represents the intrinsic rate constant for transport of material via vesicles from the endosome back to the cell surface and is assumed independent of the sorting process (Ward *et al.*, 1989). The contents of these vesicles are denoted by $(1 - f_R)$ and $(1 - f_L)$, the fraction of endocytosed receptors and ligands, respectively, routed along the recycle pathway. k_{deg} (time^{-1}) represents a lumped rate constant for the routing of material from the endosome to the lysosome, degradation in the lysosome, and the release of fragments in the extracellular medium. However, this simplification is likely to be ultimately found inadequate because enzymatic degradation rate constants are probably different for the biochemically-distinct receptor and ligand molecules. The fractions of endocytosed receptors and ligands following this degradation pathway are f_R and f_L respectively. The sorting mechanism is assumed to determine the fractions f_R and f_L, which will in general be dependent on ligand concentration and receptor/ligand binding properties. Mechanistic models for sorting which predict the values of f_R and f_L will be considered later in this chapter. Such models will include intracellular receptor/ligand binding and dissociation reactions. These details are not incorporated into this present base model, but will appear in the extensions to be discussed shortly.

Notice that most of the rate processes have been assumed to be simple first-order processes. This relationship has been demonstrated for internalization, recycling, and degradation in a few systems (Carpenter and Cohen, 1976; Bauman and Doyle, 1978; Zigmond et al., 1982; Tomoda et al., 1989), though use of a time-delay for movement between the endosomal compartment and the lysosomal compartment before degradation has been proposed (Bridges et al., 1982; Wiley and Cunningham, 1981).

Kinetic balance equations can now be written for the relevant species: R_s, the number of free receptors on the cell surface; C_s, the number of receptor/ligand complexes on the cell surface; R_{Ti}, the total number of free plus bound receptors in endosomes; $L_i^\#$, the amount of intracellular ligand ($\#$/cell); and L, the ligand concentration in the medium (M). Because we distinguish only receptors and ligands but not between free or bound receptors and ligands in the quantities f_R and f_L describing the outcome of the endosomal sorting process, we do not follow intracellular free and bound receptors separately. In our discussion of sorting later in this chapter, we will examine this step of the cycle more thoroughly.

With this assumption, the kinetic equations describing the whole-cell endocytic trafficking cycle are:

$$\frac{dR_s}{dt} = -k_f L R_s + k_r C_s - k_{eR} R_s + k_{rec}(1 - f_R) R_{Ti} + V_s \qquad (3\text{–}1a)$$

$$\frac{dC_s}{dt} = k_f L R_s - k_r C_s - k_{eC} C_s \qquad (3\text{–}1b)$$

$$\frac{dR_{Ti}}{dt} = k_{eR} R_s + k_{eC} C_s - [k_{rec}(1 - f_R) + k_{deg} f_R] R_{Ti} \qquad (3\text{–}1c)$$

$$\frac{dL_i^\#}{dt} = k_{eC} C_s - [k_{rec}(1 - f_L) + k_{deg} f_L] L_i^\# + k_{fp} N_{Av} L \qquad (3\text{–}1d)$$

Non-specific, or *"fluid-phase"*, *uptake* of ligand is described by the term $k_{fp} N_{Av} L$ in Eqn. (3–1d), where k_{fp} is the rate constant for fluid internalization (μm^3/min-cell) and N_{Av} is Avogadro's number. Fluid-phase uptake typically contributes significantly to ligand uptake only at high ligand concentrations and is therefore generally neglected in analyses of RME. Examination of Eqn. (3–1d) shows that this assumption requires that the quantity ($k_{fp} N_{Av} L / k_{eC} R_T$), which is the ratio of fluid-phase ligand uptake to maximal receptor-mediated uptake, be insignificant compared to 1. Typically, for ligand concentrations less than 10^{-8} M neglect of non-specific uptake is quantitatively justifiable.

When fluid-phase uptake is neglected, Eqn. (3–1d) for internalized ligand, $L_i^{\#}$, is similar to what the equation for intracellular complexes would be if written independently:

$$\frac{dC_i}{dt} = k_{eC}C_s - [k_{rec}(1 - f_R) + k_{deg}f_R]C_i \qquad (3\text{–}1e)$$

where C_i is the number of intracellular receptor/ligand complexes per cell. Aside from replacement of C_i for $L_i^{\#}$, the only difference between Eqns. (3–1d) and (3–1e) is the possible distinction between endosomal sorting fractions for receptor and ligand degradation, f_R and f_L, respectively. When receptor and ligand are considered to sort similarly, so that $f_R = f_L$, $C_i(t)$ will be identical to $L_i^{\#}(t)$ and intracellular complexes can be determined by measuring intracellular ligand. This is a procedure followed almost universally by experimental investigators due to the difficulty in measuring intracellular receptors independently. For systems in which f_R is not equal to f_L, however, this approach cannot be used.

We will assume for simplicity during our foundational analysis in this section that the extracellular ligand concentration L is constant. This assumption can be easily relaxed by writing a kinetic equation for the change in free ligand concentration for cells present at concentration n (#/volume):

$$\frac{dL}{dt} = (-k_f L R_s + k_r C_s - k_{fP}L)n \qquad (3\text{–}1f)$$

An additional quantity that is frequently measured is the total cell-associated ligand $L_T^{\#}$ (surface bound plus internalized but undegraded):

$$L_T^{\#} = C_s + L_i^{\#} \qquad (3\text{–}1g)$$

Some general behavior of the endocytic trafficking process can be gleaned from Eqns. (3–1a)–(3–1c) before moving to a specific application. Consider the system at *steady-state in the absence of ligand*, i.e., $L = 0$. In this case, $C_s = L_i^{\#} = 0$ and all time derivatives are equal to zero. We then solve the algebraic Eqns. (3–1a) and (3–1c) to obtain R_{s0}, the number of free surface receptors in the absence of ligand. C_{s0}, the number of surface complexes when $L = 0$, is obviously equal to zero. Thus the *total number of surface receptors*, free and bound, in the absence of ligand, R_{sT0}, is simply equal to R_{s0}, given by the expression:

$$R_{s0} = R_{sT0} = \left[1 + \frac{k_{rec}(1 - f_{R0})}{k_{deg}f_{R0}}\right]\left(\frac{V_{s0}}{k_{eR}}\right) \qquad (3\text{–}2)$$

Anticipating the possibility that the value of f_R depends on the presence of ligand, i.e. that the endosomal sorting of free and bound receptors may differ, we use the notation f_{R0} for the value of f_R when $L = 0$. Similarly, we use the notation V_{s0} for the value of V_s when $L = 0$ in case receptor synthesis also depends on ligand.

Next, consider a steady-state in the presence of ligand at concentration L. The total number of surface receptors, R_{sT}, equal to the sum of the number of free receptors, R_s, and receptor/ligand complexes, C_s, is now

$$R_{sT} = \left[1 + \frac{K_D + (k_{eC}/k_f)}{L}\right]\left\{\frac{\left[1 + \frac{k_{rec}(1 - f_{RL})}{k_{deg}f_{RL}}\right]V_{sL}}{k_{eC} + k_{eR}\left[\frac{K_D + (k_{eC}/k_f)}{L}\right]}\right\} \tag{3-3}$$

where f_{RL} and V_{sL} denote the values of f_R and V_s in the presence of ligand. The ratio of total surface receptor number in the presence of ligand to that in the absence of ligand is given by:

$$\frac{R_{sT}}{R_{s0}} = \frac{(1 + \kappa)}{(\delta + \kappa)}\left[\frac{1 + \{[k_{rec}(1 - \sigma f_{R0})]/k_{deg}\sigma f_{R0}\}}{1 + \{[k_{rec}(1 - f_{R0})]/k_{deg}f_{R0}\}}\right]v \tag{3-4}$$

In this expression there are four dimensionless parameter groups:

1. δ, the ratio of the complex internalization rate constant to the free receptor internalization rate constant

$$\delta = \frac{k_{eC}}{k_{eR}} \tag{3-5a}$$

2. σ, the ratio of the fraction of internalized receptors degraded in the presence of ligand to that degraded in the absence of ligand

$$\sigma = \frac{f_{RL}}{f_{R0}} \tag{3-5b}$$

3. v, the ratio of the receptor synthesis rate in the presence of ligand to the rate in the absence of ligand

$$v = \frac{V_{sL}}{V_{s0}} \tag{3-5c}$$

4. κ, a dimensionless ligand binding constant

$$\kappa = \frac{K_D + (k_{eC}/k_f)}{L} \tag{3-5d}$$

When $\delta = \sigma = v = 1$, the key parameters are identical whether ligand is absent or present. In this case $R_{sT}/R_{s0} = 1$ and there is no change in the number of surface receptors when ligand is added to the system.

If $\delta > 1$, $\sigma > 1$, and/or $v < 1$, then $(R_{sT}/R_{s0}) < 1$ so the ratio of surface receptor number in the presence of ligand to the surface receptor number in the absence of ligand is less than 1. This loss of cell surface receptors is known as *receptor downregulation*. Its extent depends on the four quantities δ, σ, κ, and v. Figure 3–4a,b illustrates the effects of δ, σ, and κ on (R_{sT}/R_{s0}).

Three major types of receptor downregulation are predicted by this model. When $\sigma = v = 1$, the only effect of ligand is to alter the receptor internalization

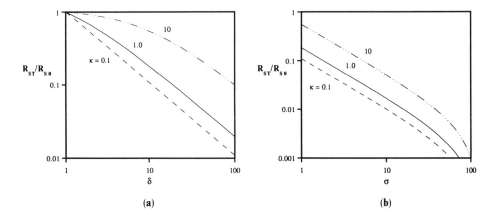

Figure 3–4 Receptor downregulation predicted by the base model for endocytosis, Eqns. (3–1a)–(3–1e). Plotted is the ratio of surface receptor number in the presence of a constant ligand concentration, L, at steady-state, to that in the absence of ligand, R_{sT}/R_{s0}, according to Eqn. (3–4). (a) Downregulation as a function of the dimensionless group δ, which characterizes the ratio of the internalization rate constant for bound versus free receptors (Eqn. (3–5a)). $v = 1$, $\sigma = 1$, $f_{R0} = 1$. (b) Downregulation as a function of the dimensionless group σ, which characterizes the ratio of receptor sorting to degradation for bound versus free receptors (Eqn. (3–5b)). $v = 1$, $\delta = 10$, $f_{R0} = 0.01$. In both plots, $k_{rec} = 5.8 \times 10^{-2}\,\mathrm{min}^{-1}$ and $k_{deg} = 1.0 \times 10^{-2}\,\mathrm{min}^{-1}$. κ is a modified dimensionless ligand binding constant, equal to $(K_D + (k_{eC}/k_f))/L$.

rate constant. Eqn. (3–4) simplifies to

$$\frac{R_{sT}}{R_{s0}} = \frac{1 + \kappa}{\delta + \kappa} \tag{3-6}$$

This important result was first obtained by Wiley (1985). The quantity R_{sT}/R_{s0} is less than 1 when $\delta > 1$, i.e., when $k_{eC} > k_{eR}$. The resulting downregulation is termed *endocytic downregulation*, implying that the root of the observed behavior lies in the internalization (*i.e.*, endocytosis) step. Eqn. (3–6) implies that receptors which show prelocalization in coated pits without ligand (class 2-type), *i.e.* $k_{eC} = k_{eR}$, will not exhibit significant downregulation of this origin. Class 1-type receptors, such as the EGF receptor, should in fact be substantially downregulated due to an enhanced internalization rate constant in the presence of ligand.

For the condition $\delta = v = 1$, the only effect of ligand is that endosomal sorting is affected. Here Eqn. (3–4) simplifies to:

$$\frac{R_{sT}}{R_{s0}} = \frac{1 + \{[k_{rec}(1 - \sigma f_{R0})]/k_{deg}\sigma f_{R0}\}}{1 + \{[k_{rec}(1 - f_{R0})]/k_{deg}f_{R0}\}} \tag{3-7a}$$

Under the simplifying assumption that the vesicular transport rate constants are roughly equal, $k_{rec} \sim k_{deg}$, this reduces to

$$\frac{R_{sT}}{R_{s0}} = \frac{1}{\sigma} = \frac{f_{R0}}{f_{RL}} \tag{3-7b}$$

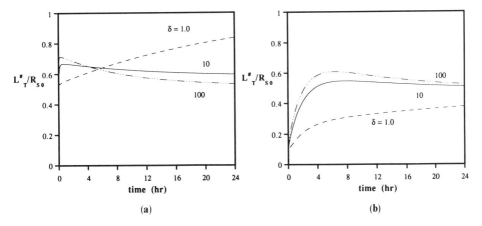

Figure 3–5 Total cell-associated ligand, $L_T^{\#}$ (Eqn. (3–1g)), as a function of time predicted by the base model for endocytosis, Eqns. (3–1a)–(3–1e) in the presence of constant ligand concentration, L. Shown is the effect of the dimensionless group δ, which characterizes the ratio of the internalization rate constant for bound versus free receptors (Eqn. (3–5a)). (a) Results when ligand causes no change in sorting, $\sigma = 1$. Also, $k_{\text{deg}} = 5 \times 10^{-2}\,\text{min}^{-1}$. (b) Results when ligand causes a significant increase in sorting to degradation, $\sigma = 5$. Also, $k_{\text{deg}} = 1 \times 10^{-2}\,\text{min}^{-1}$. For both plots, $k_f = 7.2 \times 10^{7}\,\text{M}^{-1}\,\text{min}^{-1}$, $k_r = 3.4 \times 10^{-1}\,\text{min}^{-1}$, $k_{\text{eR}} = 3.0 \times 10^{-2}\,\text{min}$, $k_{\text{rec}} = 5.8 \times 10^{-2}\,\text{min}^{-1}$, $V_s = 130\,\#/\text{min}$.

Thus, if the fraction of internalized receptors that are degraded in the absence of ligand is less than the fraction degraded in the presence of ligand, what we will term *sorting downregulation* can occur. This may occur in the neutrophil chemotactic peptide receptor system (Zigmond *et al.*, 1982).

A third possible basis for receptor downregulation is the case of $\delta = \sigma = 1$, so that influence of ligand is on receptor synthesis. In this case,

$$\frac{R_{\text{sT}}}{R_{\text{s0}}} = v \tag{3–8}$$

When $v < 1$, *i.e.* there is ligand-induced inhibition of receptor synthesis, *synthetic downregulation* is the predicted result. The LDL receptor appears to exhibit this type of behavior (Goldstein *et al.*, 1985).

The above discussion centered on the steady-state behavior of our base model for endocytosis. For a constant extracellular ligand concentration L, Eqns. (3–1a)–(3–1e) are linear and may be solved analytically using matrix methods or numerically to obtain the transient behavior of the model.

Example transient plots of total cell-associated ligand, $L_T^{\#}(t)$, are provided in Figure 3–5a,b. This figure shows the effect of the dimensionless ligand-enhanced endocytosis rate constant δ, for small and large values of the dimensionless ligand-enhanced degraded fraction σ. Total cell-associated ligand is a quantity that is relatively easy to measure, so it is useful to have some insight concerning the effects on this due to underlying trafficking properties.

3.2.1.b Use of Whole-Cell Kinetic Models in Data Analysis
Besides elucidating general trafficking effects, this whole-cell kinetic model of the endocytic cycle allows data on receptor/ligand dynamics to be analyzed

for specific experimental systems. The rate constants corresponding to each step can be determined, and quantitative information on the various trafficking processes thereby obtained. This information can be insightful for its own sake as well as useful for incorporating into larger models for cell behavioral functions.

Two rather distinct approaches to *parameter determination* exist. One is to estimate a complete set of model parameter values simultaneously from an entire compilation of experimental data, typically utilizing a numerical parameter fitting routine. The other is to design specific experimental procedures dedicated to isolating particular features of the overall system, so that smaller subsets of parameter values are deduced independently from separate data. An example of each procedure is given here but we must express a prejudice for the latter approach. Our primary reason relates to validation of the model itself, which is frequently of uncertain foundation. Analysis of an experiment that focuses on an individual step permits one to decide how representative the mathematical term in the model for that step actually is. This sort of validation is comparatively difficult for a procedure that examines the whole model at once; the overall fitness of the model can be assessed but the contributions of specific terms can be elusive. A secondary reason is that parameters can often be essentially "measured" from suitable data transformations, replots, or simple mathematical formulae, rather than "fit". Once all the individual terms are validated and their parameter values determined, then *a priori* predictions can be made from computations with the overall model for additional data.

On the other hand, arguments against reliance on the isolation approach can be made. A substantial one is that the methods used to gain such separation of individual steps may often perturb the processes under investigation. This can often be the case when temperature alterations or pharmacological agents are used. Restriction to experimental time scales appropriate for a subset of events is safer, but not universally applicable.

An excellent example of the *overall parameter fitting procedure* is the work by Waters *et al.* (1990) for EGF binding and trafficking in fibroblasts. They used two types of experimental protocols. In protocol 1, cells were preincubated in radiolabeled ligand at $4\,°C$, so that internalization was negligible, for sufficient time that surface receptor binding reached equilibrium. At time $t = 0$, the cells were placed in medium lacking ligand and warmed to $37\,°C$ to allow membrane trafficking to begin. For a sequence of time points over a 15-minute period, measurements were made of free ligand in the medium, surface-bound ligand, and intracellular ligand (the latter two distinguished by acid wash). The same three quantities were measured in protocol 2 but by an alternative procedure. Here, the cells were placed in $37\,°C$ medium containing labeled ligand at time $t = 0$ and measurements were made at time points up to 60 min. Hence, the only difference between these two protocols is the initial conditions appropriate to the model equations under investigation. At $t = 0$ all labeled EGF is surface-bound in protocol 1, whereas in protocol 2 it is all in the medium.

Examples of the experimental data are given in Figure 3–6a,b. For both protocols, transient values of free ligand in the medium $L(t)$, surface-bound ligand $C_s(t)$, and internalized ligand $L_i^{\#}(t)$ are measured and normalized to the total

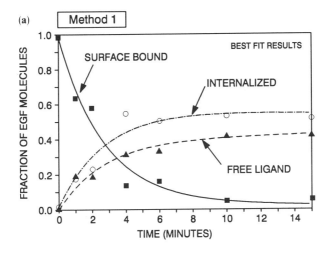

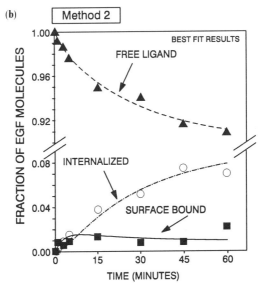

Figure 3–6 Experimental data of Waters *et al.* (1990) for the internalization of radiolabeled EGF by fibroblasts, along with computer fit of their mathematical model. Shown are the fractions of EGF molecules which are surface bound (C_s), internalized ($L_i^\#$), and free in the medium (L). The initial condition for (a) is that all ligand is bound to surface receptors; the initial condition for (b) is that all ligand is free in solution. Reprinted with permission from Waters *et al.* (1990). Copyright 1990 American Chemical Society.

amount of ligand (the sum of these three quantities) after correction for non-specific binding.

For data analysis a set of mathematical model equations quite similar to Eqns. (3–1) are applied. Because the free ligand concentration, L, varied in the experimental protocols used, Eqn. (3–1f) is needed. Receptor and ligand recycling is neglected because in the cell type used (rat epithelial cells) previous work indicated little evidence of EGF receptor recycling. Fluid-phase ligand uptake is ignored due to the low ligand concentrations present. Finally, free intracellular receptors are ignored, because both internalization of free receptors and dissociation of intracellular complexes are neglected. Thus, the model equations applied by Waters *et al.* are equivalent to Eqns. (3–1a,b,e and f) for $R_s(t)$, $C_s(t)$, $C_i(t)$, and $L(t)$, but with $f_R = f_L = 1$. It should be noted that the measurements for internalized ligand $L_i^{\#}(t)$ are used to represent intracellular complexes $C_i(t)$ because of the neglect of fluid-phase ligand uptake and the assumption of identical endosomal sorting for receptors and ligands (all going the degradation route). Calculations of $L(t)$ were also corrected for the radioactivity that appears at rate $k_{deg}C_i(t)$ in the medium due to degraded ligand.

The ordinary differential equations (Eqns. (3–1a, b, e, f) were integrated numerically using a commercial software package. Initial conditions required the amount of labeled ligand in the three forms: free, surface-bound, and intracellular, along with the total number of cell receptors. The Simplex method was used to determine the "best fit" set of values for the six model parameters (k_f, k_r, k_{eR}, k_{eC}, k_{deg}, and V_s) simultaneously by minimizing the error between computer-generated transient curves and the data points for $L(t)$, $C_s(t)$, and $C_i(t)$. This was accomplished for data from both protocols, and a sensitivity analysis performed to decide which parameter values could be determined with confidence from each.

Figures 3–6a,b show the excellent fit between model and data for a particular ligand concentration. It must be emphasized that this does not represent a successful model prediction *per se*, but instead a successful model fit. If the model, with parameter values now determined, could be used to generate computations applied to further independent sets and/or types of experimental observations, that would make it truly predictive.

Waters *et al.* conclude that protocol 1 is better for determining the dissociation rate constant k_r whereas protocol 2 is better for determining the association rate constant k_f. This is not surprising, since protocol 1 is essentially a dissociation experiment ($L(0) = 0$) whereas protocol 2 is essentially an association experiment ($C_s(0) = 0$). Both protocols work equally well for the endocytic rate constant k_{eC}, because the time scale of both experiments allows for significant receptor internalization. Neither protocol was found to yield confident estimates for the other three parameters: k_{eR}, k_{deg}, V_s. This is also expected, because the actual quantities measured have little to do with these aspects of receptor trafficking on the time scale observed.

As an example of how this sort of approach can be used to gain significant biological insights, this group of investigators has successfully applied its whole-cell model to examination of a central issue concerning regulation of EGF-receptor (EGF-R) function: the role of receptor autophosphorylation sites in EGF-R internalization (Sorkin *et al.*, 1991b, 1992). Values of k_{eC} were determined for a series of site-directed mutations in EGF-R cytoplasmic tail tyrosine residues. Results indicated that phosphorylation of these sites is necessary for efficient coated-pit internalization of this receptor, suggesting a potentially-important regulatory mechanism.

A philosophically different approach, though one with the same ultimate aim of elucidating receptor regulatory mechanisms, is that of employing *focused experiments* (also mostly for EGF-R), designed to isolate particular aspects of trafficking individually (Wiley and Cunningham, 1981, 1982; Lund *et al.*, 1990; Wiley *et al.*, 1991). This is accomplished with the use of pharmacologic agents and by considerations of experimental time scale. We will outline key procedures for each of the central classes of parameters in the base model: k_f, k_r, k_{eR}, k_{eC}, k_{rec}, k_{deg}, and V_s, though variations of these can certainly be applied.

First, the *binding rate constants* k_f and k_r can be determined from measurements of cell surface ligand binding as detailed in section 2.2.1.d. This requires using lower temperatures or pharmacological inhibitors of internalization such as phenylarsineoxide (Low *et al.*, 1981). Here, the latter is probably to be preferred when significant effects of temperature on binding rate constants are anticipated.

Second, the *internalization rate constants* k_{eC} and k_{eR} can be determined from measurements of cell surface-bound and intracellular label over sufficiently short time periods that recycling and degradation have relatively little effect on these data. k_{eC} is found when the ligand is labeled, while k_{eR} is obtained when the receptor is labeled, both by the same analysis procedure. Consider Eqn. (3–1e) for C_i. When recycling and degradation are neglected, this simplifies to:

$$\frac{dC_i}{dt} = k_{eC} C_s \tag{3–9}$$

This ordinary differential equation can be integrated to obtain a transient solution for intracellular complexes as a function of surface complexes:

$$C_i(t) = C_i(0) + \int_0^t k_{eC} C_s(t') \, dt' \tag{3–10a}$$

where t' is merely a dummy time variable for carrying out the integration. When there are no intracellular complexes at $t = 0$, the start of the experiment (as in most experimental protocols), and the internalization rate constant does not vary with time, this reduces to the expression:

$$C_i(t) = k_{eC} \int_0^t C_s(t') \, dt' \tag{3–10b}$$

Thus, measurements for $C_i(t)$ and $C_s(t)$ can be graphed in the form $C_i(t)$ ("Internalized") versus the integral of $C_s(t)$ ("Surface") – the "In/Sur" plot originally developed by Wiley and Cunningham (1982). A linear slope with value k_{eC} should be obtained if the analysis assumptions hold. An example of this result is shown in Figure 3–7. For EGF and fibroblasts, it is typically found that linear "In/Sur" plots are obtained for time periods below about 10 min; recycling generally begins to show up at that point and degradation about 20 min later. These effects are observable on In/Sur plots for measurements taken over longer time periods, manifested by a temporally decreasing slope. Wiley *et al.* (1991) have, in fact, derived a formula correcting k_e for any such effects of degradation and recycling. This correction can be used in conjunction with estimates of the recycling and degradation rate constants obtained from independent experiments.

Using the In/Sur plot procedure, k_{eC} is determined from measurements of surface-bound and internalized ligand, labeled with either a radioactive or

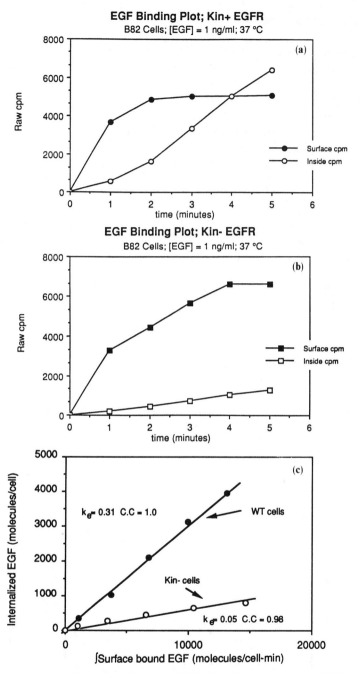

Figure 3–7 Experimental data from Starbuck (1991) for internalization of EGF by B82 cells transfected with EGF receptors. Data are shown plotted in two forms. (a,b) The amount of surface bound ("Sur", or $C_s(t)$) and intracellular ligand ("In", or $C_i(t)$) are shown as a function of time: (a) results for cells transfected with wild-type (WT) EGF receptor; (b) results for cells transfected with a mutant EGF receptor lacking tyrosine kinase activity (Kin–) due to an amino acid substitution at residue 721. Cells were incubated with 1 ng/ml ^{125}I-labeled EGF at 37 °C. (c) Data replotted as $C_i(t)$ versus the integral of $C_s(t)$. The slope of the line gives the value of the complex internalization rate constant k_{eC} (see Eqn. 3–10b). In this experiment, $k_{eC} = 0.31$ min^{-1} for the WT receptor, and $k_{eC} = 0.05$ min^{-1} for the Kin– mutant receptor for the ligand concentration used here.

fluorescent tag. k_{eR} can be determined in similar fashion using measurements of surface and internalized receptor, most commonly by means of a labeled anti-receptor antibody (*e.g.*, Wiley *et al.*, 1991).

Recycling rate constants for intracellular receptors can be determined by pulse/chase experiments in which $C_i(t)$ is measured using labeled ligand, or $R_i(t)$ is measured using labeled antireceptor antibody. Recall that our base model lumps all intracellular receptors. It is possible that the measured parameters might vary with ligand concentration, however; sorting may be affected by the proportions of receptors free and bound. Indeed, when ligand and receptor have dissimilar sorting fractions (*i.e.*, f_R not equal to f_L), data on intracellular labeled ligand will not accurately reflect receptor recycling behavior. For EGF and EGF-R, the ligand and receptor are believed to remain associated within the endosomal sorting compartment (Sorkin *et al.*, 1991a) so that f_R and f_L are equal (as long as fluid-phase uptake is negligible).

Cells are preincubated with labeled moiety to get a substantial intracellular pool. After a mild surface stripping procedure to remove surface-bound ligand, they are placed at time $t = 0$ in chase medium lacking labeled moiety but containing saturating concentrations of unlabeled moiety. The absence of label during the chase ensures that no additional intracellular receptors will be generated, and the presence of saturating concentrations of unlabeled moiety prevents rebinding of any label which dissociates after recycling. The resulting decrease in $C_i(t)$ or $R_i(t)$ monitored is due to recycling if the experiments are carried out for a short enough time period that degradation is negligible. Under these conditions, Eqn. (3–1e) for intracellular complexes simplifies to:

$$\frac{dC_i}{dt} = -k_{rec}(1 - f_R)C_i \tag{3–11}$$

The solution to this can again be found by integration:

$$\frac{C_i(t)}{C_i(0)} = \exp\{-k_{rec}(1 - f_R)t\} \tag{3–12a}$$

or, rearranging:

$$\ln\frac{C_i(t)}{C_i(0)} = -k_{rec}(1 - f_R)t \tag{3–12b}$$

Hence, a plot of the logarithm of the fraction of intracellular label remaining versus time t after the unlabeled chase period begins should yield a linear plot with slope equal to $-k_{rec}(1 - f_R)$. Unfortunately, this product cannot be resolved, so only the effective recycling rate constant $k'_{recR} = k_{rec}(1 - f_R)$ can be obtained. Example results are plotted in Figure 3–8. A similar procedure can be used for $R_i(t)$ data.

To obtain the receptor *recycling fraction* $(1 - f_R)$, one approach might be to measure k'_{recR} for a different receptor in the same cell type, one for which it is a good approximation that $f_R \ll 1$ and thus $k'_{recR} = k_{rec}$. One such candidate is, in fact, the transferrin receptor (Tf-R) (Hanover and Dickson, 1985). Assuming that k_{rec} characterizes the trafficking organelle movement process, and is independent of the organelle's contents, the unknown f_R for the original receptor can be

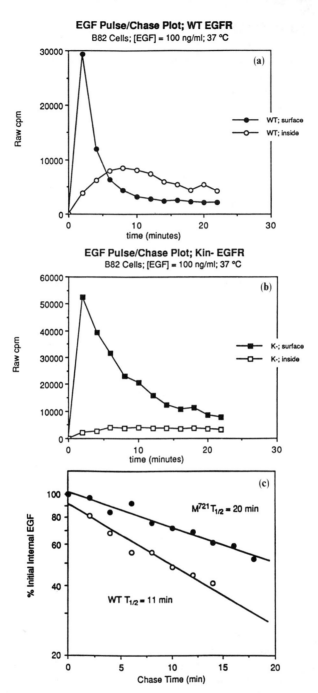

Figure 3–8 Experimental data from Starbuck (1991) for recycling of EGF by B82 cells transfected with EGF receptors. Cells were pulsed for 2 min with 100 ng/ml ^{125}I-labelled EGF at 37 °C, then chased with 100 ng/ml unlabeled EGF. (a,b) The amounts of surface bound and intracellular labeled ligand are shown as a function of chase time: (a) results for cells transfected with WT EGF receptors; (b) results for cells transfected with K– EGF receptors, as in Figure 3–7. (c) The logarithm of the fraction of intracellular labeled ligand remaining as a function of chase time is shown, for the two receptors. The slope of the plot is equal to $-k_{rec}(1 - f_R)$ (see Eqn. (3–12b), equal to 0.069 min^{-1} for WT receptor and 0.048 min^{-1} for the K– receptor.

calculated from the measured values of k'_{rec} and k_{rec}. As an example, Wiley *et al.* (1991) found that $k'_{recR} = 0.047$ min^{-1} for EGF-R and 0.088 min^{-1} for Tf-R, both in B82 fibroblasts. Assuming $f_R \sim 0$ for Tf-R, this yields $f_R = 0.47$ for EGF-R in this cell type. Because it is possible that f_R varies with endosomal ligand concentration or receptor number, this procedure should be repeated for different preincubation concentrations of labeled ligand (see, for example, Herbst *et al.*, 1993). As with internalization, recycling parameter values can be measured for receptor complexes using labeled ligand or for free receptors using labeled antireceptor antibodies.

The simple result in Eqn. (3–12a) can be corrected for the effects of receptor complex re-internalization, and an approximate formula for this has also been derived by Wiley *et al.* (1991). Such effects will arise if the receptor/ligand complex dissociation rate constant k_r is small compared to the internalization rate constant k_{eC}, so that significant numbers of recycling complexes can re-enter the cell instead of dissociating.

To obtain the overall ligand recycling rate constant $k'_{recL} = k_{rec}(1 - f_L)$, analogous analysis could be applied to pulse/chase experiments using labeled ligand. We are not aware of any experimental effort to date aimed at determining the ligand recycling fraction $(1 - f_L)$, though an approach has been offered in principle (Linderman and Lauffenburger, 1989). If the transport rate constant k_{rec} is indeed identical for ligand and receptor recycling, then

$$k'_{recL}/k'_{recR} = (1 - f_L)/(1 - f_R).$$

Thus, f_L can be estimated from this ratio if f_R has already been determined as mentioned above. This will fail, however, if diacytotic pathways exist (Besterman *et al.*, 1981; Daukas *et al.*, 1983; McKinley and Wiley, 1988) which allow for rapid ligand exocytosis without commensurate receptor recycling. Under such circumstances, k_{rec} will not be the same for ligand and receptor.

The overall *ligand degradation rate constant* $k_{deg}f_L$ is easier to measure than that for receptors. Ligand degradation products can be distinguished from intact ligand by protein precipitation assays (*e.g.*, trichloroacetic acid, or TCA) or by electrophoretic gel separation (Wiley *et al.*, 1985). Measurement of the amount of degraded product in the medium, $L_d(t)$, and the amount of intracellular ligand, $L_i^\#(t)$, is necessary. A kinetic equation for the amount of degraded product can be most simply written as:

$$\frac{dL_d}{dt} = k_{deg}f_L L_i^\# \qquad (3\text{–}13)$$

In a manner similar to the treatment of Eqn. (3–9) for internalization, this ordinary differential equation can be integrated to obtain:

$$L_d(t) = k_{deg}f_L \int_0^t L_i^\#(t')\, dt' \qquad (3\text{–}14)$$

where we have assumed that no degraded ligand is present at time $t = 0$. Hence, a plot of degraded ligand product versus the integral of intracellular ligand would be predicted to yield a straight line with slope equal to $k_{deg}f_L$. Often, a lag time (of about 30–45 min) before the onset of significant product is observed, because traverse from endosome through lysosome to extracellular product is not a simple first-order process. Therefore, $k_{deg}f_L$ may be better determined from the quasi-linear portion of the curve after an estimated lag time. Ligand degradation data can, in fact, be accounted for more accurately by inclusion of a time-lag in the model between endosome and product (Wiley and Cunningham, 1981).

The overall *receptor degradation rate constant* $k_{deg}f_R$ and *receptor synthesis rate* V_s are typically determined from labeled receptor synthesis experiments, usually using $[^{35}S]$methionine uptake in a pulse/chase protocol (Stoscheck and Carpenter, 1984). Alternatively, they can be found by measuring the total numbers of surface and intracellular receptors at steady-state in the absence of ligand, using a labeled antireceptor antibody (Turner et al., 1988). Either fluorescence labeling or radioactive labeling can be used for this approach. At steady-state conditions with $L = 0$, Eqn. (3–1c) predicts that the ratio of surface receptor number, R_{sT0}, to intracellular receptor number, R_{iT0}, should be:

$$\frac{R_{sT0}}{R_{iT0}} = \frac{k_{rec}(1 - f_R) + k_{deg}f_R}{k_{eR}} \qquad (3–15)$$

If the effective receptor recycling rate constant, $k_{rec}(1 - f_R)$, and the free receptor internalization rate constant, k_{eR}, are known, then the overall receptor degradation rate constant, $k_{deg}f_R$, can be calculated from an experimentally-determined value of this ratio. Further, the receptor synthesis rate constant can be determined from the number of intracellular receptors, combining the solutions to Eqns. (3–1a)–(3–1c) at steady-state with $L = 0$:

$$R_{iT0} = \frac{V_s}{k_{deg}f_R} \qquad (3–16)$$

This permits the receptor synthesis rate, V_s, to be calculated from measurement of R_{Ti} because $k_{deg}f_R$ has just been determined. The value of V_s found in this way is that in the absence of ligand, or V_{s0}. Receptor synthesis may be altered by the presence of ligand, in which case the value V_{sL} at a given ligand concentration, L, can be determined by $[^{35}S]$methionine uptake in a pulse/chase receptor synthesis experiment (Earp et al., 1986; Weissman et al., 1986).

As an example, Table 3–1 gives a list of these parameter values for the EGF receptor transfected into the B82 fibroblast cell line, determined using these approaches (Starbuck, 1991; Wiley et al., 1991; Starbuck and Lauffenburger, 1992; Herbst et al., 1993). An interesting and important feature of this EGF parameter determination work is that $k_{eC} > k_{eR}$, indicating that receptor downregulation in the fibroblast EGF system is due to internalization parameter change. Thus it falls into the endocytic downregulation category of section 3.2.1.a. The next section will show that real systems exhibiting receptor downregulation may also fall into other categories.

Notice that for most of these procedures in the focused experiment approach to parameter determination, model terms for particular steps in trafficking can be tested individually by the dedicated experimental data. With all parameter values specified, truly *a priori* predictions can be made of further kinetic data on ligand

Table 3–1 Receptor trafficking parameters for
EGF/EGF receptors on B82 fibroblasts

Parameter	Value
k_f	$7.2 \times 10^7 \text{ M}^{-1} \text{ min}^{-1}$
k_r	$3.4 \times 10^{-1} \text{ min}^{-1}$
k_{eR}	$3.0 \times 10^{-2} \text{ min}^{-1}$
k_{eC}	$3.0 \times 10^{-2} - 3.0 \times 10^{-1} \text{ min}^{-1}$
k_{rec}	$5.8 \times 10^{-2} \text{ min}^{-1}$
k_{deg}	$2.2 \times 10^{-3} \text{ min}^{-1}$
V_s	$1.3 \times 10^2 \text{ #/min}$
f_R	0.2–0.8
f_L	0.2–0.8

From Starbuck (1991), Wiley *et al.* (1991), Starbuck and
Lauffenburger (1992), Herbst *et al.* (1993). k_{eC}, f_R, and f_L
vary with EGF concentration; all three quantities decrease
as EGF concentration increases.

uptake. Success in this offers confidence for application of a trafficking model
when needed for studies of cell behavioral functions, as will be seen in Chapter 6.

3.2.1.c *Inclusion of Fluid-Phase Ligand Uptake and Endosomal Sorting Effects*

Zigmond *et al.* (1982) analyzed the trafficking behavior of the chemotactic peptide
receptor on neutrophil leukocytes by combining a variety of experimental measurements with mathematical modeling. Discovering the underlying basis for receptor
downregulation in this system, *i.e.*, whether it was due primarily to endocytic,
sorting, or synthetic effects of ligand, as outlined in section 3.2.1.a, was the goal
of this work.

Ligand uptake and release were examined by measuring the amount of
cell-associated ligand as a function of incubation time, and the fraction of
undegraded ligand released (recycled) as a function of incubation time as well as
chase time without ligand. These measurements are shown in Figure 3–9a, an
important feature of which is the decrease in fraction of ligand released with
increasing incubation time. The rate of recovery of surface receptors upon
removing cells from medium containing ligand at various concentrations was also
measured (Figure 3–9b). This plot demonstrates that the receptor recycling rate
constant apparently decreases for higher ligand incubation concentrations.

Based on these observations, the Zigmond's group formulated the model
illustrated in Figure 3–10. R_s and C_s are the numbers of surface free receptors and
receptor/ligand complexes, and R_i and C_i are the numbers of internalized free
receptors and complexes (all in #/cell). L_i is the concentration of free ligand
(moles/volume) within the endosomal compartment, which possesses volume v.

The central difference between this model and the base model is the explicit
inclusion of the possibility that endosomal sorting may be different between free
and bound intracellular receptors, based on the experimental observation that
receptor recycling appears to vary with ligand concentration. Hence, two distinct
sorting fractions are defined: f_R^{free} and f_R^{bound}, for free and bound intracellular
receptors, respectively. In particular, the hypothesis that $f_R^{\text{free}} = 0$ and $f_R^{\text{bound}} = 1$

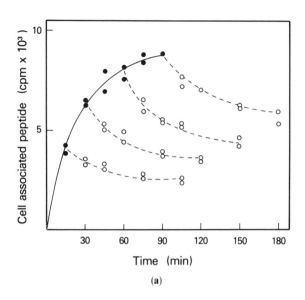

(a)

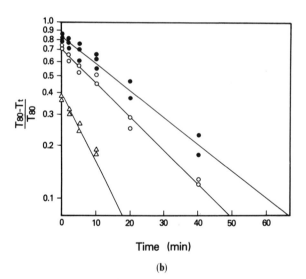

(b)

Figure 3–9 Kinetics of ligand internalization and receptor and ligand recycling. Data from Zigmond *et al.* (1982) for the uptake of the peptide ligand FNLLP (*N*-formylnorleucylleucylphenylalanine) by neutrophils. (a) Time-course of uptake and release of ligand. The solid circles mark the amount of ligand that is cell-associated after incubation of cells for various times at 37 °C with 2×10^{-8} M labeled ligand. In some experiments, cells incubating with ligand were then switched to 37 °C medium containing no ligand; these results are marked with open circles. (b) Time-course of receptor recycling. Cells were incubated with unlabeled ligand at 2×10^{-7} M (closed circles), 1×10^{-7} M (open circles), or 1×10^{-8} M (triangles). Cells were then washed and incubated in medium at 37 °C containing no ligand. The logarithm of the fraction of total receptors (T_{80}) that is unavailable for binding is plotted as a function of recovery time. Reproduced from the *Journal of Cell Biology*, 1982, Vol. 92, pp. 34–43, by copyright permission of the Rockerfeller University Press.

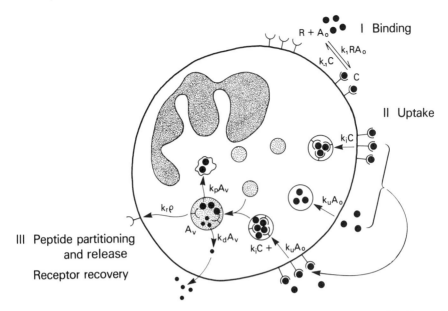

Figure 3–10 Zigmond *et al.* (1982) model for receptor and ligand uptake by neutrophils. Except for the inclusion of a storage pool for peptide ligand, the model is similar to that shown in Figure 3–3. Reproduced from the *Journal of Cell Biology*, 1982, Vol. 92, pp. 34–43, by copyright permission of the Rockefeller University Press.

is examined; *i.e.*, the theoretical postulate to be tested is that essentially all free receptors recycle whereas all bound receptors are degraded.

A major consequence of the differential sorting hypothesis in this model is the explicit distinction of intracellular free and bound receptors, R_i and C_i, respectively. Correspondingly, receptor/ligand association and dissociation events within the endosomal compartment must be included. The association and dissociation rate constants k_f and k_r are known to be fairly independent of pH for this receptor/ligand system, so they are used in both the surface and endosomal reaction terms. It is necessary to consider free ligand present in the endosomal compartment due to both complex dissociation and fluid-phase endocytosis. k_{fp} is the rate constant for fluid-phase endocytosis, which is related to k_{eC} but can be determined independently from measurements of ligand uptake versus bulk ligand concentration.

An additional simplifying assumption is that only receptor/ligand complexes are internalized ($k_{eR} = 0$), consistent with the "signaling" receptor expectation. Finally, ligand is assumed to partition from the endosome to a non-degraded storage pool with rate constant k_p, in a process parallel with degradation. No ligand is assumed to recycle, so that $f_L = 1$. Receptor degradation is neglected, however, as is receptor synthesis during the short (< 3 h) experimental time period. Intracellular bound receptors are thus considered to accumulate until dissociating, with recovery of surface receptors occurring when cells are removed from ligand-containing medium. Such receptor recovery was found to be not quite complete, though, indicating that receptor degradation probably should have been included.

The kinetic equations for free and bound surface and intracellular receptors, along with intracellular free ligand, resulting from these model assumptions are:

$$\frac{dR_s}{dt} = -k_f L R_s + k_r C_s + k_{rec}(1 - f_R^{free})R_i \tag{3-17a}$$

$$\frac{dC_s}{dt} = k_f L R_s - k_f C_s - k_{eC} C_s + k_{rec}(1 - f_R^{bound})C_i \tag{3-17b}$$

$$\frac{dR_i}{dt} = -k_f L_i R_i + k_r C_i - k_{rec}(1 - f_R^{free})R_i - k_p f_R^{free} R_i \tag{3-17c}$$

$$\frac{dC_i}{dt} = k_f L_i R_i - k_r C_i + k_{eC} C_s - k_{rec}(1 - f_R^{bound})C_i - k_p f_R^{bound} C_i \tag{3-17d}$$

$$\frac{d(v N_{Av} L_i)}{dt} = -k_f L_i R_i + k_r C_i + k_{fp} N_{Av} L - (k_{deg} + k_p) v N_{Av} L_i \tag{3-17e}$$

The Zigmond et al. model can be seen to be a variant of the base model, with the inclusion of additional intracellular detail allowing for differential sorting behavior of free and bound receptors. In terms of the base model, this permits the average sorting fraction of intracellular receptors, f_R, to depend on ligand concentration, L, since L will influence the relative proportions of internalized receptors in free versus bound state. These investigators derived an implicit expression for this dependence, applying the hypothesis $f_R^{free} = 0$ and $f_R^{bound} = 1$:

$$f_R = \frac{f_R^{free} R_i + f_R^{bound} C_i}{R_i + C_i} = \frac{C_i}{R_i + C_i} \tag{3-18}$$

The fraction of internalized receptors degraded is thus simply the ratio of endosomal complexes to total endosomal receptors at whatever ligand concentration prevails. The dependence of f_R on L, as well as on other parameters is manifest through the influence of these on R_i and C_i. A mechanistic explanation for this was offered by Linderman and Lauffenburger (1988, 1989) (see section 3.3.2b).

For a contant bulk ligand concentration L, the set of differential equations above was solved analytically by Zigmond et al. and applied to their kinetic data on ligand uptake/release and receptor recovery. This allowed straightforward evaluation of the model rate constants in a manner not requiring numerical algorithms. An estimate for the intracellular endosomal volume, $v = 10^{-11}$ ml/cell, about 2% of the total cell volume, was used. Using these independent parameter estimates, the steady-state total surface receptor number density could be predicted a priori with an analytical expression (which appears incorrectly in the original reference). A fairly clear approximate formula, most accurate when fluid-phase uptake is significant compared to receptor-mediated uptake of ligand, is:

$$\frac{C_s + R_s}{R_{s0}} = \frac{1}{\omega}\left[1 + \frac{1 + (k_{eC}/k_r)}{L_0/K_D}\right] \tag{3-19a}$$

where ω is a dimensionless quantity which can be thought of as a scaled fluid-phase ligand uptake rate constant:

$$\omega = \left(1 + \frac{K_D}{L_0}\right)\left(1 + \frac{k_{eC}}{k_r}\right) + \frac{k_{eC}}{k_{rec}}\left(1 + \frac{L_0}{K_D}\right)\left(\frac{k_{fp}}{v[k_{deg} + k_p]}\right) \quad (3\text{--}19\text{b})$$

The larger ω is, the greater the level of receptor downregulation.

Figure 3–11 shows successful qualitative agreement of calculations from this model with experimentally observed receptor downregulation behavior over a range of ligand concentration. A quantitative deviation exists, however, in which the model predicts a greater number of surface receptors remaining than is measured. This discrepancy may be due to the neglect of receptor degradation, or perhaps to surface expression of a cryptic pool of intracellular receptors upon ligand stimulation.

From these results, it can be inferred that receptor downregulation in the

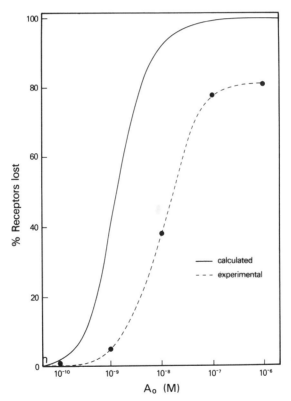

Figure 3–11 Downregulation of neutrophil surface peptide receptors. The percentage of surface receptors lost at steady-state $[(R_{sT0} - R_{sT})/R_{sT0}] \times 100$, is plotted against the ligand concentration L_0. Experimental data are shown as closed circles. Model predictions are shown as the solid line. Parameter values used in the model: $k_{deg} = 0.025 \, \text{min}^{-1}$, $k_p = 0.015 \, \text{min}^{-1}$, $k_{fp} = 4 \times 10^{-14} \, \text{l} \, \text{min}^{-1} \, \text{cell}^{-1}$, $k_{eC} = 0.15 \, \text{min}^{-1}$, $K_D = 2 \times 10^{-8} \, \text{M}$, $k_f = 2 \times 10^7 \, \text{M}^{-1} \, \text{min}^{-1}$, $k_r = 0.4 \, \text{min}^{-1}$, $k_{rec} = 0.3 \, \text{min}^{-1}$, $R_{s0} = 5 \times 10^4$ receptors cell^{-1}, $v = 10^{-14} \, \text{l} \, \text{cell}^{-1}$. From Zigmond *et al.* (1982). Reproduced from the *Journal of Cell Biology*, 1982, Vol. 92, pp. 34–43, by copyright permission of the Rockefeller University Press.

neutrophil chemotactic peptide system is due largely to a decreased receptor recycling rate constant as ligand concentration increases. Using the terminology of section 3.2.1.a, this would be categorized as sorting downregulation. Recall that, as shown above, the fibroblast EGF system, in contrast, falls into the category of endocytic downregulation.

3.2.1.d Antigen Processing

A recent mathematical model based on receptor trafficking phenomena analyzes the kinetics of *antigen processing* and *presentation* by macrophages and B lymphocytes (Singer and Linderman, 1990). T helper lymphocytes do not recognize antigen free in solution; the participation of antigen presenting cells (APC), typically macrophages and B lymphocytes, is required. APC ingest a complex antigen from the bloodstream or tissue fluid by either fluid-phase or receptor-mediated endocytosis, process it by intracellular enzymatic degradation to expose immunogenic peptide fragments (ag), and reexpress these fragments on their surface membrane in a complex with the major histocompatibility complex class II molecule (MHC). The entities recognized by the helper T-cells are surface MHC/ag complexes. Stimulation of T helper cells in this way leads to interleukin 2 (IL-2) production, cell division, and T helper cell interaction with other cells of the immune system. (In a related system, MHC class I molecules complexed with processed antigen are used for recognition by cytotoxic T-cells.) For further background information, see Unanue (1984) and Brodsky and Guagliardi (1991).

Many investigators believe that antigen is processed to the relevant peptides within endosomes and that the MHC class II molecules are present in that same compartment (Cresswell, 1985; Guagliardi *et al.*, 1990). Thus the events of antigen degradation and MHC/ag complex formation may occur simultaneously, followed by recycling of MHC/ag complexes to the cell surface. This suggests that the explicit tracking of events within endosomes, rather than the approach used earlier in this chapter, is necessary for the understanding of this system.

The model of Singer & Linderman (1990) was developed to address several quantitative questions concerning the processing of antigen. First, experiments relating an extracellular antigen concentration to the ensuing stimulation of helper T-cells (measured by IL-2 production rate or subsequent T-cell proliferation) give no information on the number of MHC/antigen complexes on the APC surface required to produce stimulation. In addition, observed differences in the rates of antigen processing between macrophages and B-cells for the case of fluid-phase antigen uptake have not been explained. B-cells require a processing time of about 8 h to stimulate effectively a population of T-cells (Eisenlohr *et al.*, 1988; Gosselin *et al.*, 1988; Lakey *et al.*, 1988) whereas macrophages need only about 1 h (Ziegler and Unanue, 1981). Further, the mode of endocytosis dramatically influences the extent of T helper cell stimulation, presumably by influencing the number of MHC/ag complexes presented by the APC. When utilizing receptor-mediated antigen endocytosis via binding to their surface immunoglobulin receptors, B-cells can effectively stimulate a population of T-cells with an antigen concentration much less than that needed when only non-specific uptake occurs (Casten and Pierce, 1988).

Figure 3–12 illustrates the model proposed by Singer and Linderman (1990

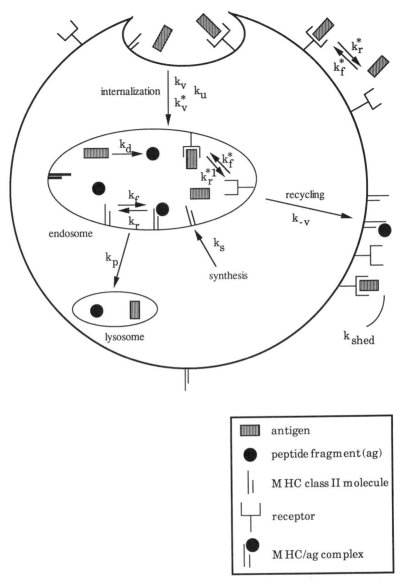

Figure 3–12 Model of antigen processing and presentation. The steps included in the model are: binding of antigen to surface receptors, if present (rate constant k_f^*), dissociation of antigen from cell surface receptors (rate constant k_r^*), internalization of free MHC class II molecules (rate constant k_v), internalization of receptor/antigen complexes (rate constant k_v^*), fluid-phase internalization of antigen (rate constant $k_u \approx k_v^*$ total endosome volume*cell surface area/total endosomal surface area), binding of antigen to receptors within endosomes (rate constant k_f^*), dissociation of antigen from receptors within endosomes (rate constant k_r^{*1}), degradation of antigen to form immunogenic peptide fragments (rate constant k_d), binding of peptide to MHC molecules (rate constant k_f), dissociation of peptide from MHC molecules (rate constant k_r), routing of unbound ligand in endosomes to lysosomes (rate constant k_p), recycling of free MHC molecules, MHC/peptide complexes, free receptors, and bound receptors to the cell surface (rate constant k_{-v}), loss of MHC/peptide complexes from the cell surface by any route (rate constant k_{shed}), and delivery of newly synthesized MHC molecules to endosomes (rate k_s). Modified from Singer and Linderman (1990, 1991).

101

to address these questions. The general model includes two routes for protein antigen uptake: non-specific and receptor-mediated. The receptor-mediated pathway is applicable only when antigen-binding receptors are present. Internalized antigen and receptors, together with both internalized and newly-synthesized MHC molecules, meet in the endosomal compartment. Within endosomes, antigen undergoes limited proteolysis to form peptide fragments (ag) which then bind to MHC molecules. In a simplification of the endosomal sorting analysis presented earlier, all ligand (undegraded antigen and unbound antigen fragments) is assumed routed to the lysosome for degradation, whereas MHC/ag complexes and free MHC molecules are recycled to the cell surface, allowing presentation of antigen to T helper cells. The assumption of perfect recycling of MHC molecules is a first approximation based on the long half-life of the similar MHC class I molecules (Tse *et al.*, 1986).

Kinetic balance equations are written for surface and endosomal numbers of each of the various species, with parameter values estimated from a variety of literature sources. The most dramatic difference in parameter values between macrophages and B-cells is that the former are of greater surface area, volume, and fluid-phase endocytosis rates (Chesnut *et al.*, 1982; Swanson *et al.*, 1985; Harding and Unanue, 1989). With all other parameters identical, the discrepancy in antigen presentation times when antigen uptake occurs only by non-specific internalization, as mentioned above, can be accounted for quantitatively with MHC internalization rate constants of 0.042 min^{-1} for macrophages and 0.0075 min^{-1} for B-cells, along with surface areas of $2200 \text{ }\mu\text{m}^2$ for macrophages and $900 \text{ }\mu\text{m}^2$ for B-cells. These parameter estimates are obtained from literature data, rather than from fits of the data to the model. We note that rate constants for non-specific uptake also differ between the two cell types because they are related to the MHC internalization rate constant. The number density of MHC/ag complexes then computed to be present on the macrophages at 1 hour and the B-cells at 8 hours is about $5 \text{ }\#/\mu\text{m}^2$, yielding a prediction for a number density of complexes resulting in maximal T-cell stimulation.

Figure 3–13 illustrates the difference in antigen presentation between fluid-phase and receptor-mediated antigen uptake by B-cells. The experimental data of Casten and Pierce (1988) are shown for comparison with model computations by Singer and Linderman. The data in Figure 3–13a imply that 50% and 100% T-cell stimulation occur at extracellular antigen concentrations of approximately 0.003 and $0.1 \text{ }\mu\text{M}$ for receptor-mediated endocytosis and approximately 1.0 and $4.0 \text{ }\mu\text{M}$ for fluid-phase endocytosis of antigen. The model computations in Figure 3–13b indicate that the numbers of MHC/ag surface complexes expressed for receptor-mediated uptake at $0.003 \text{ }\mu\text{M}$ and for fluid-phase uptake at $1.0 \text{ }\mu\text{M}$ are roughly equal, as are the numbers of receptor-mediated uptake at $0.1 \text{ }\mu\text{M}$ and fluid-phase uptake at $4.0 \text{ }\mu\text{M}$. Assuming that equal numbers of complexes on APC generate equal stimulation of T cells, this equality suggests that the model has captured the physical behavior of the system. In addition, it yields an estimate for the absolute number of complexes needed for 50% T-cell stimulation of about 500, and for 100% stimulation of about 2300. These values are consistent with experimental measurements of the threshold number of complexes needed to

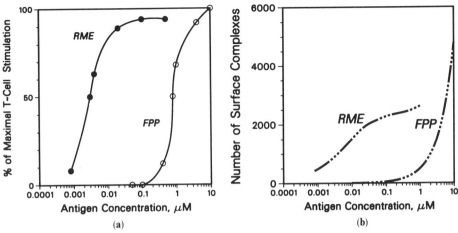

Figure 3–13 (a) T-cell stimulation as a function of antigen concentration when antigen is internalized via receptor-mediated endocytosis (RME) or fluid phase pinocytosis (FPP). The data of Casten and Pierce (1988), who used TPc9.1 T-cells, murine splenic B cells, and the antigen pigeon cytochrome c or pigeon cytochrome c coupled to Ig antibody, are plotted. The results are expressed as the percentage of the maximal T-cell response obtained for the fluid-phase case. (b) Number of MHC/ag complexes as a function of antigen concentration for the receptor-mediated endocytosis or fluid-phase pinocytosis route of antigen internalization. The predictions of the model of Singer and Linderman (1990) are shown. The complex number plotted is for a pulse time of 8 h and B-cell antigen presenting cells. Reproduced from the *Journal of Cell Biology*, 1990, Vol. 111, pp. 55–68, by copyright permission of the Rockefeller University Press.

stimulate T helper cells, approximately 300 MHC/ag complexes per APC (Demotz *et al.*, 1990; Harding and Unanue, 1990).

Finally, this model was used to examine the likely relevant *in situ* situation: the case of multiple MHC-binding antigens present simultaneously (Singer and Linderman, 1991). In this case, the mathematical model is complicated by the addition of equations to follow each antigen present in the system; however, the types of equations and analyses do not change. In Figure 3–14, the presentation of one particular type of MHC/ag complex by a macrophage is shown as a function of that antigen concentration when other antigens (denoted competing proteins) are also present and may compete for the antigen binding site on the MHC molecule. Competition experiments suggest that this effect is observable (Lorenz and Allen, 1988; Adorini *et al.*, 1989) and may suggest a way to manipulate antigen presentation and thus the immune response *in vivo*.

3.2.1.e Other Considerations

A few phenomena not included in the models discussed above may be important in some systems. These include the possibility of multiple recycling pathways (Sorkin *et al.*, 1991a), reversible trafficking between endosomal and lysosomal compartments (Buktenica *et al.*, 1987; Draye *et al.*, 1988) and polarized trafficking of receptors among distinct cell surface regions (Matlin, 1986). In addition, subtleties of diffusion-limited binding, crosslinking, and probabilistic phenomena may be relevant in some systems. Such aspects will be discussed in Chapter 4.

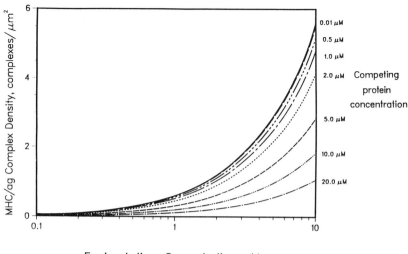

Figure 3–14 Stimulatory MHC/ag complex density (complexes/μm²) on the surface of a macrophage antigen presenting cell as predicted by the model of Singer and Linderman (1991). The concentration of foreign antigen, the antigen that gives rise to immunogenic peptide fragments and thus the stimulatory MHC/ag complexes plotted on the ordinate, is given on the abscissa. Simulations were done for 0.01–20 μM of competing proteins. These proteins can be degraded to peptide fragments which bind MHC molecules, thus competing with the binding of the immunogenic peptide fragments.

It must also be emphasized that there are mathematical models for a different type of receptor/ligand trafficking behavior, that involved in steroid hormone receptors (*e.g.*, Munck and Holbrook, 1984, 1987). In these systems there appears to be cycling between cytoplasmic and nuclear compartments which can be analyzed quantitatively in a manner similar to that seen for endocytic trafficking. Finally, a substantial amount of experimental work on the protein synthesis pathway is revealing the likely operation of very similar trafficking processes (Chung *et al.*, 1989), perhaps meeting the endocytic pathway in some common compartment near the Golgi region (see Klausner (1989) and Pugsley (1989)).

3.2.2 Mechanistic Models for Individual Trafficking Processes

The last section discussed models for whole-cell receptor/ligand trafficking dynamics. In those models, the goal was to develop equations describing the locations and state (*i.e.*, free or bound) of receptors and ligands in whole cells, with the aim of elucidating the dynamics of the endocytic cycle and in preparation for future work linking trafficking models with the production of cell behavioral responses. Although several different variations of our base model have been applied, the models are similar in that each of the rate processes is characterized by a single kinetic rate constant, or, in the case of separation of receptors and ligands in endosomes, by a single fraction or sorting efficiency. In many cases, these constants are known to be lumped parameters describing several different steps. For example, the rate constant k_{eC} for the internalization of receptor/ligand complexes includes not only the budding off of smooth or coated pits but also the movement of complexes to these specialized regions of the cell surface. As a second example, the sorting fractions f_R and f_L actually represent a variety of kinetic and transport events occurring within the endosome. In such cases, the parameter values measured should reflect more fundamental properties of the cell and the particular molecules involved in underlying mechanisms.

In this section, we will describe the use of mathematical modeling to investigate *individual steps* of the endocytic cycle in more detail. The aim here is to develop mechanistic models which can aid in the interpretation of biological data. Because these models incorporate details that are, in general, less well-known and more difficult to measure, few such models have been developed. We will discuss models for two particular situations: (1) the receptor internalization step, examining details of receptor interactions with coated- and smooth pits; and (2) the endosomal receptor sorting step, investigating factors affecting receptor destination.

3.2.2.a Saturation of the Coated-Pit Pathway

A key assumption of the models examined in section 3.2.1.b was that the rate constant for receptor internalization is independent of the number of receptors present. However, some experimental data suggest a saturation effect at high receptor densities. This was observed early on for the LDL receptor (Goldstein *et al.*, 1977), but the first clear quantitative demonstration was published by Wiley (1988) for the EGF receptor (EGF-R). Figure 3–15 shows the value of the receptor/ligand complex internalization rate constant, k_{eC}, as a function of the

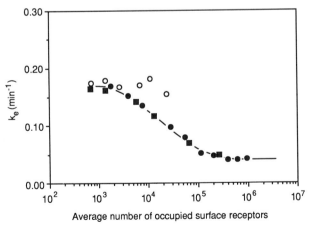

Average number of occupied surface receptors

Figure 3–15 Saturation of the EGF receptor internalization rate constant at high receptor occupancy. The receptor/ligand complex internalization rate constant k_{eC} (denoted k_e on graph) is plotted as a function of the average number of occupied surface receptors. Results from normal human fibroblasts are shown with open symbols; results from A431 cells are shown with closed symbols. From Wiley (1988). Reproduced from the *Journal of Cell Biology*, 1988, Vol. 107, pp. 801–810, by copyright permission of the Rockerfeller University Press.

number of surface receptor/ligand complexes, C_s. Increases in C_s are achieved by increasing the extracellular ligand concentration, L. The open symbols are data for normal human fibroblasts possessing about 5×10^4 receptors/cell, yielding a fairly constant value of k_{eC} with C_s of about 0.18 min^{-1}. Data for the transformed A431 cells, possessing about 1×10^6 receptors/cell, are represented by the closed symbols; here, k_{eC} decreases from its initially high value of about 0.17 min^{-1} to a lower limit of roughly 0.05 min^{-1} as C_s increases. Wiley inferred that the coated-pit apparatus is somehow saturated by EGF/EGF-R complexes on the A431 cells at high EGF concentrations, but not saturated on the normal fibroblasts. After saturation of the coated-pit pathway, any additional receptor complexes taken in must be internalized by an alternative pathway. Similar results have been obtained for internalization of the EGF-R on B82 mouse L cells (Starbuck et al., 1990), the insulin receptor on Chinese Hamster ovary cells (Backer et al., 1991), and the vitellogenin receptor on Xenopus oocytes (Opresko and Wiley, 1987). Plots of k_e versus number of surface complexes (e.g., Figure 3–15) are termed "SatIn plots", for "Saturation of Internalization" (Lund et al., 1990).

An additional endocytic pathway is offered by long-standing observations of so-called smooth pits on cell surfaces (van Deurs et al., 1989). Experiments have suggested use of this alternative pathway for the insulin receptor (Smith and Jarett, 1983). Electron microscopy studies have shown that the majority of EGF-R clusters on A431 cells are found in smooth, rather than coated, pits (McKanna et al., 1979; Willingham et al., 1983; Hopkins et al., 1985).

These considerations led Lund et al. (1990) to propose the Dual Pathway Internalization model (see Figure 3–16) and to test the ability of this model to explain their data. In this model the coated-pit adaptor molecules are termed internalization components. Coated pits are assumed to have an intrinsic internalization rate, λ (time^{-1}), but the only receptors internalized with these pits are those coupled with the internalization components. For the EGF system, only receptor/ligand complexes are assumed able to couple with the internalization components to form ternary complexes, with coupling and uncoupling rate constants k_c and k_u. Smooth pits are also modeled as internalizing at a constant rate, and are assumed to be indiscriminate about their contents; that is, both free receptors and receptor/ligand complexes may be internalized within them. Thus, there is a uniform rate constant k_t (time^{-1}) for free receptor and receptor/ligand complex endocytosis via smooth pits. (This rate constant also implicitly combines the rate of smooth pit internalization and the fraction of cell surface area covered by smooth pits, since receptors and complexes are considered to locate within these membrane structures purely by random distribution.) The coated-pit and smooth-pit pathways are also termed the induced (i.e., through the presence of ligand) and constitutive internalization pathways, respectively. Combining these two mechanisms gives the overall endocytic rate constant per receptor/ligand complex, k_{eC}.

What might govern the ability of a receptor to couple with an internalization component? For some receptors, such as the LDL and Tf receptors, the binding ability must be present in the receptor cytoplasmic tail independent of ligand binding, since these receptors are clustered in coated pits regardless of whether or not they are complexed with ligand. The EGF receptor, on the other hand,

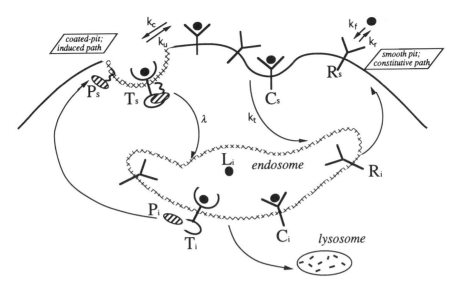

Figure 3–16 Dual pathway internalization model proposed by Lund *et al.* (1990). Steps involved in this model are: binding of ligand to cell surface receptors R_s (rate constant k_f), dissociation of ligand from cell surface complexes C_s (rate constant k_r), interaction of surface complexes and cell surface internalization components P_s to form surface ternary complexes T_s (rate constants k_c), uncoupling of ternary complexes (rate constant k_u), internalization of free and bound receptors via the smooth pit pathway (rate constant k_t), internalization of ternary complexes via the coated pit pathway (rate constant λ), and recycling of free endosomal internalization components P_i. The overall endocytic rate constant is then $k_e = (\lambda T_s + k_t C_s)/(T_s + C_s)$, following Eqn. (3–22). The coated-pit affinity is $K_{cp} = k_c/(k_u + \lambda)$, following Eqn. (3–24). Schematic is taken from Starbuck (1992).

only clusters in coated pits when occupied with ligand, suggesting that ligand binding causes a conformational change making the cytoplasmic tail region available for coupling. Wiley *et al.* (1991) have shown that activation of the EGF-R tyrosine kinase domain is necessary for these receptors to enter the induced internalization pathway; mutant receptors lacking the ability to activate the tyrosine kinase exhibit a much lower endocytic rate, presumably the rate characteristic of the constitutive pathway.

In order to determine the value of the overall complex endocytic rate constant k_{eC}, the contribution of the induced and constitutive pathways to k_{eC}, and dependence of k_{eC} on the number of surface receptor/ligand complexes (and thus on ligand concentration), Lund *et al.* (1990) pose the sequence of interactions shown schematically in Figure 3–16. This scheme neglects endosomal sorting, recycling, and degradation, for it is directed toward short time scales (approximately 5 min) during which these processes take place to minimal extent. Ligand concentration L is assumed constant over the time course of the experiments, and non-specific uptake of ligand is assumed negligible. Note that k_{eC} is not given in the above scheme as one of the intrinsic rate constants for internalization. This is because k_{eC} is the overall observed rate constant for internalization which can be related to the intrinsic rate constants for induced and constitutive internalization, λ and k_t, by the procedure described below.

The net rate of ligand internalization (disregarding fluid-phase endocytosis, which is negligible at low ligand concentrations) is

$$\frac{dL_i^\#}{dt} = \lambda T_s + k_t C_s \qquad (3-20)$$

Recall that $L_i^\#$ is the total amount of intracellular ligand ($\#$/cell). This is what can be measured most easily using labeled ligand, and as mentioned earlier is equivalent to the number of intracellular receptor/ligand complexes when the receptor and ligand sort together. C_s is the number of surface receptor/ligand complexes and T_s is the number of surface ternary complexes. Using the overall internalization rate constant k_{eC}, to represent the endocytic rate constant averaged over all surface receptor/ligand complexes, we can write:

$$\frac{dL_i^\#}{dt} = k_{eC}(C_s + T_s) \qquad (3-21)$$

It is important to note that the intrinsic internalization rate constants describing the induced and constitutive pathways, λ and k_t, are assumed to be constants, likely dependent on cell type. On the other hand, the overall or observed endocytic rate constant, k_{eC}, is assumed to be a lumped parameter which will vary with the number of receptor/ligand complexes. Its value should reflect biochemical properties of the receptor and coated-pit internalization components. Equating the right-hand sides of Eqns. (3–20) and (3–21) gives

$$k_{eC} = \frac{\lambda T_s + k_t C_s}{T_s + C_s} \qquad (3-22)$$

In a sense, this is our desired result: it tells us the dependence of the overall complex internalization rate constant k_{eC} on the individual pathway rate constants λ and k_t. However, this expression for k_{eC} is difficult to use, for the number of surface ternary complexes, T_s, is difficult to measure directly. Thus the next step is to express T_s and thus k_{eC} in terms of more fundamental parameters of the system.

A population balance on the number of ternary complexes per cell gives

$$\frac{dT_s}{dt} = k_c C_s P_s - k_u T_s - \lambda T_s \qquad (3-23)$$

where P_s is the number of available coated-pit internalization components. To simplify the analysis, we assume a quasi-steady-state for the number of surface ternary complexes, so that setting $dT_s/dt = 0$ in Eqn. (3–23) yields a relationship between ternary and binary receptor complexes on the cell surface:

$$T_s = K_{cp} C_s P_s \qquad (3-24)$$

where

$$K_{cp} = \frac{k_c}{k_u + \lambda} \qquad (3-25)$$

is a coated-pit association constant ($[\#/\text{cell}]^{-1}$), incorporating both molecular (k_c, k_u) and cellular (λ) properties). The quasi-steady-state assumption is not strictly valid, because the time constants for endocytosis are typically the same order as the experimental observation period.

Assuming that the total number of surface internalization components is constant due to continued coated-pit recycling, we can invoke conservation of the total surface number of these molecules: $P_T = P_s + T_s$. Utilizing the difference $P_s = P_T - T_s$ in Eqn. (3–24) and substituting the resulting relation into Eqn. (3–22), we obtain an expression for k_{eC} in terms of C_s:

$$k_{eC} = \frac{(k_t + \lambda K_{cp}P_T) + k_t K_{cp}C_s}{(1 + K_{cp}P_T) + K_{cp}C_s} \tag{3-26}$$

If ternary complexes are a relatively small fraction of total surface complexes, this expression is a good approximation for the effects of surface complexes on the endocytic rate constant.

Eqn. (3–26) attempts to explain how the overall endocytic rate constant for receptor/ligand complexes might vary with ligand concentration via changes in the number of surface complexes. It predicts that k_{eC} should take on its largest value for small numbers of surface complexes, with the possibility of decreasing to a minimal value as the number of surface complexes increases if coated-pit internalization components are saturated. The condition for saturation at high complex levels can be derived from requiring that C_s in the denominator of Eqn. (3–26) has influence: $C_s > (P_T + K_{cp}^{-1})$. Since the maximum value of C_s is R_T, we are roughly requiring that $R_T > (P_T + K_{cp}^{-1})$. Notice that this criterion involves both stoichiometry (the ratio of R_T to P_T) and affinity (K_{cp}). So, "saturation of internalization" can be expected for cells with unusually great receptor numbers or for receptors with unusually strong affinity for coated-pit internalization components. This behavior is qualitatively consistent with Figure 3–15. A comparison of Eqn. (3–26) and data for the B82 fibroblast cell line is shown in Figure 3–17.

Explicit formulae for limiting cases of very few and very many surface complexes can be obtained:

$$\left.\begin{array}{ll} C_s \to 0, & k_{eC} \to \lambda\left(1 + \dfrac{1}{K_{cp}P_T}\right)^{-1} \\[3mm] C_s \to \infty, & k_{eC} \to k_t \end{array}\right\} \tag{3-27}$$

assuming the condition $k_t \ll \lambda K_{cp}P_T$; this implies that at low surface complex number complexes will be predominantly found in coated pits. When the number of surface complexes is so large that the coated-pit adaptors are saturated, the overall endocytic rate constant will appear to be that characteristic of the constitutive pathway. When the number of surface complexes is very low, the overall endocytic rate constant will be a fraction of the coated-pit internalization rate, depending on the number and affinity of the coated-pit adaptors for the receptor.

An especially exciting prospect, in terms of relating molecular properties to trafficking dynamics and ultimately to cell function, is the ability of the model

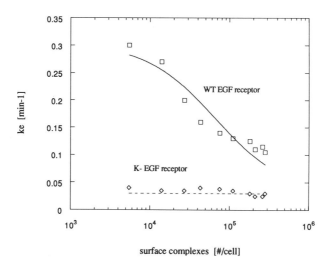

Figure 3-17 Saturation of the EGF receptor internalization rate constant, k_{eC}, as the number of surface complexes, C_s, increases. Data and model fits of Eqn. (3-26) for the internalization of wild-type (squares) and kinase-negative (diamonds) EGF receptors in transfected B82 fibroblasts are shown. Parameter values are $K_{cp} = 1.8 \times 10^{-5}$ $(\#/\text{cell})^{-1}$ for wild-type receptors and $K_{cp} = 8.8 \times 10^{-10}$ $(\#/\text{cell})^{-1}$ for kinase-negative receptors, with $P_T = 1.0 \times 10^4$ #/cell for both types; also, $\lambda = 1.8$ min^{-1}, $k_t = 0.03$ min^{-1}. From Starbuck (1991).

to account for differences in a variety of experimental measurements between receptor forms. When wild-type EGF-R are transfected into B82 cells (which lack endogenous EGF-R), receptor downregulation is observed along with substantial ligand degradation. In contrast, when certain site-directed mutants of EGF-R are transfected into the same cell type no downregulation is seen, nor is ligand degradation significant (Chen *et al.*, 1987). An example of such mutants is the substitution mutant M721, in which methionine replaces lysine at residue 721. This mutant receptor lacks tyrosine kinase activity ordinarily induced by EGF binding (thus preventing transduction of a mitogenic signal) so it can be denoted Kin−. It is useful to ask what model parameter is altered by these mutations to cause the differences in trafficking behavior.

Starbuck and coworkers (Starbuck *et al.*, 1990; Starbuck and Lauffenburger, 1992) have applied the dual pathway model, along with the other parts of the base whole-cell trafficking model, to these B82 cell data. The key finding is that the only notable difference in model parameter values between the wild-type EGF-R and the Kin− receptor M721 is a greatly reduced coated-pit affinity constant, K_{cp}, for Kin− EGF-R. Figure 3-17 shows SatIn plots for these two receptors, along with model calculations based on Eqn. (3-26). With $\lambda = 1.8$ min^{-1} set independently (Wiley *et al.*, 1985), a least-squares computer fit gives $k_t = 0.03$ min^{-1} and $P_T = 1.0 \times 10^4$ #/cell for both receptor types with $K_{cp} = 1.8 \times 10^5$ $(\#/\text{cell})^{-1}$ for WT and 8.8×10^{-10} $(\#/\text{cell})^{-1}$ (effectively zero) for Kin− receptor. Therefore, the mutant receptor does not couple to any significant extent with coated pits. As a result, it cannot effectively utilize the induced, coated-pit endocytic pathway but instead is internalized only through the constitutive, smooth-pit pathway at a much lower rate. This comparative behavior can be seen in Figure 3–7 as well as in Figure 3–17. Coated-pit adaptor molecules are stoichiometrically limiting in this cell type, since R_T for these transfected B82 cells is approximately 2×10^5 #/cell so that $P_T/R_T = 0.05$. This limitation accounts for the dramatic decrease of k_e with C_s for the WT receptors which couple strongly to the internalization components.

Using this single parameter value alteration, $K_{cp}P_T \sim 0$ (actually, 8.8×10^{-6}) for Kin$-$ instead of 0.18 as for WT, *a priori* predictions can be successfully made for EGF-R downregulation and EGF degradation data, accounting for the differences found between the mutant and wild-type receptors (Figure 3–18a,b).

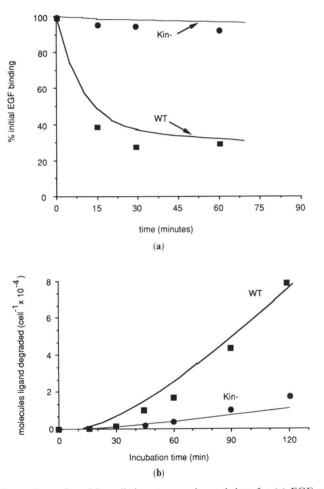

Figure 3–18 Comparison of model predictions to experimental data for (a) EGF receptor down-regulation, and (b) EGF depletion from medium, for B82 fibroblasts transfected with wild-type (squares) or kinase-negative mutant (circles) EGF receptors. From Starbuck and Lauffenburger (1992). The wild-type receptor data in (b) were used to help determine the ligand degradation rate constant in the model. All other curves are *a priori* model predictions, with all parameter values previously specified by independent experiments. Reprinted with permission from Starbuck and Lauffenburger (1992). Copyright 1992 American Chemical Society.

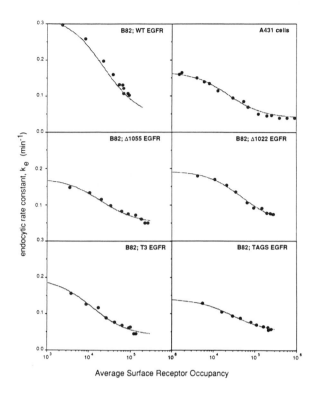

Figure 3–19 Saturation of internalization in B82 fibroblasts transfected with various forms of the EGF receptor. Data for A431 cells are also shown. The overall observed endocytic rate constant for receptor/ligand complexes, k_{eC}, is plotted as a function of surface complex number C_s. Model fits to the data are shown as solid curves. From Starbuck (1991).

In Chapter 6 this approach will be applied to cell proliferation studies by Wells *et al.* (1990) to explain altered mitogenic effects in terms of a trafficking parameter change.

More subtle changes in K_{cp} values have also been measured for various truncation and substitution mutants of the EGF receptor (see Figure 3–19, from Starbuck (1991)), demonstrating that variations of structural determinants can be manifested as quantitative changes in trafficking model parameters. This offers hope for relating receptor structural properties in a clear manner to trafficking dynamics, and ultimately to cell behavioral responses.

3.2.2.b *Endosomal Sorting*

In the endocytosis models described earlier in this chapter, the sorting or separation of molecules slated for recycling from molecules destined for the lysosome was largely uncharacterized. We defined the sorting fractions f_R and f_L, the fractions of internalized receptors and ligands routed from the endosome to the lysosome, but did not offer any mechanism by which the values of the fractions might be determined. We now turn to the development of a mechanistic model of the sorting process, in the hopes of offering a physical explanation for the outcome of sorting. A detailed exposition of theoretical analyses applied to this problem is given by Linderman and Lauffenburger (1988).

The mechanism of the endosomal sorting process is not known. Electron micrographs of endosomes show that they consist of a central vesicular chamber, approximately 0.2–0.8 µm in diameter, and one or more attached thin tubules, about 0.01–0.06 µm in diameter (Geuze *et al.*, 1983). Marsh and coworkers (Marsh *et al.*, 1986; Griffiths *et al.*, 1989) have performed a detailed study on the morphology of endosomes in BHK-21 cells. In these cells, the endosome volume is about 0.04 µm³/endosome, with 60–70% of this volume contained within the vesicle. Tubules account for the majority of the surface area, about 60–70% of the total surface area of 1.5 µm²/endosome.

The unusual structure of the endosome is related to the way that it is believed to function. The tubules are believed to be intermediates in the recycling of molecules back to the cell surface (Hatae *et al.*, 1986; Geuze *et al.*, 1987). Thus, receptors or ligands found in tubules when they break their connection with the vesicle will be recycled. Molecules remaining behind in the vesicle are assumed to be degraded when the vesicle matures into a lysosome (Murphy, 1991; Dunn and Maxfield, 1992). The primary support for this hypothesis comes from observations by Geuze *et al.* using immunogold electron microscopy to locate the asialoglyco-protein receptor and its ligand, asialoorosomucoid, in the endosomes of rat hepatocytes. The receptors, which recycle efficiently in this system, were found concentrated in the tubular extensions of the endosome, whereas most ligand molecules, which in this system are typically degraded, were found free in the lumen of the vesicle. Hatae *et al.* followed a fluid-phase marker, horseradish peroxidase, during transient endocytosis. They found it first in small vesicles, then in larger vesicles, and finally within tubular extensions of the large vesicles, suggesting a time sequence for molecular movement from the vesicular portion to the tubular portion of the endosome.

The acidic environment of the endosome, measured in the range 5.0–5.5 (Tycko *et al.*, 1983; Roederer *et al.*, 1987; Yamashiro and Maxfield, 1988) may also play a role in the sorting process. It has been shown that the affinity of receptors for their ligands may be substantially reduced at this pH for many, but not all, receptor/ligand systems (Dunn and Hubbard, 1984; Mellman *et al.*, 1984; Wileman *et al.*, 1985a).

The outcome of the sorting process, in terms of the fractions of internalized receptors and ligands which are routed along the recycling or degradative pathways, depends on the receptor for any given cell type, and may vary with cell type for the same receptor. Marshall (1985) found that insulin receptors on rat adipocytes are efficiently recycled to the cell surface and that even after 4 h of continuous ligand uptake, during which the receptors cycle through the sorting process about once every 6 min, there is no significant loss of receptors to the degradative pathway. During this time, about 75% of the internalized ligand is degraded and the remaining 25% is exocytosed. Yet the outcome of sorting is very different in the Tf receptor system. Tf and its receptor are not separated and both apparently recycle to the cell surface after releasing iron into the cytoplasm, with Tf still bound to its receptor (Ciechanover *et al.*, 1983). In contrast, the opposite seems to occur in the EGF receptor system, with both the receptor and its ligand predominantly degraded in normal fibroblasts (Stoscheck and Carpenter, 1984); in other cell types, however, there may be significant EGF receptor recycling (Wiley *et al.*, 1991).

Further, some properties of the ligand may influence the sorting outcome, including ligand concentration (Zigmond *et al.*, 1982), ligand valency (Hopkins and Trowbridge, 1983; Mellman *et al.*, 1984; Ukkonen *et al.*, 1986; Weissman *et al.*, 1986), and ligand affinity for receptor binding (Anderson *et al.*, 1982; Townsend *et al.*, 1984; Schwartz *et al.*, 1986). For the most part, these findings are qualitative and do not actually permit estimation of the values of the sorting fractions, f_R and f_L. The studies by Zigmond *et al.* (1982) and Mellman and colleagues are the two exceptions, at this time.

Zigmond *et al.* (1982) applied a whole-cell kinetic model (section 3.2.1.c) to the dynamics of endocytic trafficking of the chemotactic peptide receptor in neutrophil leukocytes. Although not able to estimate the sorting fractions separately, these investigators were able to determine the overall rate constant for receptor recycling, equal to the combined quantity $k_{rec}(1 - f_R)$, as a function of chemotactic peptide concentration in the extracellular medium, L_0. For $L_0 = 1 \times 10^{-8}$ M, 3×10^{-8} M, and 2×10^{-7} M, they obtained $k_{rec}(1 - f_R) = 0.11$ min^{-1}, 0.086 min^{-1}, and 0.038 min^{-1}, respectively (see Figure 3–9). Clearly, as the ligand concentration increases the overall rate constant for receptor recycling decreases. With the assumption that the rate constants for organellar movement k_{rec} and k_{deg} are independent of the ligand concentration, this implies a corresponding increase in the fraction of internalized receptors degraded, f_R, as ligand concentration concentration increases. Experimental evidence for this independence of the compartment transfer rate constant k_{rec} from compartment composition comes from observations of cohort movement of different ligands and receptors in aveolar macrophages (Ward *et al.*, 1989).

Similarly, the data of Mellman and coworkers on the dynamics of trafficking of the Fc$_\gamma$ receptor in the mouse macrophage line J774 was analyzed in terms of a whole-cell kinetic model by Linderman and Lauffenburger (1989). This receptor physiologically provides for clearance of immunoglobulin G (IgG) complexes with recognized antigens by binding, internalization, and degradation. The experiments provided kinetic binding and receptor recycling data in response to a monovalent ligand (an Fab antibody fragment directed against the receptor) and a multivalent ligand (IgG complexes). Again, the values of the sorting fractions could not be estimated separately, but the overall rate constant for receptor degradation expressed in the combined quantity $k_{deg}f_R$ could be determined for the two ligands. For the monovalent ligand, the best-fit value of $k_{deg}f_R$ was 3×10^{-3} min^{-1}, whereas for the multivalent ligand it was 8×10^{-3} min^{-1}. Again assuming that organellar transport rate constants are not affected, this implies a factor of almost three increase in f_R for the ligand with higher valency. Thus, an increase in valency appears to lead to an increase in the fraction of internalized receptors sorted to the degradative pathway. We note in passing that the internalization rate constant was found to be significantly greater for the IgG complexes than for the Fab fragment, approximately 0.5 min^{-1} as compared to 0.05 min^{-1}, and that the analysis of the whole-cell kinetic data with a mathematical model of endocytosis similar to those described in section 3.2.1 enabled the dissection of the different rate constants for the various steps.

Linderman and Lauffenburger (1988) proposed a mechanism for endosomal sorting, and their analysis of the corresponding mathematical model demonstrated that this mechanism could account for all the qualitative effects mentioned earlier (*i.e.*, the effects of ligand concentration, affinity, and valency on the outcome of the sorting process) as well as the two quantitative results presented just above. This mechanism is based on an affinity-binding interaction between a receptor's cytoplasmic tail and a membrane-associated accessory molecule (Figure 3–20), analogous to that described in the models of section 2.2.4.c for ternary complex

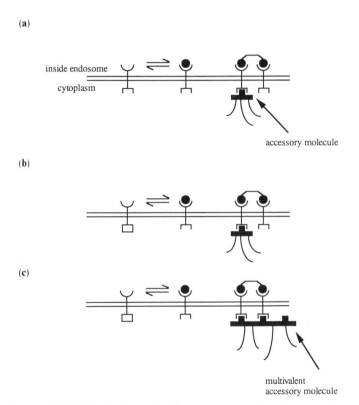

Figure 3–20 Proposed affinity binding interaction between receptors in endosomes and a membrane-associated accessory molecule, modified from Linderman and Lauffenburger (1988). The mobility of receptors in the endosome membrane, and hence their ability to move into the recycling endosome tubules, may be influenced by their binding to immobile accessory molecules. (a) In this scenario, receptors have a low affinity for a monovalent accessory molecule. Ligand binding produces no conformational change in the receptor, but a multivalent binding of ligand to receptors increases the probability that at least one of the receptors in the aggregate, and therefore the entire aggregate, will be immobilized. (b) In this scenario, free receptors have no affinity for the accessory molecule. Ligand binding produces a conformational change in the receptor so that it may bind the interaction molecule. The probability of receptor/ligand immobilization depends on the affinity of the accessory molecule for the complex and on the degree of receptor crosslinking by ligand. (c) This scenario is as in (b) except that the accessory molecule is multivalent and thus has a higher avidity for receptor aggregates than for single receptors. All of these possibilities are consistent with the proposed mathematical model for receptor/ligand sorting in endosomes.

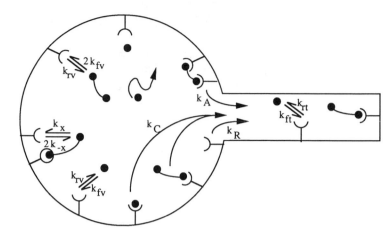

Figure 3–21 Single endosome model, adapted from Linderman and Lauffenburger (1988). Receptors and ligand can exist in bound or free states; multivalent ligands (here a bivalent ligand is shown) can crosslink receptors. The binding and dissociation rate constants may be different in the vesicle and tubule. Receptors may move by passive diffusion or with the aid of a membrane current into the tubule, where they are trapped and cannot return to the vesicle. The rate constant for the movement of receptors from the vesicle into tubules depends on the state of the receptor. Ligand molecules may diffuse between the vesicle and tubule compartments.

formation during cell surface ligand binding and in the model of section 3.2.2.a for receptor aggregation in coated pits by coated-pit internalization components.

The model defines the temporal beginning of the sorting process as the time when internalized receptor/ligand complexes arrive in an endosome, a low pH environment with the apparent geometry of a central spherical vesicle connected to thin tubular extensions (Figure 3–21). All internalized receptors and ligands are assumed to be initially within the central vesicle, but following diffusive transport during a sorting time period t_{sort}, some fraction of each will move into the attached tubules. The fractions of receptors and ligands remaining in the central vesicle at t_{sort} are postulated to then be routed to lysosomes for degradation, and so are equal to f_R and f_L, respectively. The fractions moved into tubules by t_{sort}, $(1 - f_R)$ and $(1 - f_L)$, are assumed to be recycled/exocytosed to the cell surface. Estimates for t_{sort} of the order of 5–10 min have been obtained (Schwartz et al., 1982; Marshall, 1985; Evans and Flint, 1985).

In the low pH conditions typically found in the endosomal compartment, the rate of ligand dissociation from receptors is generally accelerated over that of the rate at typical cell surface pH (about 7.3), although there are few clear quantitative measurements of rate constants as functions of pH. Systems for which such data are currently available include the leukocyte chemotactic peptide receptor (Zigmond et al., 1982), the macrophage Fc_γ receptor (Mellman et al., 1984), and the Tf receptor (Klausner et al., 1983). The Linderman & Lauffenburger model allows for the transient dissociation and reassociation of receptor/ligand complexes within the endosome vesicle and tubules, following the same binding scheme present at the cell surface but with pH-affected rate constants.

Free ligand, either dissociated from a receptor complex or originally endocytosed in the bulk fluid phase trapped by an internalizing coated pit, is assumed to diffuse freely in three dimensions within the endosome vesicle and tubules with diffusion coefficient D_L. Receptors, receptor/ligand complexes, and aggregated clusters of multivalent receptor/ligand complexes may diffuse in two dimensions within the vesicle and tubule membrane with diffusion coefficients D_R, D_C, and D_A, respectively. The crucial postulate of the model is that D_R, D_C, and D_A may be unequal, due to differing interactions with the membrane-associated accessory component by receptors in the various states of binding (free, bound, aggregated), as illustrated in Figure 3–20. As the simplest example, consider a monovalent ligand and receptor, such as for EGF or chemotactic peptide. As indicated by internalization studies of EGF (Lund et al., 1990), bound receptors have an affinity for a membrane-associated accessory molecule which may immobilize the complexes in coated pits; in such a case, one could easily postulate that $D_C \sim 0$ even while $D_R > 0$. Fc_ε receptors on RBL cells serve as an example for a multivalent ligand/monovalent receptor case. Menon et al. (1986) found that Fc_ε receptors bound by monomeric IgE molecules have a cell surface diffusivity of 3×10^{-10} cm^2/s, while the diffusivity of receptors crosslinked by IgE trimers and higher oligomers is reduced by a factor of about 100. Robertson et al. (1986) have further shown that this change in receptor mobility is likely the result of an interaction between the crosslinked complexes and cytoskeletal elements. In this case, $D_A \sim 0$ while D_R and $D_C > 0$.

Given diffusivities for the free ligand and the receptor in its various states of binding, rate constants for transport of these molecules from the central endosomal vesicle to attached tubules can be calculated (Linderman and Lauffenburger, 1986). These calculations show that ligand transport is very fast compared to receptor transport (because D_L is likely to be a couple to orders of magnitude greater than D_R); in fact, on the time scale of t_{sort} free ligand can be assumed to be distributed in an equilibrium fashion between the vesicle and tubule lumen fluid volumes. The receptor transport rate constants corresponding to D_R, D_C, and D_A are defined as k_R, k_C, and k_A, respectively. Combining these rate constants with the receptor/ligand association and dissociation rate constants as shown in Figure 3–21 provides the core of the mathematical model. Actual interactions of the receptor with the postulated accessory molecule are not included explicitly; they are assumed to occur instantaneously depending on the receptor binding state in this model. A further simplifying assumption is that receptors do not diffuse back from the tubules into the vesicles at an appreciable rate, perhaps due to convective membrane currents from the vesicle into the tubule or to a trapping mechanism present in the tubule (Linderman and Lauffenburger, 1989).

Mathematical formulation of the model consists of kinetic balance equations for the various species in the endosome vesicle and tubule. For the case of monovalent ligand and monovalent receptor, the following equations completely govern the numbers of free receptors in vesicle and tubules, R_v and R_t, receptor/ligand complexes in vesicle and tubules, C_v and C_t, and concentration of free ligand in vesicle and tubules, L_v and L_t:

$$\frac{dR_v}{dt} = -k_{fv}L_v R_v + k_{rv}C_v - k_R R_v \tag{3-28a}$$

$$\frac{dC_v}{dt} = k_{fv}L_v R_v - k_{rv}C_v - k_c C_v \tag{3-28b}$$

$$\frac{dC_t}{dt} = k_{ft}L_t R_t - k_{rt}C_t + k_c C_v \tag{3-28c}$$

$$R_{eT} = R_v + R_t + C_v + C_t \tag{3-28d}$$

$$(V_v + V_t)L_T = (V_v L_v + V_t L_t) + \frac{(C_v + C_t)}{N_{Av}} \tag{3-28e}$$

$$L_t = \kappa_e L_v \tag{3-28f}$$

Subscripts v and t on the rate constants k_f and k_r signify values appropriate to the pH values present in the vesicle and tubules, respectively (there is scant information, but the pH may be higher in the tubules if proton pumps are excluded from their membranes). V_v and V_t are the volumes of the endosomal vesicular and tubular compartments κ_e is a partition coefficient for the equilibrium distribution of ligand between vesicular and tubular lumen volumes; κ_e values less than one account for steric hindrance when ligands are of substantial size relative to the tubule diameter). Eqns. (3-28d,e) dictate that the total receptor number and ligand concentration within the endosome, R_{eT} and L_T, remain constant during the sorting time.

These equations are then integrated numerically from $t = 0$ to $t = t_{sort}$, using initial conditions $R_v(0) = 0$, $C_v(0) = R_{eT}$, and $C_t(0) = 0$ with $L_T = L_0 + (C_v + C_t)/[N_{Av}(V_v + V_t)]$. The desired result, the value of the sorting fractions, are computed from $f_R = [R_v(t_{sort}) + C_v(t_{sort})]/R_{eT}$, $f_L = [V_v L_v(t_{sort})]/[(V_v + V_t)L_T]$. Similar equations and procedures arise for multivalent receptors and/or ligands (see Linderman and Lauffenburger, 1988). In such cases, the crosslinking events are included (as in the models of section 4.3).

Parameter estimates are discussed by Linderman and Lauffenburger (1988, 1989). For the illustrations discussed here, the following apply: endosome vesicle radius $a_v = 0.3$ μm, endosome tubule radius $a_t = 0.02$ μm, $D_R = 1 \times 10^{-10}$ cm^2/s, $\rho_e = 0.75$, $V_v = 3 \times 10^{-14}$ cm^3, $V_t = 1 \times 10^{-14}$ cm^3. Receptor/ligand binding rate constants depend on the particular example system chosen. Using a theoretical expression for the rate constant for receptor transport from the vesicle of the endosome into a tubule, k_R, in terms of the receptor diffusion coefficient D_R (Linderman and Lauffenburger, 1986):

$$k_R = \frac{D_R}{a_v^2}\left\{ \frac{2\ln\left[\dfrac{2}{1 - \cos\left(\dfrac{a_t}{a_v}\right)}\right]}{1 + \cos\left(\dfrac{a_t}{a_v}\right)} - 1 \right\}^{-1} \tag{3-29}$$

we find that $k_R \sim 1 \, \text{min}^{-1}$, indicating that for $t_{\text{sort}} \sim 5{-}10 \, \text{min}$ there should be sufficient time for substantial movement of receptors into the tubules for recycling. This is an important result, given experimental observations that in some systems the recycling fraction can be as high as 0.99 (*i.e.*, $f_R \leq 0.01$).

Figure 3–22 illustrates model predictions of the dynamic behavior of the various receptor species during the sorting process. Figure 3–22a shows transient curves for a monovalent receptor and ligand system, for free and bound receptors

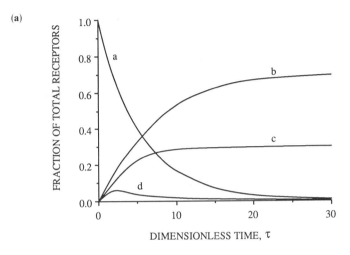

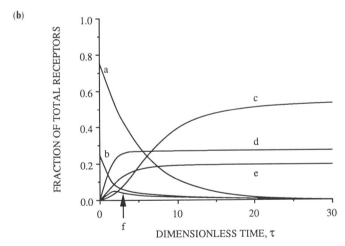

Figure 3–22 General behavior of the single endosome model. (a) Monovalent receptor, monovalent ligand system. Parameter values: $\varepsilon = 0.1$, $\eta = 10.0$, $(V_v + V_t)N_{\text{Av}}L_{\text{Tot}}R_{\text{eT}} = 1.001$, $\kappa_e = 0.75$, $k_C/k_R = 0.1$. Initial condition: $C_v(0) = R_{\text{eT}}$. Curve a: bound receptors in vesicle; curve b: bound receptors in tubules; curve c: free receptors in tubules; curve d: free receptors in vesicle. (b) Monovalent receptor, bivalent ligand system. Parameter values: $\varepsilon = 0.1$, $\eta = 10.0$, $\chi = 10.0$, $k_{-x}/k_r = 1.0$, $(V_v + V_t)N_{\text{Av}}L_{\text{Tot}}/R_{\text{eT}} = 0.63$, $\kappa_e = 0.75$, $k_C/k_R = 1.0$, $k_A/k_R = 0.0$. Initial condition: 75% of all receptors are crosslinked and found in the vesicle; 25% of all receptors are bound (uncrosslinked) and found in the vesicle. Curve a: crosslinked receptors in vesicle; curve b; bound (uncrosslinked) receptors in vesicle; curve c: crosslinked receptors in tubules; curve d: bound (uncrosslinked) receptors in tubules, curve e: free receptors in tubule; curve f: free receptors in vesicles.

in the vesicle and tubules; Figure 3–22b shows curves for a monovalent receptor bivalent ligand system, for free, bound, and crosslinked receptors in the vesicle and tubules. Time is made dimensionless by scaling with the rate constant for receptor transport from vesicle into tubule, so that $\tau = k_R t$. In the former plot bound receptors are assumed to be nearly immobilized in the vesicle ($k_C/k_R = 0.1$), whereas in the latter plot only crosslinked receptors are assumed to be immobilized ($k_C/k_R = 1.0$, $k_A = 0$). Further, it is assumed that the binding kinetics within the vesicle and tubule are identical, i.e. $k_f = k_{fv} = k_{ft}$ and $k_r = k_{rv} = k_{rv}$. Notice that

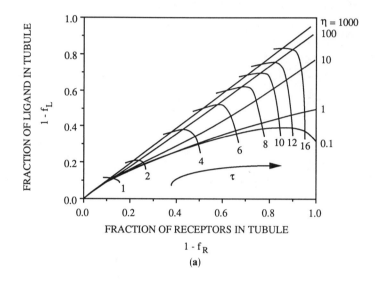

(a)

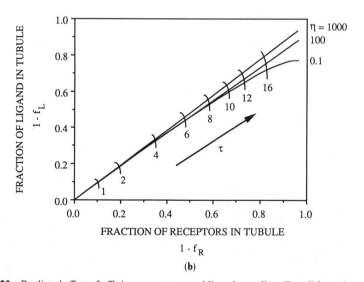

(b)

Figure 3–23 Predicted effect of affinity on receptor and ligand recycling. For all four plots, $\kappa_e = 0.75$, $V_v/V_t = 7/3$. (a) Monovalent receptor, monovalent ligand system. Parameter values: $\varepsilon = 0.1$, $(V_v + V_t)N_{Av}L_{Tot}/R_{eT} = 1.001$, $k_C/k_R = 0.1$. Initial condition: $C_v(0) = R_{eT}$. (b) Monovalent receptor, monovalent ligand system. Parameter values and initial condition as in (a) except that $\varepsilon = 0.01$.

receptor accumulation in the tubules (and thus recycling), is essentially rate-limited by the kinetics of receptor/ligand complex dissociation or receptor crosslink dissociation, which may be regulated by the endosomal pH.

Computational predictions of the sorting fractions f_R and f_L can be presented most compactly as shown in Figures 3–23a–d, plots of the fraction of internalized ligand that is exocytosed $(1 - f_L)$ versus the fraction of internalized receptors that are recycled $(1 - f_R)$. Figures 3–23a and 3–23b are for the completely monovalent

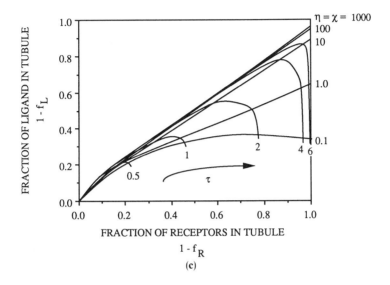

(c)

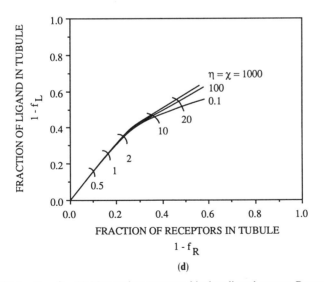

(d)

Figure 3–23 (*continued*)   (c) Monovalent receptor, bivalent ligand system. Parameter values: $\varepsilon = 1.0$, $k_{-x}k_r = 1.0$, $(V_v + V_t)N_{Av}L_{Tot}/R_{eT} = 0.63$, $k_C/k_R = 1.0$, $k_A/k_R = 0.0$. Initial condition: 75% of all receptors are crosslinked and found in the vesicle; 25% of all receptors are bound (uncrosslinked) and found in the vesicle. (d) Monovalent receptor, bivalent ligand system. Parameter values and initial condition as in (c) except that $\varepsilon = 0.01$.

system with ε, the ratio of the receptor/ligand complex dissociation rate to the transport rate constant for receptors entering tubules, k_r/k_R, equal to 0.1 and 0.01, respectively. The curves are parameterized in dimensionless time, $\tau = k_R t$, and dimensionless binding affinity $\eta = R_{eT}/[N_{Av}(V_v + kV_t)K_D]$, where $K_D = k_r/k_f$.

Figures 3–23c and 3–23d show the monovalent receptor with bivalent ligand system, with an additional parameter, the dimensionless crosslinking equilibrium constant, $\chi = R_{eT}(k_x/k_r)$, where k_x is the receptor crosslinking rate constant. In both figures, $k_C/k_R = 1.0$ and $k_A = 0$. In all four plots of Figure 3–23, increasing binding and crosslinking affinities are shown to lead to decreased receptor recycling and variable effect on ligand exocytosis, with greatest sensitivity to these parameters at shorter sorting times.

This type of analysis can be used to make quantitative predictions for comparison with the data of Zigmond et al. and Mellman and coworkers. Using known values for the receptor/ligand association and dissociation rate constants, along with the estimates given above for endosomal geometric parameters and free receptor diffusivity, the experimentally observed variation of $(1 - f_R)$ with L_0 for the neutrophil/chemotactic peptide–receptor system can be explained with this model for a sorting time of $t_{sort} = 4$ min. Similarly, the experimentally observed dependence of f_R on ligand valency for the macrophage/Fc$_\gamma$-receptor system can be accounted for by this model for a sorting time of about 3 min. The key postulate in each case was that bound neutrophil peptide receptors and crosslinked macrophage Fc$_\gamma$ receptors are immobile or significantly hindered in diffusing within the endosome. Other than this, all parameters were estimated independently. It is interesting to recognize that these postulates are consistent with internalization behavior of the receptors; bound neutrophil peptide receptors and crosslinked macrophage Fc$_\gamma$ receptors seem to be preferentially endocytosed (recall the results of the Zigmond et al. and Linderman–Lauffenburger whole-cell kinetic models mentioned earlier).

In addition to these phenomena, this model predicts a number of additional effects of system variables and parameters that should be amenable to experimental investigation. Application of molecular biology techniques to alter parameter values by genetic modification of receptor and ligand properties now offers a promising avenue for testing this sort of mechanistic model (for example, see Herbst et al., 1993).

NOMENCLATURE

Symbol	Definition	Typical Units
a_t	endosome tubule radius	μm
a_v	endosome vesicle radius	μm
B_i	total number of internalized bound ligand molecules	#/cell
B_s	total number of surface-associated ligand molecules	#/cell
C_i	number of intracellular receptor/ligand complexes	#/cell
C_s	number of receptor/ligand complexes on the surface	#/cell
C_{s0}	number of surface complexes in the absence of ligand	#/cell
C_t	number of receptor/ligand complexes in endosome tubules	#/cell
C_v	number of receptor/ligand complexes in endosome vesicles	#/cell

Symbol	Definition	Typical Units
D_A	diffusion coefficient of receptors crosslinked by ligand	cm^2/s
D_C	receptor/ligand complex diffusion coefficient	cm^2/s
D_L	free ligand diffusion coefficient	cm^2/s
D_R	free receptor diffusion coefficient	cm^2/s
f_L	fraction of endocytosed ligands degraded	
f_R	fraction of endocytosed receptors degraded	
f_{RL}	value of f_R in presence of ligand	
f_{R0}	value of f_R in the absence of ligand	
f_R^{bound}	fraction of bound receptors in endosomes that are not recycled	
f_R^{free}	fraction of free receptors in endosomes that are not recycled	
k_A	rate constant for the transport of receptors crosslinked by ligand from endosome vesicles into endosome tubules	min^{-1}
k_c	rate constant for coupling of complexes with internalization components	$(\#/cell)^{-1}\,min^{-1}$
k_C	rate constant for the transport of receptor/ligand complexes from endosome vesicles into endosome tubules	min^{-1}
k_{deg}	overall rate constant for the routing of material from the endosome to the lysosome, degradation in the lysosome, and the release of fragments in the extracellular medium	min^{-1}
k_{eC}	rate constant for internalization of receptor/ligand complexes	min^{-1}
k_{eR}	rate constant for internalization of free receptors	min^{-1}
k_f	rate constant for association of receptor/ligand complexes	$M^{-1}\,min^{-1}$
k_{fp}	rate constant for fluid internalization	$\mu m^3/min\text{-}cell$
k_{ft}	rate constant for association of receptor/ligand complexes within endosome tubules	$M^{-1}\,min^{-1}$
k_{fv}	rate constant for association of receptor/ligand complexes within endosome vesicles	$M^{-1}\,min^{-1}$
k_p	rate constant for movement of ligand from endosomes to a partitioned pool	min^{-1}
k_r	rate constant for dissociation of receptor/ligand complexes	min^{-1}
k_{rt}	rate constant for dissociation of receptor/ligand complexes within endosome tubules	min^{-1}
k_{rv}	rate constant for dissociation of receptor/ligand complexes within endosome vesicles	min^{-1}
k_R	rate constant for the transport of free receptors from endosome vesicles into endosome tubules	min^{-1}
k_{rec}	rate constant for transport of material via vesicles from the endosome back to the cell surface	min^{-1}
k'_{recL}	effective ligand recycling rate constant, equal to $k_{rec}(1 - f_L)$	min^{-1}
k'_{recR}	effective receptor recycling rate constant, equal to $k_{rec}(1 - f_R)$	min^{-1}
k_t	rate constant for free receptor and receptor/ligand complex endocytosis via smooth pits	min^{-1}
k_u	rate constant for uncoupling of complex from internalization component	min^{-1}
k_x	rate constant for receptor crosslinking	$(\#/cell)^{-1}\,min^{-1}$
K_{cp}	coated-pit associated constant	$(\#/cell)^{-1}$
K_D	equilibrium dissociation constant	M
L	ligand concentration in the medium	M
L_d	number of degraded ligand molecules in the medium	$\#/cell$
L_i	concentration of free ligand molecules in endosomes	M
$L_i^\#$	number of intracellular ligand molecules	$\#/cell$
L_t	concentration of free ligand in endosome tubules	M
L_T	total concentration of ligand (free plus bound) in endosomes	M
L_v	concentration of free ligand in endosome vesicles	M
L_0	ligand concentration in the medium	M

Symbol	Definition	Typical Units
n	cell concentration	#/volume
N_{Av}	Avogadro's number ($=6.02 \times 10^{23}$ #/mole)	
P_s	free cell surface internalization component	#/cell
P_T	total number of surface internalization components	#/cell
R_i	number of free receptors in endosomes	#/cell
R_{Ti}	number of receptors (free plus bound) in endosomes	#/cell
R_s	number of free receptors on the cell surface	#/cell
R_{sT}	total number of surface receptors	#/cell
R_{s0}	number of free surface receptors in the absence of ligand	#/cell
R_{sT0}	total number of surface receptors in the absence of ligand	#/cell
R_t	number of free receptors in endosome tubules	#/cell
R_T	total number of receptors	#/cell
R_{eT}	total number of receptors in endosomes	#/cell
R_v	number of free receptors in endosome vesicles	#/cell
t	time	min
t_{sort}	endosome sorting time	min
T_s	number of surface receptor/ligand/internalization component ternary complexes	#/cell
T_i	number of intracellular receptor/ligand/internalization component ternary complexes	#/cell
v	total endosomal volume	μm^3/cell
V_s	rate of new receptor synthesis and expression	(#/cell)-time^{-1}
V_t	total endosome tubule volume	μm^3/cell
V_{sL}	value of V_s in the presence of ligand	(#/cell)-time^{-1}
V_{s0}	value of V_s in the absence of ligand	(#/cell)-time^{-1}
V_v	total endosome vesicle volume	μm^3/cell
δ	ratio of complex internalization rate constant to free receptor internalization rate constant	
ω	scaled fluid-phase ligand uptake rate constant	
λ	rate constant for internalization of ternary complexes in coated pits	min^{-1}
σ	ratio of the fraction of endocytosed receptors degraded in the presence of ligand to the fraction degraded in the absence of ligand	
κ	modified ligand binding constant	
κ_e	partition coefficient for endosome tubules	
τ	dimensionless time	
ε	ratio of the receptor/ligand complex dissociation rate constant to the rate constant for receptors entering tubules	
η	dimensionless binding affinity	
χ	crosslinking equilibrium constant	
v	ratio of the receptor synthesis rate in presence of ligand to the rate in the absence of ligand	

REFERENCES

Adorini, L., Appella, E., Doris, G., Cardinaux, F. and Nagy, Z. A. (1989). Competition for antigen presentation in living cells involves exchange of peptides bound by Class II MHC molecules. *Nature*, **342**:800–803.

Anderson, R. G. W., Brown, M. S., Beisiegel, U. and Goldstein, J. L. (1982). Surface distribution and recycling of the low density lipoprotein receptor as visualized with antireceptor antibodies. *J. Cell Biol.*, **93**:523–531.

Anderson, R. G. W., Vasile, E., Mello, R. J., Brown, M. S. and Goldstein, J. L. (1978). Immunocytochemical visualization of coated pits and vesicles in human fibroblasts: relation to low density lipoprotein receptor distribution. *Cell*, **15**:919–933.

Backer, J. M., Shoelson, S. E., Haring, E. and White, M. F. (1991). Insulin receptors internalize by a rapid, saturable pathway requiring receptor autophosphorylation and an intact juxtamembrane region. *J. Cell Biol.*, **115**:1535–1545.

Bajzer, Z., Myers, A. C. and Vuk-Pavlovic, S. (1989). Binding, internalization, and intracellular processing of proteins interacting with recycling receptors: a kinetic analysis. *J. Biol. Chem.*, **264**:13623–13631.

Bauman, H. and Doyle, D. (1978). Turnover of plasma membrane glycoproteins and glycolipids of hepatoma tissue culture cells. *J. Biol. Chem.*, **253**:4408–4418.

Beaumelle, B. D., Gibson, A. and Hopkins, C. R. (1990). Isolation and preliminary characterization of the endocytic pathway in lymphocytes. *J. Cell Biol.*, **111**:1811–1823.

Beck, J. S. (1988). On internalization of hormone–receptor complexes and receptor recycling. *J. Theor. Biol.*, **132**:263–276.

Beck, J. S. and Goren, H. J. (1983). Simulation of association curves and 'Scatchard' plots of binding reactions where ligand and receptor are degraded or internalized. *J. Receptor Res.*, **3**:561–577.

Beck, J. S. and Goren, H. J. (1985). Determination of binding parameters in the presence of coupled reactions. *Cell Biophys.*, **7**:31–42.

Besterman, J. M., Airhart, J. A., Woodworth, R. C. and Low, R. B. (1981) Exocytosis of pinocytosed fluid in cultured cells: kinetic evidence for rapid turnover and compartmentation. *J. Cell Biol.*, **91**:716–727.

Bretcher, M. S. (1988). Fibroblasts on the move. *J. Cell Biol.*, **106**:235–237.

Bridges, K., Harford, J., Ashwell, G. and Klausner, R. D. (1982). Fate of receptor and ligand during endocytosis of asialoglycoproteins by isolated hepatocytes. *Proc. Natl. Acad. Sci. USA*, **79**:350–354.

Brodsky, F. M. and Guagliardi, L. E. (1991). The cell biology of antigen processing and presentation. *Annu. Rev. Immunol.*, **9**:707–744.

Brown, M. S., Anderson, R. G. W. and Goldstein, J. L. (1983). Recycling receptors: the round-trip itinerary of migrant membrane proteins. *Cell*, **32**:663–667.

Buktenica, S., Olenick, S. J., Salgia, R. and Frankfater, A. (1987). Degradation and regurgitation of extracellular proteins by cultured mouse peritoneal macrophages and baby hamster kidney fibroblasts: kinetic evidence that the transfer of proteins to lysosomes is not irreversible. *J. Biol. Chem.*, **262**:9469–9476.

Carpenter, G. and Cohen, S. (1976). ^{125}I-labelled human epidermal growth factor: binding, internalization, and degradation in human fibroblasts. *J. Cell Biol.*, **71**:159–171.

Casten, L. A. and Pierce, S. K. (1988). Receptor-mediated B cell antigen processing. *J. Immunol.*, **140**:404–410.

Chen, W., Goldstein, J. L. and Brown, M. S. (1990). NPXY, a sequence often found in cytoplasmic tails, is required for coated pit-mediated internalization of the low density lipoprotein receptor. *J. Biol. Chem.*, **265**:3116–3123.

Chen, W. S., Lazar, C. S., Poenie, M., Tsien, R. Y., Gill, G. N. and Rosenfeld, M. G. (1987). Requirement for intrinsic protein tyrosine kinase in the immediate and late actions of the EGF receptor. *Nature*, **328**:820–823.

Chesnut, R. W., Colon, S. M. and Grey, H. M. (1982). Antigen presentation by normal B cells, B cell tumors, and macrophages: functional and biochemical comparison. *J. Immunol.*, **128**:1764–1768.

Chung, K., Walter, P., Aponte, G. W. and Moore, H.-P. (1989). Molecular sorting in the secretory pathway. *Science*, **243**:192–197.

Ciechanover, A., Schwartz, A. L. and Lodish, H. F. (1983). The asialoglycoprotein receptor internalizes and recycles independently of the transferrin and insulin receptors. *Cell*, **32**:267–275.

Collawn, J. F., Stangel, M., Kuhn, L. A., Esekogwu, V., Jing, S., Trowbridge, I. S. and Tainer, J. A. (1990). Transferrin receptor internalization sequence YXRF implicates a tight turn as the structural recognition motif for endocytosis. *Cell*, **63**:1061–1072.

Cresswell, P. (1985). Intracellular class II HLA antigens are accessible to transferrin–neuraminidase conjugates internalized by receptor-mediated endocytosis. *Proc. Natl. Acad. Sci. USA*, **82**:8188–8192.

Daukas, G., Lauffenburger, D. A. and Zigmond, S. (1983). Reversible pinocytosis in polymorphonuclear leukocytes. *J. Cell Biol.*, **96**:1642–1650.

Demotz, S., Grey, H. M. and Sette, A. (1990). The minimum number of Class II MHC-antigen complexes needed for T cell activation. *Science*, **249**:1028–1030.

Draye, J., Courtoy, P. J., Quintart, J. and Baudhuin, P. (1988). A quantitative model of traffic between plasma membrane and secondary lysosomes: evaluation of inflow, lateral diffusion, and degradation. *J. Cell Biol.*, **107**:2109–2115.

Dunn, K. W. and Maxfield, F. R. (1992). Delivery of ligands from sorting endosomes to late endosomes occurs by maturation of sorting endosomes. *J. Cell Biol.*, **117**:301–310.

Dunn, W. A. and Hubbard, A. L. (1984). Receptor-mediated endocytosis of epidermal growth factor by hepatocytes in the perfused rat liver: ligand and receptor dynamics. *J. Cell Biol.*, **98**:2148–2159.

Earp, H. S., Austin, K. S., Blaisdell, J., Rubin, R. A., Nelson, K. G., Lee, L. W. and Grisham, J. W. (1986). Epidermal growth factor (EGF) stimulates EGF receptor synthesis. *J. Biol. Chem.*, **261**:4777–4780.

Eisenlohr, L. C., Gerhard, W. and Hackett, C. J. (1988). Individual class II-restricted antigenic determinants of the same protein exhibit distinct kinetics of appearance and persistence on antigen presenting cells. *J. Immunol.*, **141**:2581–2584.

Evans, W. H. and Flint, N. (1985). Subfractionation of hepatic endosomes in Nycodenz gradients and by free-flow electrophoresis: separation of ligand-transporting and receptor-enriched membranes. *Biochem. J.*, **232**:25–32.

Gennis, R. B. (1989). *Biomembranes: Molecular Structure and Function*. New York: Springer-Verlag, 533 pp.

Geuze, H. J., Slot, J. W. and Schwartz, A. L. (1987). Membranes of sorting organelles display lateral heterogeneity in receptor distribution. *J. Cell Biol.*, **104**:1715–1723.

Geuze, H. J., Slot, J. W. and Strous, G. J. (1983). Intracellular site of asialoglyco-protein receptor–ligand uncoupling: double-label immunoelectron microscopy during receptor-mediated endocytosis. *Cell*, **32**:277–287.

Gex-Fabry, M. and DeLisi, C. (1984). Model for kinetic and steady state analysis of receptor mediated endocytosis. *Math. Biosci.*, **72**:245–261.

Glickman, J. N., Conibear, E. and Pearse, B. M. F. (1989). Specificity of binding of clathrin adaptors to signals on the mannose-6-phosphate/insulin-like growth factor II receptor. *EMBO J.*, **8**:1041–1047.

Goldstein, J. L., Brown, M. S., Anderson, R. G. W., Russell, D. W. and Schneider, W. J. (1985). Receptor-mediated endocytosis: concepts emerging from the LDL receptor system. *Annu. Rev. Cell Biol.*, **1**:1–39.

Goldstein, J. L., Brown, M. S. and Stone, N. J. (1977). Genetics of the LDL receptor: evidence that the mutations affecting binding and internalization are allelic. *Cell*, **12**:629–641.

Gosselin, E. J., Tony, H. and Parker, D. C. (1988). Characterization of antigen processing and presentation by resting B lymphocytes. *J. Immunol.*, **140**:1408–1413.

Griffiths, G., Back, R. and Marsh, M. (1989). A quantitative analysis of the endocytic pathway in baby hamster kidney cells. *J. Cell Biol.*, **109**:2703–2720.

Guagliardi, L. E., Koppleman, B., Blum, J. S., Marks, M. S., Cresswell, P. and Brodsky, F. M. (1990). Co-localization of molecules involved in antigen processing and presentation in an early endocytic compartment. *Nature*, **343**:133–139.

Haigler, H. T., Maxfield, F. R., Willingham, M. C. and Pastan, I. (1980). Dansylcadaverine inhibits internalization of ^{125}I-epidermal growth factor in BALB 3T3 cells. *J. Biol. Chem.*, **255**:1239–1241.

Hanover, J. A. and Dickson, R. B. (1985). Transferrin: receptor-mediated endocytosis and iron delivery. *In* I. Pastan and M. C. Willingham (Eds.), *Endocytosis*, pp. 131–161. New York: Plenum Press.

Hansen, S. H., Sandvig, K. and van Deurs, B. (1991). The preendosomal compartment comprises distinct coated and noncoated endocytic vesicle populations. *J. Cell Biol.*, **113**:731–741.

Harding, C. V. and Unanue, E. R. (1989). Antigen processing and intracellular Ia. *J. Immunol.*, **142**:12–19.

Harding, C. V. and Unanue, E. R. (1990). Quantitation of antigen-presenting cell MHC class II/peptide complexes necessary for T-cell stimulation. *Nature*, **346**:574–576.

Hatae, T., Fujita, M., Sagara, H. and Okuyama, K. (1986). Formation of apical tubules from large endocytic vacuoles in kidney proximal tubule cells during absorption of horseradish peroxidase. *Cell Tissue Res.*, **246**:271–278.

Herbst, J. J., Opresko, L. K., Walsh, B. J., Lauffenburger, D. A. and Wiley, H. S. (1994). Regulation of post-endocytic trafficking of the epidermal growth factor receptor through endosomal retention. *J. Biol. Chem.*, **269**:12865–12873.

Holtzman, E. (1989). *Lysosomes.* New York: Plenum Press, 439 pp.

Hopkins, C. R., Miller, K. and Beardmore, J. M. (1985). Receptor-mediated endocytosis of transferrin and epidermal growth factor receptors: a comparison of constitutive and ligand-induced uptake. *J. Cell Sci. Suppl.*, **3**:173–186.

Hopkins, C. R. and Trowbridge, I. S. (1983). Internalization and processing of transferrin and the transferrin receptor in human carcinoma A431 cells. *J. Cell Biol.*, **97**:508–521.

Kaplan, J. and Keogh, E. A. (1982). Temperature shifts induce the selective loss of alveolar macrophage plasma membrane components. *J. Cell Biol.*, **94**:12–19.

Keen, J. H. (1990). Clathrin and associated assembly and disassembly proteins. *Annu. Rev. Biochem.*, **59**:415–438.

Klausner, R. D. (1989). Sorting and traffic in the central vacuolar system. *Cell*, **57**:703–706.

Klausner, R. D., Van Renswoude, J., Ashwell, G., Kempf, C., Schechter, A. N., Dean, A. and Bridges, K. R. (1983). Receptor-mediated endocytosis of transferrin in K562 cells. *J. Biol. Chem.*, **258**:4715–4724.

Krangel, M. S. (1987). Endocytosis and recycling of the T3-T cell receptor complex: the role of T3 phosphorylation. *J. Exp. Med.*, **165**:1141–1159.

Lakey, E. K., Casten, L. A., Niebling, W. L., Margoliash, E. and Pierce, S. K. (1988). Time dependence of B cell processing and presentation of peptide and native protein antigens. *J. Immunol.*, **140**:3309–3314.

Lauffenburger, D. A., Linderman, J. and Berkowitz, L. (1987). Analysis of mammalian cell growth factor receptor dynamics. *Ann. NY Acad. Sci.*, **506**:147–162.

Limbird, L. E. (1986). *Cell Surface Receptors: A Short Course on Theory and Methods.* Boston: Martinus Nijhoff Publishing, 196 pp.

Linderman, J. J. and Lauffenburger, D. A. (1986). Analysis of intracellular receptor/ligand sorting: calculation of mean surface and bulk diffusion times within a sphere. *Biophys. J.*, **50**:295–305.

Linderman, J. J. and Lauffenburger, D. A. (1988). Analysis of intracellular receptor/ligand sorting in endosomes. *J. Theor. Biol.*, **132**:203–245.

Linderman, J. J. and Lauffenburger, D. A. (1989). *Receptor/Ligand Sorting Along the Endocytic Pathway*. Berlin: Springer-Verlag, 164 pp.

Lorenz, R. G. and Allen, P. M. (1988). Direct evidence for functional self-protein/Ia-molecule complexes in vivo. *Proc. Natl. Acad. Sci. USA*, **85**:5220–5223.

Low, D. A., Baker, J. B., Koonce, W. C. and Cunningham, D. D. (1981). Released protease-nexin regulates cellular binding, internalization, and degradation of serine proteases. *Proc. Natl. Acad. Sci. USA*, **78**:2340–2344.

Lund, K. A., Opresko, L. K., Starbuck, C., Walsh, B. J. and Wiley, H. (1990). Quantitative analysis of the endocytic system involved in hormone-induced receptor internalization. *J. Biol. Chem.*, **265**:15713–15723.

Marsh, M., Griffiths, G., Dean, G. E., Mellman, I. and Helenius, A. (1986). Three-dimensional structure of endosomes in BHK-21 cells. *Proc. Natl. Acad. Sci. USA*, **83**:2899–2903.

Marsh, M., Schmid, S., Kern, H., Harms, E., Male, P., Mellman, I. and Helenius, A. (1987). Rapid analytical and preparative isolation of functional endosomes by free flow electrophoresis. *J. Cell Biol.*, **104**:875–886.

Marshall, S. (1985). Degradative processing of internalized insulin in isolated adipocytes. *J. Biol. Chem.*, **260**:13517–13523.

Matlin, K. S. (1986). The sorting of proteins to the plasma membrane in epithelial cells. *J. Cell Biol.*, **103**:2565–2568.

McKanna, J. A., Haigler, H. T. and Cohen, S. (1979). Hormone receptor topology and dynamics: morphological analysis using ferritin-labeled epidermal growth factor. *Proc. Natl. Acad. Sci. USA*, **76**:5689–5693.

McKinley, D. N. and Wiley, H. (1988). Reassessment of fluid-phase endocytosis and diacytosis in monolayer cultures of human fibroblasts. *J. Cell. Physiol.*, **136**:389–397.

Mellman, I., Fuchs, R. and Helenius, A. (1986). Acidification of the endocytic and exocytic pathways. *Annu. Rev. Biochem.*, **55**:663–700.

Mellman, I. and Plutner, H. (1984). Internalization and degradation of macrophage Fc receptors bound to polyvalent immune complexes. *J. Cell Biol.*, **98**:1170–1177.

Mellman, I., Plutner, H. and Ukkonen, P. (1984). Internalization and rapid recycling of macrophage Fc receptors tagged with monovalent antireceptor antibody: possible role of a prelysosomal compartment. *J. Cell Biol.*, **98**:1163–1169.

Menon, A. K., Holowka, D., Webb, W. W. and Baird, B. (1986). Clustering, mobility, and triggering activity of small oligomers of immunoglobulin E on rat basophilic leukemia cells. *J. Cell Biol.*, **102**:534–540.

Munck, A. and Holbrook, N. J. (1984). Glucocorticoid–receptor complexes in rat thymus cells: rapid kinetic behavior and a cyclic model. *J. Biol. Chem.*, **259**:820–831.

Munck, A. and Holbrook, N. J. (1987). Steroid hormone antagonism and a cyclic model of receptor kinetics. *J. Steroid Biochem.*, **26**:173–179.

Murphy, R. F. (1991). Maturation models for endosome and lysosome biogenesis. *Trends Cell Biol.*, **1**:77–82.

Myers, A. C., Kovach, J. S. and Vuk-Pavlovic, S. (1987). Binding, internalization, and intracellular processing of protein ligands: derivation of rate constants by computer modeling. *J. Biol. Chem.*, **262**:6494–6499.

Opresko, L. K. and Wiley, H. S. (1987). Receptor-mediated endocytosis in *Xenopus* oocytes: evidence for two novel mechanisms of hormonal regulation. *J. Biol. Chem.*, **262**:4116–4123.

Orci, L., Perrelet, A. and Gorden, P. (1978). Less-understood aspects of the morphology of insulin secretion and binding. *Recent Prog. Horm. Res.*, **34**:95–121.

Pastan, I. and Willingham, M. C. (1985). *Endocytosis*. New York: Plenum Press, 326 pp.

Pearse, B. M. F. (1988). Receptors compete for adaptors found in plasma membrane coated pits. *EMBO J.*, **7**:3331–3336.

Pearse, B. M. F. and Crowther, R. A. (1987). Structure and assembly of coated vesicles. *Annu. Rev. Biophys. Biophys. Chem.*, **16**:49–68.

Pearse, B. M. F. and Robinson, M. S. (1990). Clathrin, adaptors, and sorting. *Annu. Rev. Cell. Biol.*, **6**:151–171.

Pugsley, A. P. (1989). *Protein Targeting*. San Diego: Academic Press, Inc. 279 pp.

Robertson, D., Holowka, D. and Baird, B. (1986). Cross-linking of immunoglobulin E-receptor complexes induces their interaction with the cytoskeleton of rat basophilic leukemia cells. *J. Immunol.*, **136**:4565–4572.

Roederer, M., Bowser, R. and Murphy, R. F. (1987). Kinetics and temperature dependence of exposure of endocytosed material to proteolytic enzymes and low pH: evidence for a maturation model for the formation of lysosomes. *J. Cell. Physiol.*, **131**:200–209.

Salzman, N. H. and Maxfield, F. R. (1988). Intracellular fusion of sequentially formed endocytic compartments. *J. Cell Biol.*, **106**:1083–1091.

Salzman, N. H. and Maxfield, F. R. (1989). Fusion accessibility of endocytic compartments along the recycling and lysosomal endocytic pathways in intact cells. *J. Cell Biol.*, **109**:2097–2104.

Schwartz, A. L. (1990). Cell biology of intracellular protein trafficking. *Annu. Rev. Immunol.*, **8**:195–229.

Schwartz, A. L., Ciechanover, A., Merritt, S. and Turkewitz, A. (1986). Antibody-induced receptor loss: different rates for asialoglycoproteins and the asialoglycoprotein receptor in HEPG2 cells. *J. Biol. Chem.*, **261**:15225–15232.

Schwartz, A. L., Fridovich, S. E. and Lodish, H. F. (1982). Kinetics of internalization and recycling of the asialoglycoprotein receptor in a hepatoma cell line. *J. Biol. Chem.*, **257**:4230–4237.

Singer, D. F. and Linderman, J. J. (1990). The relationship between antigen concentration, antigen internalization, and antigenic complexes: modeling insights into antigen processing and presentation. *J. Cell Biol.*, **111**:55–68.

Singer, D. F. and Linderman, J. J. (1991). Antigen processing and presentation: how can a foreign antigen be recognized in a sea of self proteins? *J. Theor. Biol.*, **151**:385–404.

Sklar, L. A., Finney, D. A., Oades, Z. G., Jesaitis, A. J., Painter, R. G. and Cochrane, C. G. (1984). The dynamics of ligand–receptor interactions: real-time analyses of association, dissociation, and internalization of an *N*-formyl peptide and its receptors on the human neutrophil. *J. Biol. Chem.*, **259**:5661–5669.

Smith, K. A. and Cantrell, D. A. (1985). Interleukin 2 regulates its own receptors. *Proc. Natl. Acad. Sci. USA*, **82**:864–868.

Smith, R. M. and Jarett, L. (1983). Quantitative ultrastructural analysis of receptor-mediated insulin uptake into adipocytes. *J. Cell. Physiol.*, **115**:199–207.

Sorkin, A., Krolenko, S., Kudrjavtceva, N., Lazebnik, J., Teslenko, L., Soderquist, A. M. and Nikolsky, N. (1991a). Recycling of epidermal growth factor-receptor complexes in A431 cells: identification of dual pathways. *J. Cell Biol.*, **112**:55–63.

Sorkin, A., Waters, C., Overholser, K. A. and Carpenter, G. (1991b). Multiple auto-phosphorylation site mutations of the epidermal growth factor receptor. *J. Biol. Chem.*, **266**:8355–8362.

Sorkin, A., Helin, K., Waters, C., Carpenter, G. and Beguinot, L. (1992). Multiple autophosphorylation sites of the epidermal growth factor receptor are essential for receptor kinase activity and internalization. *J. Biol. Chem.*, **267**:8672–8678.

Springer, T. A. (1990). Adhesion receptors of the immune system. *Nature*, **346**:425–434.

Starbuck, C. (1991). *Quantitative Studies of Epidermal Growth Factor Receptor Binding and Trafficking Dynamics in Fibroblasts: Relationship to Cell Proliferation.* University of Pennsylvania, Philadelphia, PA.

Starbuck, C. and Lauffenburger, D. A. (1992). Mathematical model for the effects of epidermal growth factor receptor trafficking dynamics on fibroblast proliferation responses. *Biotech. Prog.*, **8**:132–143.

Starbuck, C., Wiley, H. S. and Lauffenburger, D. A. (1990). Epidermal growth factor binding and trafficking dynamics in fibroblasts: relationship to cell proliferation. *Chem. Eng. Sci.*, **45**:2367–2373.

Steinman, R. M., Mellman, I. S., Muller, W. A. and Cohn, Z. A. (1983). Endocytosis and the recycling of plasma membrane. *J. Cell Biol.*, **96**:1–27.

Stoorvogel, W., Geuze, H., Griffith, J. M. and Strous, G. J. (1988). The pathways of endocytosed transferrin and secretory protein are connected in the trans-Golgi recticulum. *J. Cell Biol.*, **106**:1821–1829.

Stoscheck, C. M. and Carpenter, G. (1984). Down regulation of epidermal growth factor receptors: direct demonstration of receptor degradation in human fibroblasts. *J. Cell Biol.*, **98**:1048–1053.

Swanson, J. A., Yirinec, B. D. and Silverstein, S. C. (1985). Phorbol esters and horseradish peroxidase stimulate pinocytosis and redirect the flow of pinocytosed fluid in macrophages. *J. Cell Biol.*, **100**:851–859.

Thilo, L. (1985). Quantification of endocytosis-derived membrane traffic. *Biochim. Biophys. Acta*, **822**:243–266.

Thilo, L. and Vogel, G. (1980). Kinetics of membrane internalization and recycling during pinocytosis in *Dictyostelium discoideum*. *Proc. Natl. Acad. Sci. USA*, **77**:1015–1019.

Tomoda, H., Kishimoto, Y. and Lee, Y. C. (1989). Temperature effect on endocytosis and exocytosis by rabbit alveolar macrophages. *J. Biol. Chem.*, **264**:15445–15450.

Townsend, R. R., Wall, D. A., Hubbard, A. L. and Lee, Y. C. (1984). Rapid release of galactose-terminated ligands after endocytosis by hepatic parenchymal cells: evidence for a role of carbohydrate structure in the release of internalized ligand from receptor. *Proc. Natl. Acad. Sci. USA*, **81**:466–470.

Tran, D., Carpentier, J.-L., Sawano, F., Gorden, P. and Orci, L. (1987). Ligands internalized through coated or noncoated invaginations follow a common intracellular pathway. *Proc. Natl. Acad. Sci. USA*, **84**:7957–7961.

Trowbridge, I. S. (1991). Endocytosis and signals for internalization. *Curr. Opin. Cell Biol.*, **3**:634–641.

Tse, D. B., Cantor, C. R., McDowell, J. and Pernis, B. (1986). Recycling class I MHC antigens: dynamics of internalization, acidification, and ligand-degradation in murine T-lymphoblasts. *J. Mol. Cell. Immunol.*, **2**:315–329.

Turner, J. R., Tartakoff, A. M. and Berger, M. (1988). Intracellular degradation of the complement C3b/C4b receptor in the absence of ligand. *J. Biol. Chem.*, **263**:4914–4920.

Tycko, B., Keith, C. H. and Maxfield, F. R. (1983). Rapid acidification of endocytic vesicles containing asialoglycoprotein in cells of a human hepatoma line. *J. Cell Biol.*, **97**:1762–1776.

Ukkonen, P., Lewis, V., Marsh, M., Helenius, A. and Mellman, I. (1986). Transport of macrophage Fc receptors and Fc receptor-bound ligands to lysosomes. *J. Exp. Med.*, **163**:952–971.

Unanue, E. R. (1984). Antigen presenting function of the macrophage. *Annu. Rev. Immunol.*, **2**:395–428.

Unanue, E. R. and Allen, P. M. (1987). The basis for the immunoregulatory role of macrophages and other accessory cells. *Science*, **236**:551–557.

van Deurs, B., Petersen, O. W., Olsnes, S. and Sandvig, K. (1989). The ways of endocytosis. *Int. Rev. Cytol.*, **117**:131–177.

Ward, D. M., Ajioka, R. and Kaplan, J. (1989). Cohort movement of different ligands and receptors in the intracellular endocytic pathway of alveolar macrophages. *J. Biol. Chem.*, **264**:8164–8170.

Waters, C. M., Oberg, K. C., Carpenter, G. and Overholser, K. A. (1990). Rate constants for binding, dissociation, and internalization of EGF: effect of receptor occupancy and ligand concentration. *Biochemistry*, **29**:3563–3569.

Weigel, P. H. and Oka, J. A. (1982). Endocytosis and degradation mediated by the asialoglycoprotein receptor in isolated rat hepatocytes. *J. Biol. Chem.*, **257**:1201–1207.

Weissman, A. M., Klausner, R. D., Rao, K. and Harford, J. B. (1986). Exposure of K562 cells to anti-receptor monoclonal antibody OKT9 result in rapid redistribution and enhanced degradation of the transferrin receptor. *J. Cell Biol.*, **102**:951–958.

Wells, A., Welsh, J. B., Lazar, C. S. Wiley, S., Gill, G. N. and Rosenfeld, M. G. (1990). Ligand-induced transformation by a noninternalizing epidermal growth factor receptor. *Science*, **247**:962–964.

White, J., Kielian, M. and Helenius, A. (1983). Membrane fusion proteins of enveloped animal viruses. *Q. Rev. Biophys.*, **16**:151–195.

Wileman, T., Boshans, R. and Stahl, P. (1985a). Uptake and transport of mannosylated ligands by alveolar macrophages: studies on ATP-dependent receptor-ligand dissociation. *J. Biol. Chem.*, **260**:7387–7393.

Wileman, T., Harding, C. and Stahl, P. (1985b). Receptor-mediated endocytosis. *Biochem. J.*, **232**:1–14.

Wiley, H. S. (1985). Receptors as models for the mechanisms of membrane protein turnover and dynamics. *Curr. Top. Membr. Transp.*, **24**:369–412.

Wiley, H. S. (1988). Anomalous binding of epidermal growth factor to A431 cells is due to the effect of high receptor densities and a saturable endocytic system. *J. Cell Biol.*, **107**:801–810.

Wiley, H. S. (1992). Receptors: topology, dynamics, and regulation. In *Fundamentals of Medical Cell Biology*, volume 5A, JA1 Press, pp. 113–142.

Wiley, H. S. and Cunningham, D. D. (1981). A steady state model for analyzing the cellular binding, internalization, and degradation of polypeptide ligands. *Cell*, **25**:433–440.

Wiley, H. S. and Cunningham, D. D. (1982). The endocytotic rate constant: a cellular parameter for quantitating receptor-mediated endocytosis. *J. Biol. Chem.*, **257**:4222–4229.

Wiley, H. S., van Nostrand, W., McKinley, D. and Cunningham, D. D. (1985). Intracellular processing of epidermal growth factor and its effect on ligand/receptor interactions. *J. Biol. Chem.*, **260**:5290–5295.

Wiley, H. S., Herbst, J. J., Walsh, B. J., Lauffenburger, D. A., Rosenfeld, M. G. and Gill, G. N. (1991). The role of tyrosine kinase activity in endocytotic compartmentation and down-regulation of the epidermal growth factor receptor. *J. Biol. Chem.*, **266**:11083–11094.

Willingham, M. C., Haigler, H. T., Fitzgerald, D. J. P., Gallo, M. G., Rutherford, A. V. and Pastan, I. H. (1983). The morphologic pathway of binding and internalization of epidermal growth factor in cultured cells: studies on A431, KB, and 3T3 cells, using multiple methods of labeling. *Exp. Cell Res.*, **146**:163–175.

Yamashiro, D. J. and Maxfield, F. R. (1988). Regulation of endocytic processes by pH. *Trends Pharmacol. Sci.*, **9**:190–193.

Ziegler, K. and Unanue, E. R. (1981). Identification of a macrophage antigen processing event required for I-region-restricted antigen presentation to T lymphocytes. *J. Immunol.*, **127**:1869–1875.

Zigmond, S. H. (1981). Consequences of chemotactic peptide receptor modulation for leukocyte orientation. *J. Cell Biol.*, **88**:644–647.

Zigmond, S. H., Sullivan, S. J. and Lauffenburger, D. A. (1982). Kinetic analysis of chemotactic peptide receptor modulation. *J. Cell Biol.*, **92**:34–43.

4

Physical Aspects of Receptor/Ligand
Binding and Trafficking Processes

In the previous two chapters, receptor/ligand binding and trafficking processes were treated as essentially chemical phenomena. That is, the rate constants representing the various kinetic events were considered to be characteristics of the particular molecular components involved. Although this is true to a large degree, receptor/ligand interactions are not governed totally by purely intrinsic chemical properties. Merely physical features of the systems can sometimes significantly influence observed process rates. Examples of such physical aspects include molecular transport (primarily diffusion), receptor and/or ligand multivalency (*i.e.*, the ability to bind via more than one site), and probabilistic effects due to relatively small numbers of cell receptors. Hence, the rates of receptor/ligand binding and trafficking events can depend on quantities such as diffusion coefficients, valence, and receptor number. Although these quantities certainly reflect some characteristics of the molecular components involved, they are much less intimately related to molecular structure that might be altered dramatically by small changes in protein sequence. We will consider them as physical aspects of the system, at least for purposes of classification.

Mathematical analysis of receptor/ligand interactions is made much more complicated by explicit consideration of these physical features. In the previous two chapters we were able to model phenomena in terms of a rather limited framework of first-order ordinary differential equations based on straightforward mass action kinetic species balances. When diffusion, multivalency, and probabilistic effects are taken into account, the models can easily give rise to partial differential equations, second-order ordinary differential equations, extremely large sets of first-order ordinary differential equations, and/or probabilistic differential equations. This fact is another reason for dealing with these aspects in a separate chapter. Those readers preferring to avoid relatively difficult mathematical formulations may skip this chapter without catastrophic loss of ability to develop models for most systems. However, we would be remiss to omit these concepts entirely, as they can be responsible for significant effects in some situations. Multivalent receptors and ligands are not uncommon in important systems of application, such as antibody/antigen interactions and virus binding. Diffusion effects will play a role in ligand binding to surface receptors when the receptor concentration is relatively large or when the binding rates are very fast. In two-dimensional molecular interactions taking place on a membrane, diffusion can be expected generally to play a crucial role. Probabilistic effects must be anticipated when the receptor-mediated cell behavior is sensitive to changes in small numbers of receptors. Specific instances of these various situations will be encountered both in this chapter as illustrative examples and in Chapter 6 which describes models for some receptor-mediated cell functions.

4.1 PROBABILISTIC ASPECTS OF BINDING

In our discussion of simple receptor/ligand binding, we assumed that all cells have R_T receptors and behave identically. In other words, the unique solution to Eqn. (2–8) was presumed to describe each and every cell in the system under study. In reality, of course, this solution represents the mean behavior of the system, and individual cells may show some deviations from the mean behavior. In addition to deviations from the mean behavior caused by, for example, slightly different numbers of receptors on individual cells, these deviations may be due to the *probabilistic*, or *stochastic*, nature of the system.

When there are a very large number of molecules present in a chemically-reacting system, a great number of reaction events will occur in a small amount of time, allowing us mathematically to describe the overall rate of change of complex number in an average sense. For instance, we can follow the change in the mean number per cell of receptor/ligand complexes, C, with time for simple receptor/ligand binding (*cf.* Eqn. (2–8)):

$$\frac{dC}{dt} = k_f RL - k_r C \qquad (4-1)$$

where R is the number of free receptors per cell, L is the extracellular ligand concentration, and k_f and k_r are the forward and reverse binding rate constants. There are no explicit considerations of probability in such equations, and they are known as *deterministic* differential equations. That is, their solution uniquely determines the dynamics of the receptor/ligand binding process they model, giving the exact number of complexes and free receptors and concentration of ligand at any instant in time. In truth, however, these exact numbers are only the mean behavior expected in this situation and hence will be accurate only in that sense (Figure 4–1). In principle, receptor/ligand binding experiments should show fluctuations around a mean. As the number of participating molecules decreases, the greater the relative magnitude of these probabilistic, or stochastic, deviations (*e.g.*, McQuarrie, 1963). It must be emphasized that these deviations have nothing to do with biochemical heterogeneity of a system, but rather arise inherently as a statistical process in a perfectly homogeneous system.

Although the probabilistic nature of a chemically-reacting system may be interesting in the abstract, for most traditional applications of chemical reaction models the number of molecules present is usually so great that a deterministic model is entirely adequate. However, in cell biology, when we are concerned with the rates of reactions within a given cell, the number of molecules available for

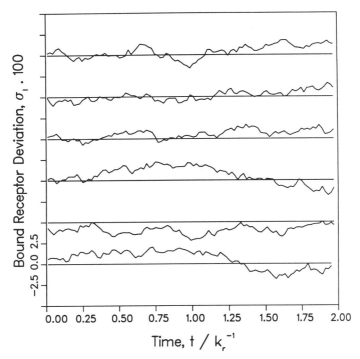

Figure 4–1 Variation in the number of receptor/ligand complexes on one cell as a function of time. The smooth curve shows the mean number of complexes, the value that would be predicted from a deterministic model. From Tranquillo (1990).

any given reaction may be comparatively small, with corresponding opportunity for relatively significant probabilistic fluctuations. Possibly relevant examples include the observation of probabilistic distributions of microtubule lengths (Mitchison and Kirschner, 1984) and bacterial chemotactic response trigger molecules (Spudich and Koshland, 1976). In considering cell receptors, we note that sometimes there are as few as 1,000–10,000 of a particular type present on a cell surface. Further, it has become apparent that cell behavior can be influenced in an exquisitely sensitive way to very small levels of signal through intracellular amplification cascades. Thus, an ability to quantitatively model and analyze receptor/ligand binding from a probabilistic viewpoint may turn out to be quite useful. In at least a few applications so far, probabilistic models of receptor/ligand binding have provided excellent interpretations of cell behavior. Examples include cell directional orientational and migration paths (Tranquillo *et al.*, 1988), detachment of cells from surfaces by distracting fluid shear flow (Cozens-Roberts *et al.*, 1990a,b), and rolling along surfaces in fluid shear flow (Hammer and Apte, 1992).

Before beginning to consider the mathematics of a system described by probabilistic modeling, it is worth noting that cell-to-cell variability is not always readily observable experimentally. For example, when ligand binding is followed by using a radioactively-labeled ligand with quantification by a gamma counter or using a fluorescently-labeled ligand with quantification by a spectrofluorometer, only the mean behavior of a group of cells is observed. However, more sophisticated techniques are now often used to follow events in individual cells; such techniques include flow cytometry and microscopic imaging of single cells.

For purposes of instruction, we wish to construct a probabilistic formulation of the simple case considered in section 2.2.1.a, that of receptor/ligand binding in the absence of significant ligand depletion and non-specific binding effects. We will consider two distinctly different sources of fluctuations: (1) fluctuations in local ligand concentration, and (2) fluctuations in receptor/ligand binding processes.

4.1.1 Fluctuations in Ligand Concentration

Fluctuating ligand concentration will be examined first, for it is the ligand concentration that cell surface receptors detect. Although the bulk ligand concentration L is typically known and may, for most purposes, be considered constant, the actual ligand concentration very near a particular cell will vary with time due to the random, thermal fluctuations of molecular diffusion (see Figure 4–2). It is, of course, this latter ligand concentration that the cell actually "samples" with surface receptors. Thus, random fluctuations in the ligand concentration near a cell may result in deviations in the number of bound receptors from the mean behavior predicted by deterministic models.

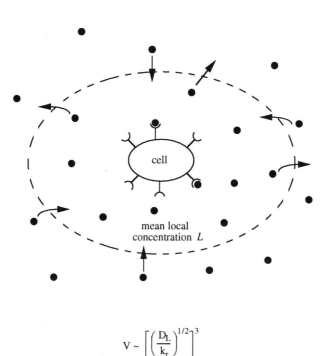

$$V \sim \left[\left(\frac{D_L}{k_r} \right)^{1/2} \right]^3$$

Figure 4–2 Variation in the ligand concentration around an individual cell. The cell samples only the ligand concentration in its local environment of approximately volume V.

For simplicity, we will examine only the effect of these fluctuations on the *equilibrium* number of bound receptors. Consider the equilibrium solution of the kinetic equation for simple ligand binding to cell surface receptors (Eqn. (4–1)), with the mean equilibrium number of receptor/ligand complexes C_{eq} determined as a function of the steady-state ligand concentration L:

$$C_{eq} = \frac{R_T L}{K_D + L} \qquad (4\text{–}2)$$

where R_T is the total number of cell surface receptors and K_D is the equilibrium dissociation constant k_r/k_f. If there are random, thermal fluctuations δL of ligand concentration in the volume of medium accessible to receptor binding, these will lead to corresponding fluctuations in the equilibrium number of complexes according to:

$$\delta C_{eq} = \frac{dC_{eq}}{dL} \delta L = \frac{R_T K_D}{(K_D + L)^2} \delta L \qquad (4\text{–}3)$$

where δC_{eq} is the standard deviation in C_{eq} as a result of a δL, the standard deviation in the value of L (Tranquillo, 1990). The relative magnitude of these fluctuations in receptor binding is given by

$$\frac{\delta C_{eq}}{C_{eq}} = \left[1 + \left(\frac{L}{K_D} \right) \right]^{-1} \frac{\delta L}{L} \qquad (4\text{–}4)$$

Eqn. (4–4) dictates, for example, that at a mean or bulk ligand concentration of $L = K_D$, a 10% fluctuation in local ligand concentration (*i.e.*, the ligand concentration that the cell detects) translates into a 5% fluctuation in equilibrium receptor occupancy.

In order to use Eqn. (4–4) to calculate the expected effect of fluctuations in ligand concentration on equilibrium receptor binding, we need to estimate a value for the term $\delta L/L$. Estimates for thermal fluctuations in local ligand concentration can be easily obtained (Tranquillo and Lauffenburger, 1987a). The magnitude of $\delta L/L$ is approximately

$$\delta L/L = (N_{Av} L V)^{-1/2} \qquad (4\text{–}5)$$

where V is the "sampling volume" of medium accessible to ligand binding and the product $N_{Av} L V$ is simply the expected number of ligand molecules in the sampling volume if there were no such fluctuations. To specify the volume V, we assume that $V \sim l^3$, where l is a characteristic system dimension. The average distance or length that a diffusing molecule will travel in a time period t_{sample} is $l \sim (D_L t_{sample})^{1/2}$, where D_L is the translational diffusion coefficient of the ligand. A reasonable estimate for t_{sample} is given by k_r^{-1} for receptor/ligand binding, since that is the mean time period between receptor binding events at steady state when $L = K_D$. Hence, $V \sim (D_L/k_r)^{3/2}$. To determine the value of $\delta L/L$ at $L = K_D$, then, we simply substitute this estimate for V into Eqn. (4–5) to find $\delta L/L = [(k_r^{3/2})/(N_{Av} K_D D_L^{3/2})]^{1/2}$. For the parameter value ranges of $D_L = 10^{-6}\text{–}10^{-5}$ cm^2/s, $k_r = 10^{-4}\text{–}10^{-1}$ s^{-1}, and $K_D = 10^{-10}\text{–}10^{-6}$ M, we find that $\delta L/L \sim 10^{-7}\text{–}10^{-2}$, or 10^{-5}% to 1%. The high end of this range should be achieved only rarely, for ligands with high affinity but large dissociation rate constant. Most commonly, the lower, negligible end of the range will be applicable. Thus fluctuations in ligand concentration are unlikely to contribute significantly to deviations from mean or deterministic equilibrium receptor binding. In other words, the "signal" of extracellular ligand concentration that the cell must detect is unlikely to vary significantly due to thermal fluctuations.

4.1.2 Fluctuations in Kinetic Binding Processes

The second and more likely source of deviations in the number of bound receptors from the mean behavior predicted by deterministic models is fluctuating reaction kinetics. Stated another way, the "signal" of ligand concentration analyzed above may be relatively constant, but the "detector" of surface receptors may contribute random errors caused by the probabilistic nature of the binding event. The rate constant for a chemical reaction represents the time-probability that the reaction will occur for any particular reactant molecule. For instance, when we write k_r with units of time^{-1} to characterize the rate constant for dissociation of a receptor/ligand complex into its component receptor and ligand molecules, that rate constant can be alternatively interpreted as the probability per unit time that a single complex will, in fact, dissociate. In other words, as the time interval Δt gets small, at most one dissociation event will occur on a particular receptor and the probability of that event is approximately $k_r \Delta t$. A similar situation exists for the association step, where $k_f L$ actually represents the probability per unit time (at a given ligand concentration) that a receptor will bind a ligand to form a complex. Figure 4–3 illustrates this concept. These fluctuations can be analyzed using a *probabilistic*, so-called "population balance", model for receptor/ligand complexes in place of the deterministic model Eqn. (4–1):

$$\frac{dC}{dt} = k_f RL - k_r C \tag{4–1}$$

An excellent background text for this approach is Gardiner (1983), and a clear derivation of a relevant example case is provided by Tranquillo (1990).

Consider the association and dissociation events that might occur during a short time interval Δt. If we choose this time interval to be small enough, at most one event of any type can take place. Let $P_j(t)$ represent the probability that there are j complexes on a cell at time t. We can then write a kinetic equation describing changes in the number of complexes during Δt, given that there are C complexes present at time t:

$$P_C(t + \Delta t) - P_C(t) = k_f L[R_T - (C - 1)]P_{C-1}(t)\Delta t$$
$$- k_f L[R_T - C]P_C(t)\Delta t - k_r C P_C(t)\Delta t$$
$$+ k_r(C + 1)P_{C+1}(t)\Delta t \tag{4–6}$$

In this equation, the first and second terms on the right-hand side represent the probability that there were $C - 1$ and C complexes, respectively, present at time t and one binding event occurred during Δt; the third and fourth terms represent the probability that there were C and $C + 1$ complexes, respectively, present at

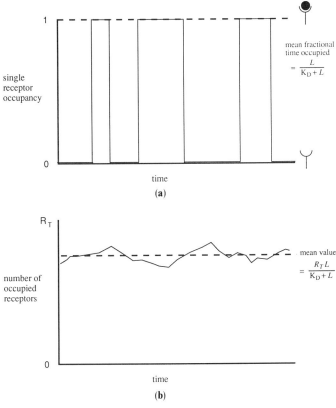

Figure 4–3   Variation in equilibrium receptor occupancy. (a) A fictional history of the occupancy of a single receptor is shown. The state of the receptor jumps between 0, or unbound, and 1, or bound. The mean fraction of time that a particular receptor is occupied is given by $L/(K_D + L)$. (b) The actual number of receptors bound on a single cell following a long exposure to a constant ligand concentration is plotted as a function of time. The mean or deterministic prediction for the equilibrium number of bound receptors is $R_T L/(K_D + L)$.

time t and one dissociation event occurred. (Terms of higher order in Δt are neglected based on our assumption that the time interval is sufficiently short that at most one event can happen.) In the limit $\Delta t \sim 0$, this discrete-time equation can be rewritten as a differential equation:

$$\frac{dP_C}{dt} = k_f L[R_T - (C - 1)]P_{C-1} + k_r(C + 1)P_{C+1}$$

$$- \{k_f L[R_T - C] + k_r C\}P_C \quad C = 1, 2, 3, \ldots, (R_T - 1) \quad (4\text{–}7a)$$

Note that there are $(R_T - 1)$ such equations, for $C = 1, 2, 3, \ldots, (R_T - 1)$. For $C = 0$ and $C = R_T$, the corresponding equations are slightly different:

$$dP_0/dt = -k_f L R_T P_0 + k_r P_1 \quad\quad\quad\quad\quad\quad (4\text{–}7b)$$

$$dP_{R_T}/dt = k_f L P_{R_T - 1} - k_r R_T P_{R_T} \quad\quad\quad\quad (4\text{–}7c)$$

This set of equations is termed the *master equation*. When the ligand concentration

L remains constant, Eqns. (4–7a–c) form a system of $(R_T + 1)$ coupled linear ordinary differential equations, and as such can be solved analytically for the various transient probabilities, $P_C(t)$. A possible set of initial conditions, those describing the case when no receptors are bound at time 0, is $P_C(0) = 0$ for $C \neq 0$ and $P_C(0) = 1$ for $C = 0$.

One method of solving Eqns. (4–7a–c) is to transform the system of ordinary differential equations into a single partial differential equation which can be more easily solved. An approach for doing this is to define a *generating function*, $G(s, t)$:

$$G(s, t) = \sum_{C=0}^{R_T} s^C P_C(t) \tag{4–8}$$

where s is a dummy variable (Bharuca-Reid, 1960). We next multiply each of our original equations (Eqns. (4–7a–c)) by s^C, sum the equations, and then recognize that the terms containing P_C can be written as functions of G and its derivatives. This procedure yields the single partial differential equation

$$\frac{\partial G}{\partial t} = (1 - s) \left\{ (k_f Ls + k_r) \frac{\partial G}{\partial s} - (k_f LR_T)G \right\} \tag{4–9}$$

The initial condition on P_C given above ($P_C(0) = 0$ for $C \neq 0$ and $P_C(0) = 1$ for $C = 0$) gives rise to the *initial condition* on G:

$$G(s, 0) = \sum_{C=0}^{R_T} s^C P_C(0) = 1 \tag{4–10a}$$

and the requirement that all probabilities must sum to one can be shown to correspond to the *boundary condition* on G:

$$G(1, t) = \sum_{C=0}^{R_T} P_C(t) = 1 \tag{4–10b}$$

An alternative method for converting the large set of ordinary differential equations as in Eqns. (4–7a–c) into a single partial differential equation involves a Taylor series expansion of P_C (see, for example, Gardiner, 1983). An example of this procedure relevant to receptor/ligand phenomena is provided by Tranquillo and Lauffenburger (1987b) for simulation of cell migration paths. This work will be described in Chapter 6.

The solution $G(s, t)$ of Eqns. (4–9) and (4–10), once found, can then be used to recover the individual probabilities from the formulae:

$$P_0(t) = G(0, t) \tag{4–11a}$$

$$P_C(t) = \frac{1}{C!} \left[\frac{d^C G}{ds^C} \right]_{s=0} \tag{4–11b}$$

Typically, one is most interested in the expected or mean value of C, denoted $\langle C \rangle$, and the variance σ_C^2. Expressions for these quantities can be found from the generating function $G(s, t)$, once the solution to Eqn. (4–9) is known. $\langle C \rangle$ and σ_C^2

are defined and related to the generating function by:

$$\langle C \rangle = \sum_{C=0}^{R_T} C P_C = \left[\frac{\partial G}{\partial s} \right]_{s=1} \tag{4-12a}$$

$$\sigma_C^2 = \sum_{C=0}^{R_T} (C - \langle C \rangle)^2 P_C = \left[\frac{\partial^2 G}{\partial s^2} + \frac{\partial G}{\partial s} - \left(\frac{\partial G}{\partial s} \right)^2 \right]_{s=1} \tag{4-12b}$$

Using the approach outlined above, we examine first the *steady-state solution* to Eqn. (4–9), the same as equilibrium in this case, arising from setting $\partial G / \partial t = 0$. This solution, obtained by direct integration, is given by

$$G(s) = \left[\frac{s + (k_r/k_f L)}{1 + (k_r/k_f L)} \right]^{R_T} \tag{4-13}$$

Using this solution together with Eqns. (4–12a,b), we can find that at equilibrium

$$\langle C_{eq} \rangle = \frac{R_T L}{K_D + L} \tag{4-14a}$$

$$(\sigma_C^2)_{eq} = \frac{R_T L K_D}{(K_D + L)^2} \tag{4-14b}$$

Notice that the mean value $\langle C_{eq} \rangle$ is identical to that given by the solution to the deterministic model (Eqn. (4–2)); this is a general result for linear equations. The interesting and important observation is that the statistical variance expected at equilibrium binding is proportional to the total number of cell receptors, R_T. The root-mean-square deviation, $\delta C_{eq} = (\sigma_C)_{eq}$, is equal to $[(R_T L K_D)^{1/2}/(K_D + L)]$.

Using this expression for δC_{eq} together with Eqn. (4–2) or (4–14a) permits us to write a simple expression for the expected relative root mean square fluctuation in equilibrium complex number due to stochastic effects in binding:

$$\frac{\delta C_{eq}}{C_{eq}} = \left(\frac{K_D}{L R_T} \right)^{1/2} \tag{4-15}$$

Calculated values of $\delta C_{eq}/C_{eq}$ are plotted in Figure 4–4 as a function of L/K_D and R_T. At $L = K_D$ the expected mean fluctuation in receptor occupancy at equilibrium will be $\delta C_{eq}/C_{eq} = R_T^{-1/2}$. When $R_T = 10^4$ receptors/cell, then, statistical fluctuations with average magnitude equal to 1% of equilibrium receptor/ligand complex number can be expected around the mean complex number. Such small fluctuations are difficult to measure experimentally, but they may have significant consequences for cell behavior. A central example is chemotaxis, in which cells can bias the spatial orientation of their movement in the direction of chemotactic attractant concentration gradients. Zigmond (1977) has shown experimentally that relative concentration gradients of about 1% across cell dimensions are sufficient to induce noticeable directional bias in neutrophil leukocytes when $L \sim K_D$ for a peptide attractant and with $R_T \sim 10^4$ peptide receptors/cell. Thus, statistical fluctuations in receptor binding may be of the same order of magnitude as the apparent gradient signal. In other situations, such as cell adhesion receptor

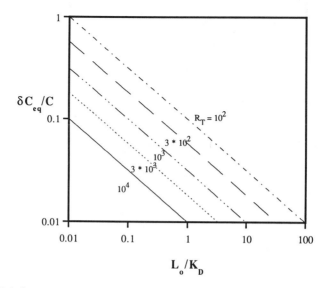

Figure 4–4 Relative root mean square fluctuation in the equilibrium number of complexes on a cell. The plot shows the prediction of the probabilistic model for homogeneous receptor/ligand binding. Note that fluctuations become more significant as the number of receptors or the ratio of L to K_D decreases.

binding in small interfacial contact zones, the number of receptors involved may be even smaller so that the significance of statistical fluctuations can be expected to be greater still.

Next, consider the *transient solution of* Eqn. (4–9) with initial and boundary conditions given by Eqns. (4–10a,b). The solution for $G(s, t)$ can be obtained by the method of characteristics (see Rhee *et al.*, 1986). The mean and variance of C can be found again by using Eqns. (4–12a,b) (McQuarrie, 1963):

$$\langle C(t) \rangle = \frac{R_T L}{K_D + L} [1 - e^{-(k_f L + k_r)t}] \tag{4–16a}$$

$$(\sigma_C^2)_{eq} = \frac{R_T L}{(K_D + L)^2} [L e^{-(k_f L + k_r)t} + K_D][1 - e^{-(k_f L + k_r)t}] \tag{4–16b}$$

Note that Eqn. (4–16a) is the same as the transient solution to the deterministic binding equation (Eqn. (4–1)) with $C_0 = 0$ (see Eqn. (2–12)).

This result was obtained by DeLisi and Marchetti (1983) in their analysis of receptor binding fluctuations in chemotactic responses. They use it to conclude that leukocytes must possess an intracellular means for averaging receptor binding events over some period of time, roughly a few minutes, so that these cells can orient their movement direction in small spatial gradients of attractant ligand concentration as observed experimentally. That is, the predicted variance in bound receptors at any instant in time is greater than the spatial difference in bound receptors across cell dimensions which apparently serves as the orienting signal. Therefore, some mechanism must exist for averaging these fluctuations over time, so that the variance becomes smaller than the orienting signal. This suggestion forms the basis for some probabilistic modeling work (Tranquillo *et al.*, 1987b, 1988) to be examined in section 6.3.3.b.

A similar result for a slightly modified model is obtained in a probabilistic analysis of cell adhesion (Cozens-Roberts *et al.*, 1990a,b). The predicted variance in the mean number of receptor/ligand bonds formed during attachment of cells to surfaces may account for transient cell detachment data in which cells continuously detached from the surface under uniform distracting fluid shear flow conditions. While heterogeneity of the cell population can also explain this observation, quantitative estimates suggest that the probabilistic effects may be at least as significant as heterogeneity in the system studied by these investigators. This cell adhesion model is discussed in section 6.2.5.b.

It should be clear that treatment of probabilistic aspects of receptor/ligand binding is mathematically much more complicated than the deterministic, or mean, behavior analyses of Chapter 2. Thus, a probabilistic modeling approach ought to be applied only when it is strongly suspected that interpretation of experimental behavior requires it. On the other hand, it is our belief that probabilistic models will prove to be of increasing utility in the near future as experimental assays for cell function become more commonly designed to generate data on individual cells. Useful texts for more information on probabilistic models include Bharuca-Reid (1960), Goel and Richter-Dyn (1974), and Doraiswamy and Kulkarni (1987).

4.2   DIFFUSION EFFECTS

Simple binding of a ligand molecule to a receptor is generally treated as a reversible, one-step process:

$$R + L \underset{k_r}{\overset{k_f}{\rightleftharpoons}} C$$

where k_f is the per receptor association rate constant and k_r is the per receptor dissociation rate constant. However, it is well-known in chemical reaction kinetics that the binding of two molecules is really a *two-step process*, requiring first the molecular transport of the individual molecular species, R and L, with *transport rate constant* k_+ before intrinsic chemical reactions or binding interactions can occur (Eigen, 1974) (see Figure 4–5). We will consider the transport mechanism to be molecular diffusion, because diffusive transport dominates convective transport at cellular and subcellular length scales (Weisz, 1973). The chemical reaction step itself is then characterized by the *intrinsic association rate constant* k_{on} and *intrinsic dissociation rate constant* k_{off}. Thus, the values of k_f and k_r that are measured in kinetic experiments are really combination rate constants which include both the transport and reaction effects, and our analysis will allow the determination of the relative contributions of each. By the analyses presented in this section, we can examine how the individual rate constants k_+, k_{on}, and k_{off} contribute to the overall observed rate constants k_f and k_r. In this way, we can gain a quantitative understanding of how the diffusion coefficient D can affect binding kinetics. We will also learn that diffusion effects can cause the binding

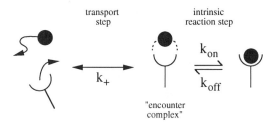

transport   intrinsic
step   reaction step

k_+

"encounter
complex"

k_{on}
k_{off}

Figure 4–5 Separation of an overall binding or dissociation event into two steps. The intrinsic binding step is characterized by rate constants k_{on} and k_{off} which are determined by receptor and ligand molecular properties. The transport step is characterized by rate constant k_+ and is influenced by diffusion and geometric considerations. The state in which receptor and ligand are close enough to bind but have not yet done so is sometimes termed the encounter complex.

parameters k_f and k_r to be not constants, but instead functions of the number of available free receptors, R; this phenomenon can arise because diffusion is highly affected by geometry.

We will consider the interplay of diffusion and intrinsic binding interactions in three types of situations: (1) binding of ligand to receptor when both are free in solution; (2) binding of ligand to receptor when the former is free in solution and the latter is on a cell surface; and, (3) interaction between a receptor and an accessory molecule when both are on a cell surface (in this latter case, the binding parameter k_f and k_r are replaced by the coupling parameters k_c and k_u, following the nomenclature of Chapter 2). These cases are illustrated in Figure 4–6, and will now be examined in order.

4.2.1 Ligand Binding to Solution Receptors

We will begin by looking at the binding of ligand to receptor when both are free in solution (Figure 4–6a) and will follow the approach used by Shoup and Szabo (1982). Our aim is to obtain expressions for the overall forward and reverse rate constants, k_f and k_r, as they depend on diffusion coefficients and free receptor number. In order to accomplish this, we will calculate the rate at which a ligand molecule would be expected to bind to a receptor molecule by the two-step diffusion and intrinsic reaction process.

A steady-state diffusion equation is written for the concentration of ligand molecules, $L(r)$, around a single receptor molecule placed at the origin of a spherical coordinate system:

$$D \frac{1}{r^2} \frac{d}{dr}\left(r^2 \frac{dL}{dr}\right) = 0 \qquad (4\text{–}17)$$

where D is the sum of the ligand and receptor diffusivities, $D = (D_L + D_R)$. The ligand concentration L varies with distance from the receptor, but very far away it is just equal to the bulk ligand concentration L_0. This provides one *boundary condition*:

$$r \to \infty \qquad L \to L_0 \qquad (4\text{–}18a)$$

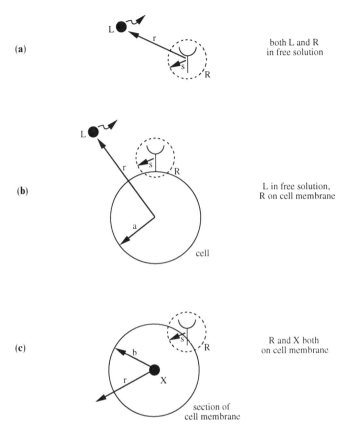

Figure 4–6 Receptor binding in three geometric situations. (a) Binding of ligand to receptors free in solution. (b) Binding of ligand to cell surface receptors. (c) Binding of cell surface molecules (X) to cell surface receptors. The radius of an encounter complex is s, the cell radius is a, a typical distance between molecule X and a receptor R is b, and r is the radial position.

The rate at which ligand molecules are bound by a receptor is equal to the intrinsic reaction rate constant k_{on} times the ligand concentration at the receptor surface, L evaluated at $r = s$; s is often termed the *encounter radius*. At steady-state this must be equal to the ligand diffusion rate (moles/time) to the receptor surface (the diffusive flux multiplied by the surface area, again evaluated at $r = s$). This provides the second boundary condition needed for Eqn. (4–17):

$$4\pi s^2 D \left. \frac{dL}{dr} \right|_{r=s} = k_{on} L(s) \tag{4–18b}$$

The solution of Eqn. (4–17) with these boundary conditions (Eqns. (4–18a,b)) is

$$L(r) = \frac{-k_{on} s L_0}{4\pi D s + k_{on}} \left(\frac{1}{r} \right) + L_0 \tag{4–19}$$

The overall flux of molecules to the receptor is given by $k_f L_0$ and is equal to

$4\pi s^2 D (\mathrm{d}L/\mathrm{d}r)_{r=s}$. Thus the overall or observable association rate constant k_f is determined by

$$k_f = L_0^{-1} 4\pi s^2 D \left[\frac{\mathrm{d}L}{\mathrm{d}r} \right]_{r=s} = k_{on} L(s) L_0^{-1} \tag{4-20}$$

giving

$$k_f = \frac{4\pi D s k_{on}}{4\pi D s + k_{on}} \tag{4-21}$$

The quantity $4\pi Ds$ (Eqn. (4–21)) is worthy of further examination. Consider the situation in which binding events occur much more rapidly than diffusion, such that all ligand reaching the receptor can be assumed to disappear instantaneously. To describe this situation using the classical diffusion theory of Smoluchowski (1917), Eqn. (4–17) is solved with the boundary condition $L = 0$ at $r = s$ replacing the boundary condition Eqn. (4–18b). The solution can then be used in Eqn. (4–20) to find that the overall rate constant in this situation is equal to $4\pi Ds$. Because this rate constant describes the situation in which diffusion is playing the limiting role, we identify the value $4\pi Ds$ with the rate constant k_+. Thus, Eqn. (4–21) can be rewritten as

$$k_f = \frac{k_+ k_{on}}{k_+ + k_{on}} = \left(\frac{1}{k_+} + \frac{1}{k_{on}} \right)^{-1} \tag{4-22a}$$

with

$$k_+ = 4\pi Ds \tag{4-22b}$$

The form of this equation suggests an appealing interpretation: the overall "resistance" to binding, $1/k_f$, is the sum of the two individual "resistances" of diffusion ($1/k_+$) and intrinsic reaction ($1/k_{on}$) in series. This expression is the key result that enables us to assess the relative contribution of the diffusion, or transport, step to the overall forward rate constant for isolated receptors and ligands binding in solution. Note that if $k_{on} \gg k_+$, then $k_f \sim k_+ = 4\pi Ds$ and the binding is termed *diffusion-limited*. On the other hand, if $k_{on} \ll k_+$, then $k_f \sim k_{on}$ and the binding is termed *reaction-limited*.

Typical values for k_+ can be easily estimated for receptor/ligand binding in free solution. Using the parameter ranges $D = 10^{-7}$–10^{-5} cm^2/s and $s = 1$–10 nm, we obtain $k_+ \sim 1 \times 10^{-13}$–$1 \times 10^{-10}$ cm^3/s. Because the initial problem solved (Eqn. (4–17)) is based on the diffusion of a single ligand molecule to a receptor, the units on k_+ (also k_{on} and k_f) are more accurately $(\#/\text{cm}^3)^{-1}$ s^{-1}. We can thus change the units of our calculated k_+ by multiplying by Avogadro's number and converting from cm^3/mole to M^{-1} to give $k_+ \sim 6 \times 10^7$–6×10^{10} M^{-1} s^{-1}. Very few values of k_f measured in free solution are found in this range, so that pure diffusion-control of isolated receptor/ligand binding reactions is not common. Thus, for receptors and ligands in free solution, the experimentally measured value of the overall forward rate constant k_f is usually a reasonable estimate for the intrinsic binding rate constant k_{on}.

As an aside, we note that the derivation of $k_+ = 4\pi Ds$ assumes that the molecules are uniformly reactive, *i.e.*, there are no spatial orientation restrictions.

For macromolecules such as ligands and receptors, this is not really true. A number of investigators have tackled this problem (see the review by Berg and von Hippel, 1985), with the finding that if the reactive site has radius s' which is small compared to s, then a good approximation is $k_+ = 4Ds'$. This result is not surprising, for it is the rate constant for diffusion to a disk of radius s' lying on an infinite plane. A more detailed expression is $k_+ = 2\pi Ds\{\sin^2(\theta/2)[(1/2)^{1/2} + (3/8)^{1/2}]\}$, where θ is the angle circumscribing the reactive portion of the receptor. Note that when $\theta = \pi/2$, the relation $k_+ \sim 4Ds$ is obtained. These alternative expressions for k_+, as well as others depending on the specific geometric situation, may be substituted as appropriate for k_+ in the equations that follow. It should also be mentioned at this point that a more sophisticated computational approach to this issue is the application of Brownian dynamics simulations incorporating molecular structure and intermolecular interactions for the reacting species (e.g., see McCammon, 1991). This approach is discussed briefly in section 4.2.5.

Shoup and Szabo (1982) define the "capture probability" γ,

$$\gamma = \frac{k_{on}}{k_{on} + k_+} \tag{4-23}$$

which is the probability that closely-associated R and L actually bind to become a true receptor/ligand complex. With this, Eqn. (4–22a) can be rewritten in the elegant form:

$$k_f = \gamma k_+ \tag{4-24}$$

Thus, γ essentially quantifies the extent to which receptor/ligand association is rate-limited by the reaction step. As γ approaches 0, association is severely reaction-limited, whereas as γ nears 1, binding is almost purely diffusion-limited.

The overall reverse rate constant, k_r, can be found by analogy. Expressed in its clearest form, the result is

$$k_r = (1 - \gamma)k_{off} \tag{4-25}$$

Hence, the overall dissociation rate constant is the product of the complex dissociation rate constant and the "escape probability" (equal to $1 - \gamma$).

4.2.2 Ligand Binding to Cell Surface Receptors

The preceding analysis has been devoted to receptor/ligand binding in free solution. In this book, on the other hand, we are mainly interested in binding of ligands to receptors restricted to two-dimensional cell membranes. Beginning with the seminal contribution by Berg and Purcell (1977), investigators have developed various approaches for this desired extension of the analysis (e.g., DeLisi and Wiegel, 1981; Brunn, 1981; Zwanzig, 1990). In the context of our current presentation, the approach by Shoup and Szabo (1982), as discussed above, is most direct. Consider a spherical cell of radius a with receptors of radius s (Figure 4–6b). Eqn. (4–22a) can be applied after two important modifications. First, the forward transport rate constant for ligand molecules diffusing to the cell surface is

$$(k_+)_{cell} = 4\pi Da \tag{4-26}$$

because the relevant radius is now a and not s. The diffusion coefficient D is equal to $(D_L + D_{cell})$, or, because $D_{cell} \ll D_L$, $D \sim D_L$. In this analysis, the diffusion of receptors on the cell surface is not taken into account. Second, the effective forward reaction rate constant for the entire cell is

$$(k_{on})_{cell} = Rk_{on} \tag{4-27}$$

for a cell possessing R free surface receptors, where k_{on} is the individual receptor intrinsic association rate constant. Thus by Eqn. (4–22a), the rate constant for ligand binding by the entire cell is

$$(k_f)_{cell} = \frac{(k_+)_{cell}Rk_{on}}{(k_+)_{cell} + Rk_{on}} \tag{4-28}$$

As shown explicitly by Goldstein (1989), the overall forward rate constant for ligand association with a free receptor on a cell surface, on our usual per receptor basis, is simply found by dividing Eqn. (4–28) by the number of free receptors, R, to obtain

$$k_f = \frac{(k_+)_{cell}k_{on}}{(k_+)_{cell} + Rk_{on}} \tag{4-29}$$

Because the interaction of isolated receptors and ligands in solution is typically close to a reaction-limited situation (as seen in the previous section), k_{on} is essentially the association rate constant that would be experimentally measured for an isolated receptor in free solution if there were no chemical effects of removing the receptor from its cell membrane lipid environment.

Recalling from Eqn. (4–24) the relationship between k_f and the capture probability γ, Eqn. (4–28) shows that for this present case of cell surface receptors, the capture probability for the entire cell is

$$\gamma_{cell} = \frac{Rk_{on}}{(k_+)_{cell} + Rk_{on}} \tag{4-30}$$

(Compare this expression to Eqn. (4–21) for an isolated receptor in free solution). Eqns. (4–25) and (4–30) are now used to evaluate the overall reverse rate constant for dissociation of a ligand molecule from cell surface receptors (Goldstein, 1989):

$$k_r = \frac{(k_+)_{cell}k_{off}}{(k_+)_{cell} + Rk_{on}} \tag{4-31}$$

A striking consequence of Eqns. (4–29) and (4–31) is that the overall rate constants for ligand association to and dissociation from cell surface receptors, k_f and k_r, are not, in fact, constants in general, but rather may be functions of the free receptor number R. k_f, k_r, and $(k_f)_{cell}$ as functions of R are illustrated in Figure 4–7. Notice from this figure that the per receptor rate constants k_f and k_r decrease as R increases. The diminution in k_f is caused by the spatial restriction of receptors onto the cell surface, where they can compete with one another for ligand molecules. An increase in R reduces the dissociation rate constant k_r by increasing the probability that ligand dissociation from one receptor results in binding to a neighboring receptor rather than escape from the cell surface entirely

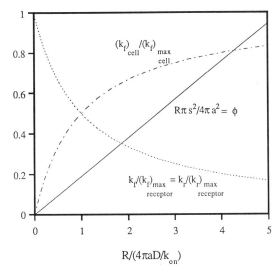

Figure 4–7   Variation of the per receptor association and dissociation rate constants, k_f and k_r, with the number of free receptors. Although the assumption is typically made that k_f and k_r are constants, this is not strictly true when diffusion effects are significant (see Eqns. (4–29) and (4–31)). $(k_f)_{\text{max rec}}$ and $(k_r)_{\text{max rec}}$, the maximum values of k_f and k_r, are given by k_{on} and k_{off}. The per cell association rate constant $(k_f)_{\text{cell}}$ also varies with the number of free receptors (see Eqn. (4–28)); the maximum value $(k_f)_{\text{max cell}}$ is given by $(k_+)_{\text{cell}}$. ϕ is the fractional surface coverage of cell area by receptors. For calculation of ϕ, $s = 10$ nm, $a = 10$ μm, $k_{\text{on}} = 10^7$ M^{-1} s^{-1}, and $D = 10^{-6}$ cm^2/s.

(Goldstein, 1989). The implications of these changing rate constants on the kinetics of receptor/ligand binding are discussed more fully in section 4.2.4.

Another consequence of these equations is that the rate constant for ligand capture by the entire cell, $(k_f)_{\text{cell}}$, approaches a significant fraction of its maximum value, $(k_+)_{\text{cell}} = 4\pi Da$, for small cell surface area coverage by receptors. This can be seen from the diagonal line on Figure 4–7, which converts the quantity $Rk_{\text{on}}/(k_+)_{\text{cell}}$ into an equivalent value of ϕ, the fractional *surface coverage* of all surface area by receptors ($\phi = R\pi s^2/4\pi a^2$), for a typical set of parameter estimates. For the estimates used in Figure 4–7, for example, a fractional surface coverage of only 0.1 (10%) gives almost 40% of the maximal rate of ligand binding to cells.

A rigorous experimental verification of these theoretical results has appeared (Erickson *et al.*, 1987). These investigators quantitatively varied cell surface receptor number by binding different amounts of monoclonal anti-DNP IgE antibodies to Fc receptors on rat basophilic leukemia (RBL) cells. These IgE molecules then served as (bivalent) receptors for the monovalent hapten ligand 2,4-dinitrophenyl (DNP)-aminocaproyl-L-tyrosine, or DCT, so that ligand binding to cells possessing a range of values for R_T could be explored. The critical result is shown in Figure 4–8, a plot comparing the value of the rate constants $(k_f)_{\text{cell}}$ and k_f as determined from their kinetic binding data to the theoretical predictions of Eqns. (4–28) and (4–29). Excellent agreement exists for realistic values of D and a. Furthermore, the value for k_{on} estimated from this comparison, $k_{\text{on}} = 1.1 \times 10^8$ M^{-1} s^{-1}, is very close to the value measured for the binding of DCT to isolated IgE molecules in free solution as expected.

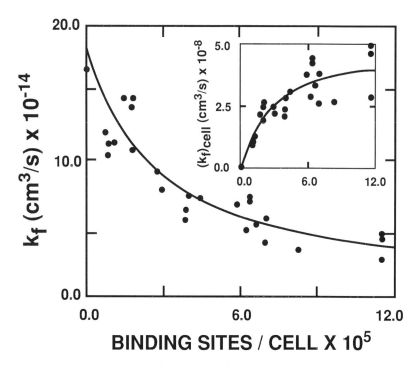

Figure 4–8 Experimental data and model fit on the variation of the per receptor and per cell association rate constants with the total number of receptors. Reproduced from the *Biophysical Journal*, 1987, Vol. 52, pp. 657–662 by copyright permission of the Biophysical Society.

A different sort of check on the diffusion-limitation theory has been performed by Northrup (1988), who reported Brownian dynamics computer simulations of ligand diffusing to cell surface receptors. His calculations were aimed at the special case in which the intrinsic binding rate constant takes on a diffusion-limited value; that is, $k_{on} = 4Ds$ (which arises for diffusion of ligand to a disk of radius s) (Berg and Purcell, 1977). In this limiting case, $(k_f)_{cell} = 4\pi DaRs/(Rs + \pi a)$ (Berg and Purcell, 1977; Shoup and Szabo, 1982). Northrup found that the computer simulations yielded a binding rate constant in close agreement with this expression at low receptor surface coverages (ϕ), but greater than this expression by about 5% when ϕ increased to about 0.25. This is certainly a minor discrepancy given uncertainty in experimental binding measurements, but it represents a real effect. At significant receptor surface coverages, competition among ligand concentration fields around neighboring receptors arises. Such competition is neglected in the theory described earlier in this section; rather, receptors are assumed to act independently, each possessing its own isolated ligand concentration field. As shown by Northrup, competition among receptor fields acts to further deplete the ligand concentration near a receptor and enhance the effective per receptor binding rate.

Zwanzig (1990) has presented a theoretical analysis of this enhancement, finding that the denominator of Eqns. (4–28) and (4–29) becomes $[(k_+)_{cell} + Rk_{on}](1 - \phi)$. Hence, as receptor coverage ϕ increases, the overall binding rate constant increases due to the interference effect. For any particular type of receptor, however, ϕ is typically much less than 1% so that this correction will usually be negligible. The Brownian dynamics approach may also be used to assess the effect of cell surface receptor distribution on the rate constants (Northrup, 1988); receptors dispersed uniformly on the cell surface will exhibit less interaction than those clustered or aggregated on the cell surface (Goldstein and Wiegel, 1983; Goldstein, 1989).

4.2.3 Receptor Coupling with Membrane Molecules

We now turn to interactions between two molecules both associated with the cell surface. Such interactions, discussed in section 2.2.4 as the formation of ternary complexes, include receptor aggregation and receptor coupling with G-proteins, cytoskeletal components, or coated-pit constituents. The effect of molecular diffusion on the rate constants for such interactions between two surface-bound species can be analyzed in a fashion analogous to that for diffusion of free solution ligand to cell surface receptors. In this new case, the coupling and uncoupling rate constants, k_c and k_u, will replace the forward and reverse binding rate constants, k_f and k_r. Thus we note that the interaction of the two species – which will be denoted as R and X – is a two-step process, requiring first diffusive transport with rate constants k_+ before coupling with rate constant k_{on} can occur. The overall coupling reaction rate constant, k_c, depends on k_+ and k_{on}. We will find that the overall uncoupling reaction rate constant, k_u, will depend on the intrinsic uncoupling rate constant k_{off} as well as the diffusion rate constant k_+.

Before we begin, a note on units is in order. For the two-dimensional reactions we are now considering, it is typical to write the rate constants k_c, k_+, and k_{on} in the units of cm^2/s, or, more accurately, $(\#/cm^2)^{-1} s^{-1}$ (recall that for the three-dimensional reactions considered in the last section, we used the units of cm^3/s, or, more accurately, $(\#/cm^3)^{-1} s^{-1}$). For comparison of the magnitudes of the rate constants for the different geometries, however, it would be convenient to be able to convert the two-dimensional rate constants to an effective three-dimensional rate constant. To do this, we will multiply the two-dimensional rate constants by the membrane thickness h to give an estimated local volume for cell surface components. In all cases, the units of $(\#/cm^3)^{-1} s^{-1}$ can be easily converted into $M^{-1} s^{-1}$. Finally, we mention that in section 2.2.4 we used the units of $(\#/cell)^{-1} s^{-1}$ for k_c. These units can easily be converted to $(\#/cm^2)^{-1} s^{-1}$ by muliplying by the surface area/cell.

Different methods for deriving the diffusion rate constant k_+ are applied by Adam and Delbruck (1968), Berg and Purcell (1977), and Keizer (1985) with similar, though not identical, results. Here we will use an approach closely following that for the case of ligand in solution binding to surface receptors (section 4.2.2) and based on the work of Shoup and Szabo (1982). A steady-state diffusion equation is first written for the concentration of receptors, $n_R(r)$ ($\#/area$), around a single membrane-associated X molecule placed at the origin of a circle (Figure 4–6c):

$$D \frac{1}{r} \frac{d}{dr} \left(r \frac{dn_R}{dr} \right) = 0 \qquad (4\text{--}32)$$

D is the sum of the diffusivities of R and X in the membrane. One boundary condition represents reaction between R and X at the encounter radius s:

$$2\pi sD \left. \frac{dn_R}{dr} \right|_{r=s} = k_{on} n_R(s) \tag{4-33a}$$

A secondary boundary condition cannot be specified at infinite r, for then Eqn. (4–32) would have no solution. (An alternative approach, analyzing a non-steady-state situation, is offered by Torney and McConnell (1983).) Instead, a "bulk" membrane receptor concentration, n_{Rb}, can be imposed at a distance $r = b$:

$$r = b \qquad n_R = n_{Rb} \tag{4-33b}$$

where b is one-half the mean distance between species X molecules. If X is the number of species X available for interaction and A is the surface area of the cell, then b can be specified from the approximate relation $X\pi b^2 = A$, so that $b = (A/\pi X)^{1/2}$. For example, if $X = 10^4$ #/cell on a cell possessing 300 μm^2 surface area, then $b \sim 0.1$ μm or 100 nm. It is essential to point out that as interactions proceed dynamically, the value of b will vary as the number of available X molecules changes.

The solution of Eqns. (4–32) and (4–33a,b) for the concentration profile of receptors, $n_R(r)$, around the species X molecule is:

$$n_R(r) = n_{Rb} \left[1 - \frac{k_{on} \ln(b/r)}{2\pi D + k_{on} \ln(b/s)} \right] \tag{4-34}$$

The overall or observable coupling rate constant, k_c, is determined from

$$k_c = n_{Rb}^{-1} 2\pi sD \left[\frac{dn_R}{dr} \right]_{r=s} = k_{on} n_R(s) n_{Rb}^{-1} \tag{4-35}$$

yielding

$$k_c = \frac{2\pi D k_{on}}{2\pi D + k_{on} \ln(b/s)} \tag{4-36}$$

Following the analogy to the case of receptor/ligand binding in solution, k_c can be expressed in terms of rate constants for the diffusive step, k_+, and the coupling step, k_{on}:

$$k_c = \left(\frac{1}{k_+} + \frac{1}{k_{on}} \right)^{-1} \tag{4-37a}$$

where the diffusive rate constant is now elucidated by comparing Eqn. (4–37a) to Eqn. (4–36):

$$k_+ = \frac{2\pi D}{\ln(b/s)} \tag{4-37b}$$

Other approaches yield forms similar except that a constant, C, is subtracted from the $\ln(b/s)$ term in the denominator of Eqn. (4–37b). Adam and Delbruck (1968) obtained $C = 1/2$, Berg and Purcell (1977) $C = 3/4$, and Keizer (1985) $C = 0.231$. For typical values of $\ln(b/s)$ of roughly 2–5 these corrections do not change the estimated value of k_+ by more than a factor of 2. Note that whatever the precise expression, k_+ varies with b and thus with the number of species X molecules available for interaction with receptors.

As before, when $k_{on} \gg k_+$, then $k_c \sim k_+$ and the coupling is *diffusion-limited*. When $k_{on} \ll k_+$, $k_c \sim k_{on}$ and coupling is *reaction-limited*. Let us examine some typical numerical values in this context. For membrane proteins, D is generally in the range 10^{-10}–10^{-9} cm^2/s. Using $s = 1$–10 nm, and $b = 100$ nm, we find that k_+ falls in the range 1×10^{-10}–3×10^{-9} cm^2/s. To convert k_+ into the same units as k_{on}, we multiply by Avogadro's number and, to give an estimated local volume for cell surface components, by a membrane thickness h of approximately 10 nm. Our value for k_+ is then approximately 6×10^4–2×10^6 M^{-1} s^{-1}. In contrast to the situation for reactions in free solution, these values – especially the lower end of the range – are much smaller than typical values for the intrinsic reaction rate constant k_{on}.

Therefore, we can expect that coupling reactions between two membrane-associated species will commonly be diffusion-limited. This result is in contrast to that for receptor/ligand binding in solution, which will not typically be diffusion-limited, and to that for ligand binding to cell surface receptors, which will be diffusion-limited only for unusually large receptor concentrations. Diffusion-limitation of membrane-associated species interactions on the one hand simplifies parameter estimation, since $k_c \sim k_+$ and knowledge of k_{on} is not necessary. On the other hand, diffusion-limitation means that k_c is not a pure constant but will vary as the concentration of reacting species changes during the process.

To obtain the overall uncoupling rate constant, k_u, we can again make use of the "capture probability" γ (see Eqn. (4–23)), such that $k_c = \gamma k_+$. Recall that γ quantifies the extent to which receptor/ligand association is rate-limited by the reaction step. As γ approaches 0, association is severely reaction-limited, while as γ nears 1, binding is almost purely diffusion-limited. k_u is thus given by

$$k_u = (1 - \gamma)k_{off} \qquad (4\text{--}38)$$

Hence, the overall uncoupling rate constant is the product of the intrinsic dissociation rate constant and the "escape probability" (equal to $1 - \gamma$).

4.2.4 Applications

Let us now explore a few applications of these theoretical treatments of diffusion effects in receptor phenomena. An important remark concerning diffusion-limitation of binding is that, when present, it affects the kinetic rate constants but not the equilibrium constant or the Scatchard plot of the equilibrium data. This can be seen by evaluating $K_D = k_r/k_f$ using Eqns. (4–29) and (4–31) for k_f and k_r, respectively, in the case of ligand binding to cell surface receptors to obtain $K_D = k_{off}/k_{on}$; the same is true for the cases of ligand binding to solution receptors or receptors coupling with membrane molecules. Diffusion limitations can, however, affect steady state, non-equilibrium processes, as well as transient phenomena.

First, let us consider possible complications in analyzing experimental receptor/ligand binding data when diffusion limitations are present. A key criterion for diffusion-limited binding of ligand to isolated receptors, according to Eqn. (4–22a), is the value of the quantity ρ_{rec} relative to 1, where

$$\rho_{rec} = \frac{k_{on}}{k_+} \tag{4–39}$$

The key criterion for diffusion-limited binding of ligand to cell surface receptors, following Eqn. (4–29), is the value of the quantity ρ_{cell} relative to 1, where

$$\rho_{cell} = \frac{R k_{on}}{(k_+)_{cell}} \tag{4–40}$$

When $\rho \ll 1$, there is no limitation of association or dissociation by ligand diffusion, so $k_f = k_{on}$ and $k_r = k_{off}$ are constants, as had been assumed in our analyses in previous sections. However, as ρ becomes comparable to 1, diffusion begins to have an influence on the rates of these processes. For the case of ligand binding to cell surface receptors when ρ is comparable to 1, the rate constants k_f and k_r are functions of R and, through $(k_+)_{cell}$, the diffusion coefficient.

As an example of the use of ρ, consider EGF binding to its fibroblast receptor, following Wiley (1988). The value of k_{on} determined for this system is about $3 \times 10^6 \ M^{-1} s^{-1}$. An estimate for k_+, using $k_+ = 4\pi Ds$, $D_L = 1.5 \times 10^{-6} \ cm^2/s$, $D = 2D_L$, and $s \sim 1$ nm, is $2 \times 10^9 \ M^{-1} s^{-1}$. Therefore, $\rho_{rec} \sim 0.001 \ll 1$. Hence, in free solution EGF binding to its receptor would not be diffusion-limited to any noticeable degree. An estimate for $(k_+)_{cell} = 4\pi Da$, using $D = D_L$ and $a = 5 \ \mu m$, is $6 \times 10^{12} \ M^{-1} s^{-1}$. For normal human fibroblasts with $R_T \sim 10^4$ receptors/cell, $\rho_{cell} \sim 0.005 \ll 1$. Clearly, this is purely reaction-limited behavior as well. In contrast, for A431 cells (an epidermoid carcinoma cell line) with $R_T > 10^6$ receptors/cell, $\rho_{cell} > 0.5$, allowing diffusion to play a significant role. For this cell type, according to Eqns. (4–29) and (4–31), the rate constants for EGF association and dissociation should be dependent on receptor occupancy through the effect of receptor occupancy on R, the free receptor number. Indeed, Wiley found this to be the case experimentally, observing an effect of receptor occupancy on EGF dissociation rates from A431 cells but not from normal fibroblasts.

When ρ_{cell} is not negligible compared to 1, then one must go back to the beginning and modify Eqn. (4–1) for simple receptor/ligand binding, using Eqns. (4–29) and (4–31) in place of the simple constants k_f and k_r, respectively. When ligand depletion can be considered negligible, the appropriately altered form of Eqn. (4–1) is:

$$\frac{dC}{dt} = \left[1 + \frac{k_{on}(R_T - C)}{4\pi aD} \right]^{-1} \{ k_{on} L_0 R_T - (k_{on} L_0 + k_{off})C \} \tag{4–41a}$$

where we have used $(k_+)_{cell} = 4\pi aD$ and the receptor conservation relation $R_T = R + C$. Alternatively, in dimensionless form using the scaled variables $\tau = k_{off} t$ and $u = C/R_T$, this equation becomes

$$\frac{du}{d\tau} = [1 + \rho_{cell}(1 - u)]^{-1} [\lambda_0 - (1 + \lambda_0)u] \tag{4–41b}$$

where $\rho_{cell} = R_T k_{on}/4\pi Da$, $\lambda_0 = L_0/K_D$, and $K_D = k_{off}/k_{on}$. This equation, though non-linear, possesses an implicit analytical solution for $u(t)$ for an association experiment with initial condition $u(0) = 0$:

$$\left(\frac{\rho_{cell}}{1+\lambda_0}\right)u + \left[\frac{\rho_{cell}\lambda_0 - (1+\rho_{cell})(1-\lambda_0)}{(1+\lambda_0)^2}\right]\ln\left|1 - \frac{(1+\lambda_0)}{\lambda_0}u\right| = \tau \quad (4\text{-}42)$$

Example computations from this expression, for parameters representing the EGF/A431 cell system, are shown in Figure 4–9a as transient $u(\tau)$ curves. Notice that increasing ρ_{cell} above unity markedly reduces the binding rate, as expected. Quantitative effects of this diffusion-limitation can best be discovered by comparison with predictions for pure reaction-limitation, following Eqn. (2–23). In that case, setting $\rho_{cell} = 0$, a plot of $\ln[u_{eq} - u(\tau)]$ versus τ gives a straight line with slope equal to $-(1 + \lambda_0)$ as in Figure 2–10a. Figure 4–9b shows the transient curves from Figure 4–9a replotted in this fashion. Clearly, values of ρ_{cell} larger than about 1 cause significant deviation of this slope from its limiting value and noticeable non-linear behavior.

This type of observation is, in fact, made by Goldstein *et al.* (1989) in experimental studies of DCT dissociation kinetics from an IgE antibody anchored as a surface receptor on RBL cells. For this system, $D = 5 \times 10^{-6}$ cm^2/s, $a = 4$ μm, $R_T = 6 \times 10^5$ #/cell, and $k_{on} = 4.8 \times 10^7$ M^{-1} min^{-1}. These values yield $\rho_{cell} \sim 2$ when all receptors are available for binding (*i.e.*, $R = R_T$), predicting that substantial diffusion-limitation effects should arise during transient association and/or dissociation experiments. Measurements showed a slowing of the effective dissociation rate of DCT from RBL cells as the number of free receptors increased during the dissociation experiment (see Figure 4–10), as predicted. Goldstein *et al.* thus quite properly used the equivalent of Eqn. (4–41) to analyze their data. We should point out that the main focus of this paper was to examine the effects of competing receptors in solution on ligand binding to cell surface receptors, a more complicated situation which will not be addressed here.

Turning to receptor processes on cell membranes (Figure 4–6c), we could infer the degree of diffusion limitation from a ratio analogous to that in Eqn. (4–39), where k_+ is given by Eqn. (4–37b). Unfortunately, values for an intrinsic reaction rate constant, k_{on}, are extremely difficult to determine for interactions between membrane-associated species. Indeed, measurements of coupling and uncoupling rate constant, k_c and k_u, are themselves rare at the present time. Mayo *et al.* (1989) determine values for overall coupling and uncoupling rate constants between EGF/EGF–receptor complexes and an hypothesized membrane-associated component on fibroblast membranes at 4 °C to be $k_c = 8 \times 10^{-5}$ (#/cell)$^{-1}$ s^{-1} and $k_u = 8 \times 10^{-5}$ s^{-1} (see section 2.2.4.c). To convert k_c into more familiar units, we multiply by Avogadro's number and by an estimated local volume for cell surface components (the product of cell surface area A and a membrane thickness h of roughly 10 nm). Using $A = 1.3 \times 10^3$ μm^2 for a cell of 10 μm radius, we obtain $k_c = 6 \times 10^5$ M^{-1} s^{-1}. An estimate for k_+ based on Eqn. (4–37b) and the estimate $b = (A/\pi X)^{1/2}$ can be calculated, using $X = 2.4 \times 10^4$ #/cell for the number of membrane-associated coupling species (as determined by Mayo *et al.* from comparison between their model and experimental binding data) on cells with 10 μm radius, $s = 3$ nm, $D = 1 \times 10^{-10}$ cm^2/s, and $h = 10$ nm.

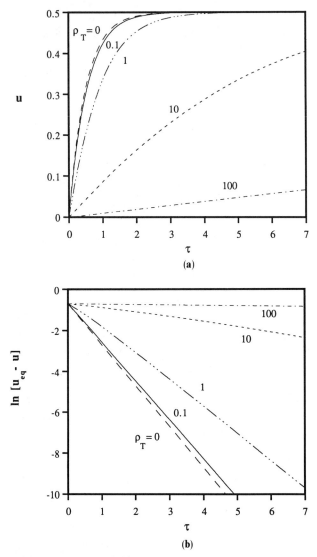

Figure 4-9 Binding of ligand to cell surface receptors. (a) The scaled number of bound receptors u is plotted as a function of the scaled time τ. (b) The logarithm of the difference between the equilibrium and transient number of bound receptors, $\ln[u_{eq} - u]$, is plotted versus τ. Note that the curves vary with the values of ρ_{cell} (ρ_T in figure), which measures the degree to which binding is diffusion-limited. When ρ_{cell} is small, the rate constants k_f and k_r are true constants. When ρ_{cell} is large, binding becomes diffusion-limited and the variation of k_f and k_r with time significantly affects the number of bound receptors.

These values yield $k_+ \sim 1 \times 10^5 \text{ M}^{-1} \text{ s}^{-1}$. Though these are clearly crude estimates, we see that k_c and k_+ are of the same order of magnitude. While k_c seems to be greater than k_+, this is not possible; our conclusion is that k_c may be very nearly approaching its diffusion-limited value, as expected.

Goldstein *et al.* (1981) used a related analysis to examine data for congregation

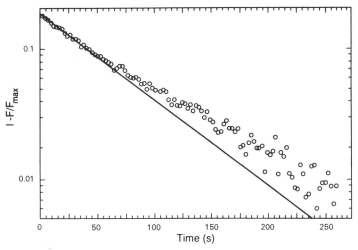

Figure 4–10 Dissociation of DCT from IgE antibodies anchored on the surfaces of rat basophilic leukemia (RBL) cells, as assayed by decline in cell-associated fluorescence, F. The rate of dissociation slows with time due to an increase in the number of free receptors. The effect is noticeable because ρ_{cell} is not $\ll 1$ and therefore k_r is not constant. Reprinted from Goldstein *et al.* (1989). Reproduced from the *Biophysical Journal*, 1989, Vol. 56, pp. 955–966, by copyright permission of the Biophysical Society.

of LDL receptors in endocytic clathrin-coated pits. Their question was whether *passive diffusion* of receptors is sufficient to account for congregation rates in coated pits or whether receptor transport by *membrane convective flow* is necessary. We can estimate a value for the diffusion rate constant, k_+, for movement of receptors to a coated pit. Using Eqn. (4–37b), but with $s = 0.05\ \mu m$ as the coated-pit radius and $b = 1\ \mu m$ the average spacing between pits, we obtain $k_+ \sim 6 \times 10^{-10}\ cm^2/s$ for $D = 3 \times 10^{-10}\ cm^2/s$. Goldstein *et al.* found that $k_c = 3 \times 10^{-10}\ cm^2/s$ was a reasonable lower bound on the rate constant for receptor/coated-pit coupling, as determined from experimental data on LDL internalization (Brown and Goldstein, 1976). Thus k_+ is of roughly the same magnitude as the value given by Goldstein *et al.* for k_c. The conclusion, then, is that passive receptor diffusion should be sufficient to provide the observed rate of receptor congregation in coated pits. Goldstein *et al.* came to the same finding with a more detailed treatment. A later analysis by Keizer (1985) using a statistical mechanics approach reinforces this conclusion. Readers interested in pursuing further analysis of effects of membrane convective flow on receptor transport are referred to a few treatments of this phenomenon (Goldstein and Wiegel, 1988; Goldstein *et al.*, 1988; Echavarria-Heras, 1988).

4.2.5 Intermolecular Effects on Diffusion

Until now, we have regarded the process of diffusion as simply characterized by a constant diffusion coefficient, D. Two complicating features merit brief mention here, both involving molecular interactions. One is the concept of "molecular crowding" for processes involving membrane-associated proteins. The other is the influence of intermolecular forces during the close encounter of two species leading to a binding interaction.

Ryan *et al.* (1988) measure spatial concentration profiles of immunoglobulin-E (IgE) receptors on the surface of rat basophilic leukemia (RBL) cells in the presence of an electric field gradient tending to congregate charged surface molecules locally at a preferred field strength. These investigators observe a clear discrepancy in the regions of greatest receptor concentration from profiles predicted by simple diffusion theory based on *ideal thermodynamic behavior* – for which molecular concentration distribution is unaffected by concentration. The measured peak concentrations are found to be lower than predicted for ideal behavior (Figure 4–11), that is, receptors appear to be negatively affected by the presence of a high concentration of cell surface molecules. Invoking a simple *steric exclusion model*, Ryan *et al.* can satisfactorily account for the observed receptor concentration profile. In this model, as cell surface molecule concentration increases, avoidance of steric overlap causes the receptors to come to an equilibrium distribution with maximal concentrations less than would be predicted if overlap was permitted. Hence, Ryan *et al.* conclude that IgE receptors on RBL cells might be crowded in a sense of steric overlap, leading to *non-ideal thermodynamic behavior.*

However, theoretical prediction of the effect of crowding on receptor transport properties, specifically receptor mobility in the cell membrane, is difficult because two competing influences exist. In reconstituted membranes, high protein-to-lipid ratios lead to slowed protein lateral diffusivities (Peters and Cherry, 1982; Tank *et al.*, 1982; Vaz *et al.*, 1984), and theoretical modeling indicates that high concentrations of mobile obstacles (*e.g.*, other receptors or proteins), or immobile

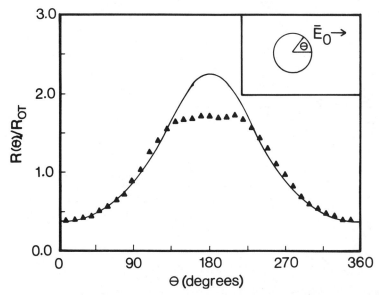

Figure 4–11 Relative numbers of IgE receptors on RBL cells as a function of the angle θ. The cells were equilibrated in an applied field of 15 V/cm to induce this redistribution of receptors. Data are shown with triangles; ideal thermodynamic behavior is shown by the solid line. Redrawn from Ryan *et al.* (1988). *Science*, vol. 239, pp. 61–64. Copyright 1988 by the AAAS.

receptors or solid membrane domains) can slow receptor diffusion by two orders of magnitude (Saxton, 1982, 1987). This would tend to prolong receptor congregation in regions of high concentration. At the same time, though, a steric exclusion effect also increases thermodynamic activity, providing an increased driving force for receptor diffusion away from regions of high concentration. From Chapter 2, we know that it is essential to understand whether the self-diffusion coefficient, D^{self} (Eqn. (2–6a)), or the mutual diffusion coefficient, D^{mutual} (Eqn. (2–6b)), is being considered. Owicki and coworkers (Scalettar *et al.*, 1988, Abney *et al.*, 1989) examine this issue theoretically for three different protein–protein interaction potentials: hard-core repulsion, soft repulsion, and soft repulsion with weak attraction. For the first two interactions, both repulsive, D^{self} is predicted to decrease with protein concentration but D^{mutual} is predicted to increase with concentration. These results would likely be found when steric exclusion predominates, as in the Ryan *et al.* model. When the weak attraction is included, both D^{self} and D^{mutual} are predicted to diminish with increasing protein concentration. Decreasing values of D^{self} with increasing concentration have indeed been experimentally obtained for bacteriorhodopsin (Peters and Cherry, 1982; Vaz *et al.*, 1984) and gramicidin (Tank *et al.*, 1982). It is generally observed that diffusion coefficients measured by FRAP (D^{self}) are smaller than those measured by PER (D^{mutual}) (Young *et al.*, 1982). The value of D appropriate for theoretical analyses of the diffusion rate constant for interactions between two membrane-associated molecules, k_+, is D^{self}, whereas for the flux of receptors into membrane structures such as coated pits, on the other hand, D^{mutual} would be more relevant.

This discussion has considered only steric interactions between molecules. However, it is well-known that other forces, such as electrostatic, van der Waals, and hydrophobic forces, become significant at close intermolecular approach. These forces can significantly influence the relative motion of two nearby molecules. Thus, attempts to analyze the role of diffusion in receptor binding or coupling events using a simple constant diffusivity approach are incomplete.

Recent advances in high-performance computer capabilities permit a more rigorous theoretical treatment of reactive encounters between two chemical species, including details of molecular structure and the intermolecular forces felt by these species as they approach one another by diffusion (an excellent review is by McCammon (1991). Diffusive trajectories of the molecules are obtained from Brownian dynamics algorithms, in which the translational and angular velocities of a molecule are governed by Newton's law of motion (Figure 4–12). Random, thermal fluctuations representing solvent molecule collisions are superimposed on deterministic molecular forces, primarily electrostatic forces arising from molecular charge. Thus, the rate of reaction between two molecules can be computed by simulating the paths of the two species in their interacting force fields and calculating the time required for them to come together in appropriate position and orientation. From a large number of such simulations, values of the diffusive encounter rate constant, k_+, can be determined. Further, given some information concerning local energy barriers to binding or coupling processes, the overall forward reaction rate constant, k_f, or coupling rate constant, k_c, can be calculated. Although until now most applications of this method have been for enzyme/substrate reactions (Northrup *et al.*, 1988; Sines *et al.*, 1990), we expect that

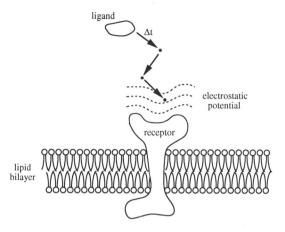

Figure 4–12 Brownian dynamics calculations for the movement of a ligand. The position of the ligand after a time step of Δt is determined by its previous position, the influence of deterministic electrostatic forces, and the random thermal displacement.

this method will become increasingly important for receptor/ligand binding studies in the very near future.

4.3 MULTIVALENT BINDING AND RECEPTOR CROSSLINKING

We now examine receptor aggregation induced by multivalent ligands. The ability of a *multivalent* ligand, *i.e.*, a ligand with more than one binding site for a receptor, to aggregate receptors is shown schematically in Figure 4–13. Receptors which are linked to other receptors via these multivalent ligands are said to be *crosslinked*; note that crosslinked receptors can result regardless of the receptor valency.

Some interesting and important cell behavioral functions involve multivalent binding phenomena. Two general categories can be defined: *multivalent ligand/ monovalent receptor* and *multivalent ligand/multivalent receptor*. A prime example of the first category is viral attachment to cell surfaces, the initial step in viral penetration of the cells. Viruses typically exhibit proteins known as viral attachment proteins (VAP) that are capable of binding to cell surface components, often physiological receptors. For instance, the HIV virus is known to bind to the CD4 receptor on T-lymphocytes, which is involved in antigen presentation to the lymphocytes for stimulation of the immune response (White and Littman, 1989). Viruses can be considered highly multivalent ligands, with the number of VAP found experimentally to range from 12 for adenovirus (Persson *et al.*, 1983) to over 1,000 for vesicular stomatitis virus (Schlegel *et al.*, 1982). The effective valency is likely to be smaller than the total number of VAP since large particles such as viruses cannot be expected to be flexible enough to permit all potential sites to access cell surface receptors simultaneously. Another example is binding of antigen/antibody complexes via the Fc domains of their antibody moieties to monovalent Fc receptors on white blood cells. These complexes typically contain

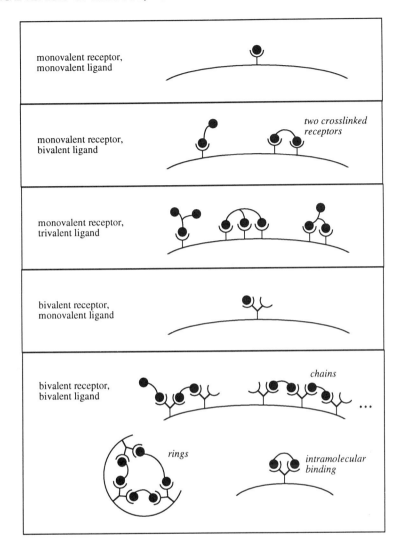

Figure 4-13 Multivalent binding. Receptors and/or ligands may have more than one binding site. Depending on the valency of the receptor and ligand, different types of receptor/ligand complexes can result. When both the ligand and the receptor are at least bivalent, chain and ring formation and intramolecular binding can occur.

many antibody molecules bound to a large antigen particle (such as a bacterium) or crosslinking numerous antigen molecules, thus the multivalency of the complex considered as a single ligand.

In the second category, multivalent ligand/multivalent receptor interactions, are many immunological stimuli. Most commonly such receptors are functionally bivalent. One example class is immunoglobulin receptors present on B-lymphocytes, for which binding of antigen ligands stimulates B-cell proliferation and antibody production during an immune response. Another example class is immunoglobulins bound with extremely high affinity (with dissociation half-time in the range of 20–50 h) to Fc receptors on circulating basophilic leukocytes and tissue mast cells, which release inflammatory mediators such as histamine when antigen ligands are bound. Immunoglobulins IgG, IgD, IgE, and the surface form of IgM have two identical ligand binding sites and thus are functionally bivalent. Substantial amounts of experimental evidence have made it clear that crosslinking of these receptors by multivalent antigen ligands is required for the signal transduction which culminates in these behavioral responses (see Baird *et al.*, 1988; Dintzis and Dintzis, 1988; Goldstein, 1988).

In our quest to describe quantitatively the formation of receptor/ligand complexes on the cell surface, the starting place for understanding cell behavioral responses, we need to examine the equilibrium and kinetic properties of multivalent ligand binding. A major new feature arising from multivalent binding is the formation of receptor aggregates, or clusters, due to crosslinking of receptors by multivalent ligand. Histamine release by basophils in response to stimulation by antigen appears to depend not only on the amount of antigen bound but also on further characteristics of the resulting receptor aggregates (Baird *et al.*, 1988). Hence, the number of receptor/ligand complexes of a given size may be an important quantity for predicting the level of histamine release. Similarly, the immunon theory (Dintzis and Dintzis, 1988) hypothesizes that a critical number of receptor/ligand aggregates of a certain size is required to activate B lymphocytes to proliferation and antibody production. Given the need for quantitative understanding along with the obviously complicated situation, it is no surprise that a great amount of effort has been devoted to developing mathematical models for multivalent receptor/ligand binding. In general, one wishes to examine the situation of f-valent ligands binding to g-valent receptors. We will not attempt to cover this body of work comprehensively, but instead will present a few fundamental cases. Readers interested in greater detail and breadth are referred to excellent reviews on this topic by Perelson (1984), Goldstein (1988), and Macken and Perelson (1985).

A key concept in this area is *avidity*, or the overall tendency of a multivalent ligand to be bound to cell receptors at equilibrium. An *avidity constant*, K_V, is basically the effective value of K_D^{-1} when multivalent ligand binding is analyzed in terms of a simple Scatchard analysis. That is, if C_{eq} is the total number of ligand molecules bound per cell at equilibrium, one might attempt to characterize this quantity using the form for monovalent ligand binding:

$$C_{eq} = \frac{R_T L}{K_D + L} \tag{4–2}$$

Of course, this expression may not capture the experimental data very well because of the additional complexities inherent in multivalent interactions. Nevertheless, an effective value of $K_V = K_D^{-1}$ can be obtained from interpretation of equilibrium binding data for multivalent ligand with this expression. It will be a combination of the equilibrium affinity constant of a single-site receptor/ligand interaction together with equilibrium constants for receptor crosslinking by the multivalent ligand. To a first approximation, it would be the product of these constants up to the maximum binding valence, with a correction factor for geometric

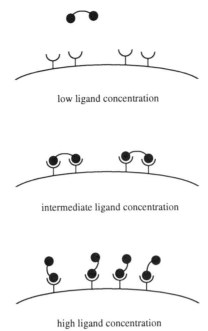

low ligand concentration

intermediate ligand concentration

high ligand concentration

Figure 4–14 Binding of multivalent ligands to monovalent receptors. At low ligand concentrations, ligands are likely to be bound by more of their binding sites than at high ligand concentrations. As a result, there is decreasing avidity with increasing ligand concentration.

considerations (Crothers and Metzger, 1972). For example, a bivalent ligand bound to cell receptors by two sites, each with equilibrium dissociation constants of $K_D = 10^{-6}$ M (a low affinity interaction), could have an effective affinity, or avidity, on the order of 10^{12} M^{-1}. Hence, a multivalent ligand, *e.g.*, a virus particle, may be attached to a cell very strongly by means of a large number of low affinity interactions.

Avidity is a difficult quantity to predict rigorously because it depends on so many factors. It is important to point out, in fact, that the effective avidity of a multivalent ligand will vary with the extent of ligand binding. At low ligand concentrations relative to the equilibrium dissociation constant, K_D, bound ligand can crosslink may receptors and so have a high avidity (Figure 4–14). At high ligand concentrations, on the other hand, ligand molecules compete with one another for available receptors and so will be typically bound to fewer receptors: hence, a lower avidity results in this situation. Another way of looking at this is to consider that the reciprocal of the effective K_D, from which the avidity constant K_V can be calculated, can be estimated from the slope of a Scatchard plot for equilibrium ligand binding. For multivalent ligands, this slope will be found to have concave-upward ("negatively cooperative") curvature. Thus, the effective K_D will be smaller at low ligand binding extent than at high, leading to a decreasing value of K_V, or avidity, as the extent of ligand binding increases.

4.3.1   Bivalent Ligand/Monovalent Receptor

We begin by considering the simplest scenario, that of a bivalent ligand and monovalent receptor, in order to provide essential insights (Figure 4–13). We further restrict our attention to surface binding, neglecting endocytic trafficking processes. Our approach will follow that of Perelson (1984). Let L_0 be the ligand concentration in free solution and assume it remains constant (*i.e.*, no significant depletion). Define C_1 as the number of ligands bound to a cell at only one site, and C_2 as the number bound to a cell at both sites. This definition for C_2 implies that the number of crosslinked receptors is simply $2C_2$. The following differential equations then govern the kinetics of ligand binding:

$$\frac{dC_1}{dt} = 2k_f L_0(R_T - C_1 - 2C_2) - k_r C_1 - k_x C_1(R_T - C_1 - 2C_2) + 2k_{-x}C_2$$

$$(4\text{--}43a)$$

$$\frac{dC_2}{dt} = k_x C_1(R_T - C_1 - 2C_2) - 2k_{-x}C_2 \qquad\qquad (4\text{--}43b)$$

Note that the *receptor conservation relation* $R_T = R + C_1 + 2C_2$ has been used to eliminate the free receptor number, R, from the equations. The factors of the integer 2 in Eqns. (4–43a,b) are present to account for statistical factors. There are two ways (either site) in which a bivalent ligand can bind to a free receptor, and two ways (again, either site) by which a doubly-bound ligand can become singly bound. k_f [concentration^{-1} time^{-1}] and k_r [time^{-1}] are the usual association and dissociation rate constants and k_x [(#/cell)$^{-1}$ time^{-1}] and k_{-x} [time^{-1}] are the receptor crosslinking and decrosslinking rate constants. As an aside, we note that k_x can also be characterized with units of [M^{-1} time^{-1}] when C_1 and C_2 are defined in terms of volumetric concentration. If we let k_x^c denote the parameter in volumetric concentration units, one way to derive a relationship is to use the cell density; *i.e.*, $k_x^c = k_x(n/N_{Av})^{-1}$ where n is the cell density (cells/volume) and N_{Av} is Avogadro's number (Perelson, 1984). An alternative procedure is to calculate an effective reaction volume for the cell-associated ligand, usually the cell surface area A multiplied by the membrane thickness h. Using this approach, $k_x^c = k_x(Ah/N_{Av})^{-1}$. Our preference is for the latter relationship, because events taking place among cell-associated molecules on an individual cell surface should not depend on the presence of other cells.

These non-linear coupled equations must, in general, be solved numerically for the *transient behavior*. An approximate solution has been obtained for the limiting case in which free solution association and dissociation steps are much faster than crosslinking steps (Perelson and DeLisi, 1980). In this case, transient behavior should exhibit two time-scales, corresponding to the ligand-binding and crosslinking steps. Dower *et al.* (1984) provide example computations for binding kinetics of bivalent antibody binding to monovalent cell surface Fc receptors, showing clearly double-exponential transient behavior and accelerated dissociation in the presence of a competing ligand. These model computations agreed favorably with their experimental data on a variety of cell types and antibody ligands.

Equilibrium behavior predicted by the model can be obtained analytically. Setting the two time derivatives in Eqns. (4–43a,b) equal to zero leaves two coupled non-linear algebraic equations for the equilibrium numbers of singly- and doubly-bound ligands, C_{1eq} and C_{2eq}, respectively. Along with the receptor conservation relation, these provide three equations for simultaneous determination of C_{1eq}, C_{2eq}, and the equilibrium free receptor number R_{eq}. Defining the equilibrium ligand/receptor dissociation constant $K_D = k_r/k_f$ and the crosslinking equilibrium constant $K_x = k_x/k_{-x}$, the following solutions can be obtained for those three quantities (Perelson and DeLisi, 1980):

$$C_{1eq} = R_T \beta \left[\frac{-1 + \sqrt{(1 + 4\delta)}}{2\delta} \right] \tag{4–44a}$$

$$C_{2eq} = \tfrac{1}{2} R_T \left[\frac{1 + 2\delta - \sqrt{(1 + 4\delta)}}{2\delta} \right] \tag{4–44b}$$

$$R_{eq} = R_T (1 - \beta) \left[\frac{-1 + \sqrt{(1 + 4\delta)}}{2\delta} \right] \tag{4–44c}$$

where $\beta = L_0/(L_0 + [K_D/2])$ is a dimensionless ligand concentration (the factor of 2 shows up due to the statistical availability of two binding sites per bivalent ligand) and $\delta = \beta(1 - \beta)K_x R_T$ is a dimensionless crosslinking parameter. This latter quantity can also be written $\beta(1 - \beta)\kappa$, where $\kappa = K_x R_T$ is the dimensionless crosslinking equilibrium constant. Example results are illustrated in Figure 4–15 by the curves marked "$f = 2$" (bivalent ligand). The scaled equilibrium number of crosslinks, C_{eq}/R_T, is plotted versus $\ln(L_0/[K_D/2])$ for several values of κ; this is termed the *crosslinking curve*. This figure shows results for higher ligand valencies as well, which we will discuss shortly.

Three features of the equilibrium crosslinking curve are important to note. First, the crosslinking curve is biphasic: at low ligand concentrations there are very few occupied receptors and consequently very few crosslinks, whereas at high ligand concentrations almost all receptors are bound singly without crosslinks. This is the same point as illustrated in Figure 4–14 in the discussion of binding avidity. Second, for bivalent ligand the curve is symmetric when plotted on the logarithmic scale shown and has a maximum at the ligand concentration $L_0 = (1/2)K_D$. Third, the height of the curve increases as the dimensionless crosslinking equilibrium constant, κ, increases. The maximum height of the curve is equal to 0.5, for this would correspond to $C_{2eq} = R_T/2$, or all receptors being crosslinked.

Now, consider properties of ligand binding itself, with avidity of primary interest. The fraction of receptors bound by ligand, whether crosslinked or not, does of course increase monotonically with ligand concentration, as can be seen in Figure 4–16a. This figure gives example plots of the equilibrium fraction of bound receptors per cell, $(f_{RB})_{eq}$:

$$(f_{RB})_{eq} = \frac{C_{1eq} + 2C_{2eq}}{R_T} \tag{4–45}$$

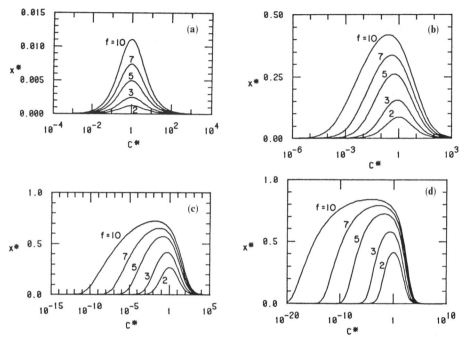

Figure 4–15 Equilibrium crosslinking curves. The scaled equilibrium number of crosslinks (denoted x^*) is plotted as a function of the scaled ligand concentration $L_0/(K_D/2)$ (denoted C^*) for several values of the ligand valency f and different values for the crosslinking equilibrium constant κ: (a) $\kappa = 0.01$; (b) $\kappa = 1$; (c) $\kappa = 10$; (d) $\kappa = 100$. Reprinted by permission of the publisher from Perelson, *Mathematical Biosciences*, vol. 53, pp. 1–39. Copyright 1981 by Elsevier Science Publishing Co., Inc.

and the total amount of bound ligand at equilibrium, C_{Beq}:

$$C_{Beq} = C_{1eq} + C_{2eq} \tag{4–46}$$

both at equilibrium conditions for a set of sample parameter values. Expressions for these two quantities can be written in a straightforward manner by substituting Eqns. (4–44a,b) into Eqns. (4–45) and (4–46). The amount of bound ligand lags behind the fraction of bound receptors as ligand concentration increases due to receptor crosslinking.

The Scatchard plot based on Eqn. (4–46) is shown in Figure 4–16b for the same parameter values as in Figure 4–16a. Note the concave-upward behavior, indicating that the effective binding affinity – or avidity – decreases as the extent of ligand binding increases at higher ligand concentration. An approximate analysis by DeLisi and Chabay (1979) (which is also relevant for bivalent ligand binding to bivalent receptors, as will be seen later) found that the avidity constant, K_V, estimated at very low and very high ligand concentrations, respectively, is related to the fundamental equilibrium parameters as follows:

$$L_0 \to 0, \qquad K_V \to \frac{8K_x R_T}{K_D} \tag{4–47a}$$

$$L_0 \to \infty, \qquad K_V \to \frac{2}{K_D} \tag{4–47b}$$

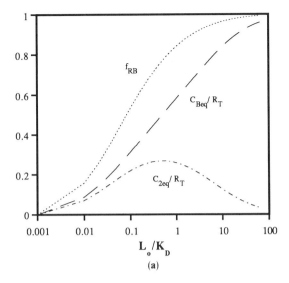

(a)

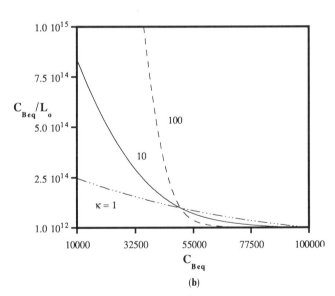

(b)

Figure 4–16 Bivalent ligand/monovalent receptor binding. (a) Equilibrium fraction of bound receptors $(f_{RB})_{eq}$, scaled total amount of bound ligand C_{Beq}/R_T, and scaled amount of doubly-bound ligand C_{2eq}/R_T as a function of the scaled ligand concentration L_0/K_D. (b) Scatchard plot showing bound ligand/free ligand (C_{Beq}/L_0) as a function of the amount of bound ligand. Parameter values: $R_T = 10^5$ #/cell, $K_D = 10^{-9}$ M, $K_x = 10^{-4}$ (#/cell)$^{-1}$.

Notice that at high ligand concentrations the avidity constant is essentially equal to the single-site binding affinity, with the statistical factor of 2 again entering. At low ligand concentrations, the avidity constant is related to the product of the fundamental equilibrium constants for ligand binding and receptor crosslinking, K_D^{-1} and K_x.

4.3.2 Multivalent Ligand/Monovalent Receptor

We next extend the bivalent ligand/monovalent receptor model to the more general case of an effectively f-valent ligand particle binding to a cell surface containing monovalent receptors. This extension has been developed by a number of investigators (Gandolfi et al., 1978; DeLisi, 1980; Perelson, 1981) and has been modified for application to viral attachment by Wickham et al. (1990). Again, we will follow the approach summarized by Perelson (1984). Here, one must define C_i, $i = 1, 2, \ldots, f$ as the number of ligands bound to the cell surface by i of its f available functional groups. Although it is not necessary to do so for writing the model formally, one can invoke the *equivalent site hypothesis*. That is, it will be assumed that the values of the forward and reverse crosslinking rate constants, k_x and k_{-x}, respectively, do not depend on the size of the cluster, or that all sites for ligand binding are identical. This may not be generally true if there are steric effects or other cooperative influences, but in the absence of quantitative information about how k_x and k_{-x} vary with bound ligand state it is a very helpful simplification.

For the case of multivalent virus binding to cell surface receptors, the total number of binding sites on the virus will be denoted as v. Thus the first binding reaction to form C_1 can occur v different ways, and a statistical factor of v must be included in the binding term. It is likely that a virus, once tethered to the cell by one of its v sites, does not express all remaining $v - 1$ sites in locations that are simultaneously available to the cell surface receptors. Thus we will use the symbol f to denote the number of sites simultaneously available, where it is expected that $f \leq v$.

The kinetic species balance equations now become (without explicitly assuming conservation of receptors or no significant depletion of ligand):

$$\frac{dL}{dt} = -vk_f LR + k_r C_1 \tag{4-48a}$$

$$\frac{dC_1}{dt} = vk_f LR - k_r C_1 - (f - 1)k_x C_1 R + 2k_{-x} C_2 \tag{4-48b}$$

$$\frac{dC_i}{dt} = (f - i + 1)k_x C_{i-1} R - ik_{-x} C_i$$
$$- (f - i)k_x C_i R + (i + 1)k_{-x} C_{i+1} \qquad i = 2, 3, \ldots, f - 1 \tag{4-48c}$$

$$\frac{dC_f}{dt} = k_x C_{f-1} R - fk_{-x} C_f \tag{4-48d}$$

If ligand is present in excess there are no depletion effects, and $L = L_0$ would replace Eqn. (4–48a). In virus attachment experiments the viruses are rarely present in excess, however, so the ligand balance equation will usually be necessary. Also, if there are no receptor trafficking phenomena such as internalization and recycling occurring, then total receptor conservation can be postulated:

$$R_T = R + \sum_{i=1}^{f} iC_i \tag{4-48e}$$

in order to complete the model system of equations. Because of the coupled, non-linear nature of this set of $f + 1$ differential equations, the kinetic behavior of this model must be found from numerical computations (Wickham *et al.*, 1990).

Equilibrium properties for this multivalent ligand/monovalent receptor model can be found analytically. With the time derivatives in Eqns. (4–48) equal to zero, and the assumption of no ligand depletion ($L = L_0$), it can be shown that the number of ligands bound by i functional groups is:

$$C_{ieq} = \left[\frac{f!}{i!(f-i)!}\right] K_x^{i-1} \frac{v}{f}\left(\frac{L_0}{K_D}\right) R_{eq}^i \tag{4-49a}$$

The number of free receptors, R_{eq}, must be determined from the implicit equation:

$$R_T = R_{eq}\left[1 + v\left(\frac{L_0}{K_D}\right)(1 + K_x R_{eq})^{f-1}\right] \tag{4-49b}$$

This polynomial equation in R_{eq} can be solved by standard numerical root-finding methods, such as the Newton–Raphson method (Press *et al.*, 1989). All the individual bound species numbers, C_{ieq}, can then be calculated using Eqn. (4–49a).

We can calculate several additional quantities. The total amount of bound ligand at equilibrium, C_{Beq}, is found from summing the numbers of ligand molecules in each bound state:

$$C_{Beq} = \sum_{i=1}^{f} C_{ieq} = \frac{v}{K_x f}\left(\frac{L_0}{K_D}\right)[(1 + K_x R_{eq})^f - 1] \tag{4-50}$$

The number of ligands in each individual bound state, C_{ieq}, is difficult to measure experimentally at the present time so it must be calculated from the model. We can further calculate the total number of crosslinks at equilibrium, C_{xTeq}, as given by

$$C_{xTeq} = \sum_{i=2}^{f} (i - 1)C_{ieq} \tag{4-51}$$

This quantity also is difficult to measure. Figure 4–15 shows example calculations of C_{xTeq} versus $(2L_0/K_D)$ for several values of κ (recall $\kappa = K_x R_T$ is the dimensionless crosslinking equilibrium constant), and a variety of valences, f. Notice that for $f = 2$, $C_{xTeq} = C_{2eq}$ from the bivalent model. As the valence of the ligand increases, the equilibrium crosslinking curves lose their symmetry and become shifted toward lower ligand concentration. As before, increasing κ leads to enhanced numbers of crosslinks.

C_{Beq}, the total number of bound ligands per cell, can be directly compared to experimental binding data, and is often converted into Scatchard plot form in order to do so. An example set of experimental data for adenovirus binding to HeLa cells (Persson *et al.*, 1983) is shown in Figure 4–17. Wickham *et al.* (1990) apply a model for multivalent ligand/monovalent receptor binding to these data. Their model is actually more complicated than that described here, taking into account non-specific attractions between the virus and cell surface, diffusion-limited virus encounter with the cell surface, and virus depletion, but we will not discuss these additional features here. A concave-upward Scatchard plot is often

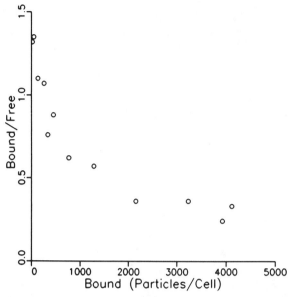

Figure 4–17 Scatchard plot of the binding of adenovirus 2 to HeLa cells at 4 °C. Data are from Persson *et al.* (1983) and are reprinted from Wickham *et al.* (1990). Reproduced from the *Biophysical Journal*, 1990, Vol. 58, pp. 1501–1516 by copyright permission of the Biophysical Society.

found for multivalent ligand binding, because as ligand concentration increases the degree of receptor crosslinking per ligand is diminished because of increased competition for receptor sites. The lower degree of crosslinking is reflected in a lower net binding affinity, since each ligand is bound by fewer receptors; hence, apparent negative cooperativity is observed.

Wickham *et al.* estimate all model parameters except two from literature information and then determine the remaining two from the Scatchard plots. Essentially, they find K_x and R_T from these equilibrium binding data. For the example shown in Figure 4–17, attachment of adenovirus to HeLa cells, they obtain $R_T = 6 \times 10^3$ #/cell and $K_x \sim 4 \times 10^7$ M^{-1} (in ligand concentration units) – a moderate affinity. Using our conversion again between (#/cell) and M, we can calculate a value for the dimensionless crosslinking constant, $\kappa = K_x R_T$; we obtain $\kappa \sim 30$, corresponding to a strong crosslinking tendency (Figure 4–15). For this system $v = 12$ and f is assumed to be 6. Despite this fairly small ligand valence, then, the strong crosslinking effect allows adenovirus to be attached by multiple moderate affinity bonds for a net high affinity interaction with the HeLa cell surface.

4.3.3 Bivalent Ligand/Bivalent Receptor

We can now turn to the case in which both receptor and ligand are multivalent. While there has recently been published some extensive and sophisticated mathematical analysis of the general case of f-valent ligands and g-valent receptors (Macken and Perelson, 1985), we will restrict this discussion to the most widely-studied case, $f = g = 2$. One reason for this is clarity of insight without requiring intense mathematical detail. Another reason, mentioned earlier in this chapter, is that most multivalent receptors of current cell behavioral relevance are bivalent.

For such systems of biological interest, a number of issues can be addressed by application of mathematical modeling and analysis. An immediate problem is the determination of quantitative values for receptor binding and crosslinking affinity constants, so that the number of receptor crosslinks and the sizes of crosslinked receptor aggregates can be calculated for correlation with resulting cell function. This is considered by DeLisi and Chabay (1979). An effort to relate receptor crosslinking to cell function is provided by MacGlashan *et al.* (1985). These investigators, using modeling results from Dembo and Goldstein (1978) and Wofsy *et al.* (1978), are able to predict the amount of histamine release from basophils as a function of antigen ligand binding to basophil cell surface antibody receptors.

Mathematical models for bivalent ligand/bivalent receptor binding and crosslinking date back to Bell (1974) and Dembo and Goldstein (1978), with a great deal of further work following. Helpful summaries are provided by Perelson (1984) and Goldstein (1988), both of which go beyond the extent of our presentation. In particular, two complicating features will be neglected here. One is the possibility of intramolecular binding reactions on a single receptor. That is, both sites of a bivalent receptor could, in principle, bind to the two functional groups on a single ligand (Figure 4–13). This can happen only when the ligand is sufficiently flexible and the functional groups are separated by an appropriate distance. A complication arises when the rate of formation of the second, intramolecular bond is influenced by molecular geometry and flexibility properties, leading to a cooperativity effect (Crothers and Metzger, 1972). The second feature we will neglect is the possibility of ring formation (Figure 4–13). Readers interested in pursuing these two related issues should consult Macken and Perelson (1985).

With intramolecular reactions and ring formation neglected, the model described earlier (Eqns. (4–43a,b)) for bivalent ligands with monovalent receptors can be applied almost directly for computations of total bound ligand and total crosslink number (though not the receptor cluster size distribution). We merely redefine R as the number of free receptor sites, instead of the total number of receptors, and replace the receptor conservation relation by the receptor site conservation relation $2R_T = R + C_1 + 2C_2$. Hence, our earlier results apply except with R_T replaced by $2R_T$. This model equivalence allows equilibrium and kinetic analyses presented by DeLisi and Chabay (1979) to apply to both bivalent and monovalent receptors for bivalent ligands.

First, consider *equilibrium behavior*. C_{1eq} and C_{2eq} can be calculated from Eqns. (4–44a,b) with R_T replaced by $2R_T$. Given that the example equilibrium results in Figure 4–15 are plotted in dimensionless terms with κ the only parameter, these results are also applicable to this present case, both qualitatively and quantitatively. The Scatchard plot for bivalent ligand binding to bivalent (as well as monovalent) receptors exhibits concave-upward curvature, as seen previously in Figure 4–16b. Here, R_T represents the total number of receptor sites per cell, so for bivalent receptors it will be twice the total number of receptors per cell. As before, avidity can be estimated at any particular ligand binding extent from the local slope of the Scatchard plot. Limiting results at very low and very high ligand concentrations are again given by Eqns. (4–47a,b).

Analyses developed by DeLisi and Chabay for *transient behavior* of bivalent receptors and ligands find that dissociation of labeled ligand in the presence of saturating concentrations of unlabeled ligand is described by two exponentials derived from the two steps necessary to dissociate a doubly-bound ligand. They also find that the model predicts that accelerated dissociation of labeled ligand in the presence of unlabeled ligand should be observed (section 2.2). This occurs because when one site of a doubly-bound labeled ligand dissociates, unlabeled ligand can bind to the freed receptor site, preventing reformation of the double bond and making complete release of the labeled ligand more likely than if no unlabeled ligand is present. These results are, as expected, similar to those previously mentioned for the bivalent ligand, monovalent receptor case.

We next turn to the specific example of *histamine release by basophils* when stimulated by antigen binding to cell surface antibodies. Dembo and Goldstein (1978) applied the model of Eqns. (4–43) at equilibrium to calculate the number of receptor crosslinks, C_{2eq}, on a basophil surface as a function of the solution concentration of a bivalent hapten stimulus; C_{2eq} is given by Eqn. (4–44b). Referring to Figure 4–15, the number of crosslinks follows a symmetric bell-shaped curve as a function of hapten concentration, L, and is maximal at the concentration $K_D/2$. Since it is known that crosslinking IgE molecules on the basophil surface is necessary to produce a histamine-release response, the hypothesis that the amount of histamine release is proportional to the number of crosslinks predicts that it should follow a similar dependence on hapten concentration. MacGlashan *et al.* (1985) found that this is, in fact, the case. They obtained a quantitative match of model predictions for histamine release to experimental data, as seen in Figure 4–18. The highest curve in this figure represents the prediction for bivalent hapten binding. The best-fit value of the equilibrium dissociation constant K_D for hapten binding to IgE was very close to an independently measured value. The other curves are further successful model predictions of the effect of a competing, non-stimulus, monovalent hapten on the histamine release curve. This monovalent competitor interferes with receptor crosslinking by the bivalent hapten and so diminishes the signal for histamine release.

The analysis described above allows determination of the total amount of bound ligand and the total crosslink number for the case of bivalent ligands binding to bivalent receptors by analogy with the bivalent ligand/monovalent receptor case. However, it does not allow determination of *cluster size distribution*. In other words, the previous analysis does not permit us to determine how the number of aggregates containing i receptors and j ligands, $C_{i,j}$, depends on the various system parameters. This distribution may have a strong influence on the ultimate cell behavioral response to ligand binding because the level of signal transduction may be a function of the aggregate size. For bivalent ligands with monovalent receptors, this problem did not arise since only $C_{1,1}$ and $C_{2,1}$ states were possible with just the latter state representing an aggregate. When the receptor is multivalent, however, a wider distribution of states can arise.

If the mathematical model is to be used to determine the cluster size distribution, it must now account explicitly for all these possible states. Kinetic

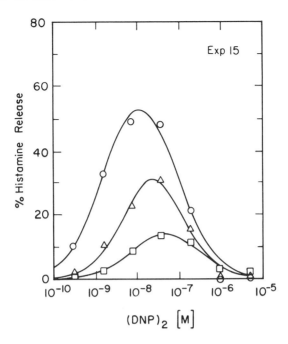

Figure 4–18 Histamine release due to receptor crosslinking. The data of MacGlashan *et al.* (1985) are plotted together with the theoretical predictions of Dembo and Goldstein (1978). The upper curve represents the case of only bivalent hapten present. The middle and lower curves represent situations in which a monovalent hapten is present at concentrations of 1×10^{-8} M and 5×10^{-8} M, respectively, in addition to the bivalent hapten. Reproduced from MacGlashan *et al.* 1985, *J. Immunol.* **135**:4129–4134, by copyright permission of the *Journal of Immunology*.

species balance equations can be written based on the various possible interactions. A variety of equilibrium analyses are available, including Dembo and Goldstein (1978), Perelson and DeLisi (1980), and Macken and Perelson (1985). We will describe only the key results, so readers interested in details of the analyses are referred to these original sources.

The equilibrium number of aggregates containing i receptors W_{ieq} (linear only, in this instance) is given by the expression:

$$W_{ieq} = \frac{R_T}{\zeta} \left[\frac{1 + 2\zeta - \sqrt{(1 + 4\zeta)}}{2\zeta} \right]^i \qquad (4\text{--}52)$$

where $\zeta = (4\lambda_0 \kappa)/(1 + 2\lambda_0)^2$; $\lambda_0 = L_0/K_D$ is the dimensionless ligand concentration and $\kappa = K_x R_T$ is the dimensionless equilibrium crosslinking constant. Example calculations (from Dembo and Goldstein, 1978) are shown plotted in Figure 4–19 for the fraction of receptors in crosslinked aggregates of various sizes, iW_{ieq}/R_T for $i = 1, 2, 3,$ and 4, versus dimensionless ligand concentration, $2\lambda_0$. The major point to note is that the higher-order aggregates are most significant when the ligand concentration is within an order of magnitude of the equilibrium dissociation constant K_D. Notice also the symmetry of these curves for this case.

Some extensions of these analyses have been published, dealing with effects such as asymmetric ligands (Wofsy *et al.*, 1978), cooperative intramolecular

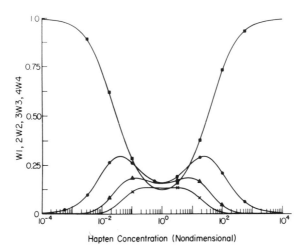

Hapten Concentration (Nondimensional)

Figure 4–19 Fraction of receptors in aggregates of 1 to 4 receptors (iW_{ieq}/R_T with $i = 1, 2, 3,$ and 4) as a function of the scaled bivalent ligand concentration $2\lambda_0$. $\kappa = 10$. Reproduced from Dembo and Goldstein 1978, *J. Immunol.* **121**:345–353, by copyright permission of the *Journal of Immunology.*

reactions (Wofsy and Goldstein, 1987), competition by monovalent ligands (Dembo and Goldstein, 1978), and more general valencies (Macken and Perelson, 1985). Readers interested in these complicating influences are encouraged to consult these references.

Many of the modeling insights described in this section may be useful to help interpret observations concerning receptor signal transduction and consequent cell function, particularly as more information about the influence of receptor cross-linking and cluster sizes on signal transduction is gained. As an example, relevant experimental data obtained by Baird and coworkers (Menon *et al.*, 1986; Baird *et al.*, 1988) indicate a quantitative influence of receptor cluster size on *serotonin release by RBL cells* in response to crosslinking of IgE receptors by multivalent DNP-protein conjugates. Figure 4–20 shows measurements of serotonin release as a function of oligomeric IgE fractions isolated by gel filtration chromatography. Decreasing fraction number corresponds to larger cluster sizes; this axis ranges from approximately hexamer to dimer receptor clusters. The key result is that the amount of serotonin release is directly correlated with the receptor cluster size. These data have not yet been analyzed in the context of a mathematical model, however.

Notice also that the mobile fraction of receptors in the cell membrane (as characterized by "% Recovery" after photobleaching in Figure 4–20) was found to be smaller for the larger receptor cluster sizes. These investigators have found that crosslinking state can affect receptor mobility by influencing interaction with cytoskeletal elements. Thus, cluster size distribution may also be an important feature to predict for receptor trafficking phenomena. Indeed, Linderman and Lauffenburger (1988) have used this hypothesis to explain differences observed for receptor recycling behavior when ligands of different valency are internalized via receptor-mediated endocytosis (section 3.2.2.b).

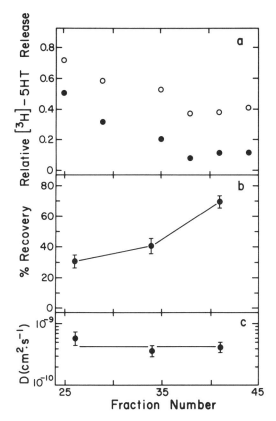

Figure 4–20 Serotonin release by rat basophilic leukemia (RBL) cells as a function of the size of receptor aggregates. Increasing cluster sizes correspond to decreasing fraction numbers. Note that as the cluster size increases, the diffusivity of receptors is essentially unchanged, the mobile fraction of receptors as measured by percentage recovery in a photobleaching experiment decreases, and the release of serotonin increases. Baird, Erickson, Goldstein, Kane, Menon, Robertson, Holowka, *Theoretical Immunology*, edited by A. S. Perelson, © 1988 by Addison-Wesley Publishing Company, Inc. Reprinted with permission of the publisher.

NOMENCLATURE

Symbol	Definition	Typical Units
a	cell radius	μm
A	cell surface area	μm^2
b	radius of a cell region	μm
C	constant	
C	receptor/ligand complex number	#/cell
C_{Beq}	equilibrium total amount of bound ligand	#/cell
C_{eq}	equilibrium receptor/ligand complex number	#/cell
C_i	number of ligands bound at i sites	#/cell
C_{ieq}	equilibrium number of ligands bound at i sites	#/cell
$C_{i,j}$	number of aggregates with i receptors and j ligands	#/cell
C_{xTeq}	total number of crosslinks at equilibrium	#/cell
C_1	number of ligands bound at one site	#/cell
C_{1eq}	equilibrium number of ligands bound at one site	#/cell

Symbol	Definition	Typical Units
C_2	number of ligands bound at two sites	#/cell
C_{2eq}	equilibrium number of ligands bound at two sites	#/cell
D	diffusion coefficient	cm^2/s
D_{cell}	cell diffusion coefficient	cm^2/s
D_L	ligand diffusion coefficient	cm^2/s
D_R	receptor diffusion coefficient	cm^2/s
f	ligand valency	
$(f_{RB})_{eq}$	equilibrium fraction of receptors bound	
g	receptor valency	
$G(s, t)$	generating function	
h	membrane thickness	nm
k_c	coupling rate constant	cm^2/s [a]
k_f	association rate constant per receptor	cm^3/s [b]
$(k_f)_{cell}$	association rate constant per cell	cm^3/s [b]
k_{off}	intrinsic dissociation rate constant	min^{-1}
k_{on}	intrinsic association rate constant per receptor	cm^3/s [b,c]
$(k_{on})_{cell}$	intrinsic association rate constant per cell	cm^3/s [b]
k_r	rate constant for dissociation of receptor/ligand complexes	min^{-1}
k_u	uncoupling rate constant	min^{-1}
k_x	receptor crosslinking rate constant	$(\#/cell)^{-1}\,min^{-1}$
k_x^c	receptor crosslinking rate constant in alternative units	$M^{-1}\,min^{-1}$
k_{-x}	receptor decrosslinking rate constant	min^{-1}
k_+	transport rate constant per receptor	cm^3/s [b,c]
$(k_+)_{cell}$	transport rate constant per cell	cm^3/s [b]
K_D	equilibrium dissociation constant	M
K_V	avidity constant	M^{-1}
K_x	crosslinking equilibrium constant	$(\#/cell)^{-1}$
l	characteristic system dimension	cm
L	ligand concentration	M
L_0	bulk ligand concentration	M
n_R	free receptor concentration on the cell surface	$\#/cm^2$
n_{Rb}	free receptor concentration at a distance b	$\#/cm^2$
N_{Av}	Avogadro's number $= 6.02 \times 10^{23}$ #/mole	
P_j	probability that there are j complexes	
R	free receptor number or site number	#/cell
R_{eq}	equilibrium number of free receptors	#/cell
R_T	total surface receptor number or receptor site number	#/cell
s	dummy variable	
s	encounter radius	nm
s'	reactive site radius	nm
t	time	min
t_{sample}	time period	min
u	scaled number of receptor ligand complexes	
v	total number of binding sites on a virus	#/cell
V	sampling volume	cm^3
W_{ieq}	equilibrium number of aggregates with i receptors	#/cell
X	number of species X	#/cell
δC_{eq}	standard deviation in C_{eq}	#/cell
δL	standard deviation in the value of L	M
Δt	time interval	min
$\langle \cdot \rangle$	expected or mean value of a quantity	
σ_c^2	variance in C	#/cell
θ	angle	radians
γ	capture probability for a receptor	

γ_{cell}	capture probability for a cell
ϕ	fractional surface coverage of a cell with receptors
ρ	ρ_{cell} or ρ_{rec}
ρ_{cell}	ratio of rate constants for binding to cell surface receptors
ρ_{rec}	ratio of rate constants for binding to solution receptors
τ	scaled time
λ_0	scaled ligand concentration
β	scaled ligand concentration
δ	crosslinking parameter
κ	scaled crosslinking equilibrium constant
ζ	crosslinking parameter

[a] More accurately, $(\#/cm^2)^{-1} s^{-1}$.

[b] More accurately, $(\#/cm^3)^{-1} s^{-1}$. Commonly converted to $M^{-1} s^{-1}$.

[c] For two-dimensional reactions the units are typically cm^2/s, or, more accurately, $(\#/cm^2)^{-1} s^{-1}$.

REFERENCES

Abney, J. R., Scalettar, B. A. and Owicki, J. C. (1989). Self diffusion of interacting membrane proteins. *Biophys. J.*, **55**:817–833.

Adam, G. and Delbruck, M. (1968). Reduction of dimensionality in biological diffusion processes. *In* A. Rich and N. Davidson (Eds), *Structural Chemistry and Molecular Biology*, pp. 198–215. San Francisco: W.H. Freeman.

Baird, B., Erickson, J., Goldstein, B., Kane, P., Menon, A. K., Robertson, D. and Holowka, D. (1988). Progress toward understanding the molecular details and consequences of IgE-receptor crosslinking. *In* A. S. Perelson (Ed.), *Theoretical Immunology, Part One*, pp. 41–59. Redwood City, CA: Addison-Wesley Publishing Company.

Bell, G. I. (1974). Model for the binding of multivalent antigen to cells. *Nature*, **248**:430–431.

Berg, H. C. and Purcell, E. M. (1977). Physics of chemoreception. *Biophys. J.*, **20**:193–219.

Berg, O. and von Hippel, P. H. (1985). Diffusion-controlled macromolecular interactions. *Annu. Rev. Biophys. Chem.*, **14**:131–160.

Bharuca-Reid, A. T. (1960). *Elements of the Theory of Markov Processes and their Application.* New York: McGraw-Hill.

Brown, M. S. and Goldstein, J. L. (1976). Analysis of a mutant strain of human fibroblasts with a defect in internalization of receptor-bound low density lipoprotein. *Cell*, **9**:663–674.

Brunn, P. O. (1981). Absorption by bacterial cells: interaction between receptor sites and the effect of fluid motion. *J. Biomech. Eng.*, **103**:32–37.

Cozens-Roberts, C., Lauffenburger, D. A. and Quinn, J. A. (1990a). Receptor-mediated cell attachment and detachment kinetics: I. Probabilistic model and analysis. *Biophys. J.*, **58**:841–856.

Cozens-Roberts, C., Quinn, J. A. and Lauffenburger, D. A. (1990b). Receptor-mediated cell attachment and detachment kinetics: II. Experimental model studies with the radial-flow detachment assay. *Biophys. J.*, **58**:857–872.

Crothers, D. M. and Metzger, H. (1972). The influence of polyvalency on the binding properties of antibodies. *Immunochemistry*, **9**:341–357.

DeLisi, C. (1980). The biophysics of ligand–receptor interactions. *Q. Rev. Biophys.*, **13**:201–230.

DeLisi, C. and Chabay, R. (1979). The influence of cell surface receptor clustering on the thermodynamics of ligand binding and the kinetics of its dissociation. *Cell Biophys.*, **1**:117–131.

DeLisi, C. and Marchetti, F. (1983). A theory of measurement error and its implications for spatial and temporal gradient sensing during chemotaxis. *Cell Biophys.*, **5**:237–253.

DeLisi, C. and Wiegel, F. W. (1981). Effect of nonspecific forces and finite receptor number on rate constants of ligand-cell bound-receptor interactions. *Proc. Natl. Acad. Sci. USA*, **78**:5569–5572.

Dembo, M. and Goldstein, B. (1978). Theory of equilibrium binding of symmetric bivalent haptens to cell surface antibody: application to histamine release from basophils. *J. Immunol.*, **121**:345–353.

Dintzis, H. M. and Dintzis, R. Z. (1988). A molecular basis for immune regulation: the immunon hypothesis. *In* A. S. Perelson (Ed.), *Theoretical Immunology*, pp. 83–103. Redwood City, CA: Addison-Wesley.

Doraiswamy, L. K. and Kulkarni, B. D. (1987). *The Analysis of Chemically Reacting Systems: A Stochastic Approach*. New York: Gordon and Breach.

Dower, S. K., Titus, J. A. and Segal, D. M. (1984). The binding of multivalent ligands to cell surface receptors. *In* A. S. Perelson, C. DeLisi and F. W. Wiegel (Eds.), *Cell Surface Dynamics: Concepts and Models*, pp. 277–328. New York: Marcel Dekker.

Echavarria-Heras, H. (1988). Convective flow effects in receptor-mediated endocytosis. *Math. Biosci.*, **89**:9–27.

Eigen, M. (1974). *In* S. L. Minz and S. M. Wiedermeyer (Eds.), *Quantum Statistical Mechanics*, pp. 37–61. New York: Plenum Press.

Erickson, J., Goldstein, B., Holowka, D. and Baird, B. (1987). The effect of receptor density on the forward rate constant for binding of ligands to cell surface receptors. *Biophys. J.*, **52**:657–662.

Gandolfi, A., Giovenco, M. A. and Strom, R. (1978). Reversible binding of multivalent antigen in the control of B-lymphocyte activation. *J. Theor. Biol.*, **74**:513–521.

Gardiner, C. W. (1983). *Handbook of Stochastic Methods for Physics, Chemistry, and the Natural Sciences*. New York: Springer-Verlag.

Goel, N. S. and Richter-Dyn, N. (1974). *Stochastic Models in Biology*. New York: Academic Press.

Goldstein, B. (1988). Densensitization, histamine release and the aggregation of IgE on human basophils. *In* A. S. Perelson (Ed.), *Theoretical Immunology, Part One. SFI Studies in the Sciences of Complexity*, pp. 3–40. Redwood City, CA: Addison-Wesley Publishing Company.

Goldstein, B. (1989). Diffusion limited effects of receptor clustering. *Comm. Theor. Biol.*, **1**:109–127.

Goldstein, B., Posner, R. G., Torney, D. C., Erickson, J., Holowka, D. and Baird, B. (1989). Competition between solution and cell surface receptors for ligand: Dissociation of hapten bound to surface antibody in the presence of solution antibody. *Biophys. J.*, **56**:955–966.

Goldstein, B. and Wiegel, F. W. (1983). The effect of receptor clustering on diffusion-limited forward rate constants. *Biophys. J.*, **43**:121–125.

Goldstein, B. and Wiegel, F. W. (1988). The distribution of cell surface proteins on spreading cells. *Biophys. J.*, **53**:175–184.

Goldstein, B., Wofsy, C. and Bell, G. (1981). Interactions of low density lipoprotein receptors with coated pits on human fibroblasts: estimate of the forward rate constant and comparison with the diffusion limit. *Proc. Natl. Acad. Sci. USA*, **78**:5695–5698.

Goldstein, B., Wofsy, C. and Echavarria-Heras, H. (1988). Effect of membrane flow on the capture of receptors by coated pits. *Biophys. J.*, **53**:405–414.

Hammer, D. A. and Apte, S. A. (1992). Simulation of cell rolling and adhesion on surfaces in shear flow: general results and analysis of selectin-mediated neutrophil adhesion. *Biophys. J.* **63**:35–57.

Keizer, J. (1985). Theory of rapid bimolecular reactions in solution and membranes. *Accounts Chem. Res.*, **18**:235–241.

Linderman, J. J. and Lauffenburger, D. A. (1988). Analysis of intracellular receptor/ligand sorting in endosomes. *J. Theor. Biol.*, **132**:203–245.

MacGlashan, D. W., Jr., Dembo, M. and Goldstein, B. (1985). Test of a theory relating to the cross-linking of IgE antibody on the surface of human basophils. *J. Immunol.*, **135**:4129–4134.

Macken, C. A. and Perelson, A. S. (1985). *Branching Processes Applied to Cell Surface Aggregation Phenomena.* Heidelberg: Springer-Verlag.

Mayo, K. H., Nunez, M., Burke, C., Starbuck, C., Lauffenburger, D. and Savage, C. R., Jr. (1989). Epidermal growth factor receptor binding is not a simple one-step process. *J. Biol. Chem.*, **264**:17838–17844.

McCammon, J. A. (1991). Computer-aided molecular design. *Science*, **238**:486–491.

McQuarrie, D. A. (1963). Kinetics of small systems. *I. J. Chem. Phys.*, **38**:433–436.

Menon, A. K., Holowka, D., Webb, W. W. and Baird, B. (1986). Clustering, mobility, and triggering activity of small oligomers of immunoglobulin E on rat basophilic leukemia cells. *J. Cell Biol.*, **102**:534–540.

Mitchison, T. and Kirschner, M. (1984). Dynamic instability of microtubule growth. *Nature*, **312**:237–242.

Northrup, S. H. (1988). Diffusion-controlled ligand binding to multiple competing cell-bound receptors. *J. Phys. Chem.*, **92**:5847–5850.

Northrup, S. H., Boles, J. O. and Reynolds, J. C. L. (1988). Brownian dynamics of cytochrome c and cytochrome c peroxidase association. *Science*, **241**:67–241.

Perelson, A. S. (1981). Receptor clustering on a cell surface. III. Theory of receptor cross-linking by multivalent ligands: description by ligand states. *Math. Biosci.*, **53**:1–39.

Perelson, A. S. (1984). Some mathematical models of receptor clustering by multivalent ligands. *In* A. S. Perelson, C. DeLisi and F. W. Wiegel (Eds.), *Cell Surface Dynamics: Concepts and Models*, pp. 223–276. New York: Marcel Dekker, Inc.

Perelson, A. S. and DeLisi, C. (1980). Receptor clustering on a cell surface. I. Theory of receptor crosslinking by ligands bearing two chemically identical functional groups. *Math. Biosci.*, **48**:71–110.

Persson, R., Svensson, U. and Everitt, E. (1983). Virus-receptor interactions in the adenovirus system. II. Capping and cooperative binding of virions on HeLa cells. *J. Virol.*, **46**:956–963.

Peters, R. and Cherry, R. J. (1982). Lateral and rotational diffusion of bacteriorhodopsin in lipid bilayers: experimental test of the Saffman-Delbruck equations. *Proc. Natl. Acad. Sci. USA*, **79**:4317–4321.

Press, W. H., Flannery, B. P., Teukolsky, S. A. and Vetterling, W. T. (1989). *Numerical Recipes: The Art of Scientific Computing.* New York: Cambridge University Press.

Rhee, H.-K., Aris, R. and Amundson, N. R. (1986). *First-Order Partial Differential Equations*, vol. 1, Englewood Cliffs, NJ: Prentice-Hall.

Ryan, T. A., Myers, J., Holowka, D., Baird, B. and Webb, W. W. (1988). Molecular crowding on the cell surface. *Science*, **239**:61–64.

Saxton, M. J. (1982). Lateral diffusion in an archipelago: effects of impermeable patches on diffusion in a cell membrane. *Biophys. J.*, **39**:165–173.

Saxton, M. J. (1987). Lateral diffusion in an archipelago: the effect of mobile obstacles. *Biophys. J.*, **52**:989–997.

Scalettar, B. A., Abney, J. R. and Owicki, J. C. (1988). Theoretical comparison of the self diffusion and mutual diffusion of interacting membrane proteins. *Proc. Natl. Acad. Sci. USA*, **85**:6726–6730.

Schlegel, R., Willingham, M. C. and Pastan, I. H. (1982). Saturable binding sites for vesicular stomatitis virus on the surface of Vero cells. *J. Virol.*, **43**:871–875.

Shoup, D. and Szabo, A. (1982). Role of diffusion in ligand binding to macromolecules and cell-bound receptors. *Biophys. J.*, **40**:33–39.

Sines, J. J., Allison, S. A. and McCammon, J. A. (1990). Point charge distributions and electrostatic steering in enzyme/substrate encounter: Brownian dynamics of modified copper/zinc superoxide dismutases. *Biochemistry*, **29**:9403–9412.

Smoluchowski, M. V. (1917). Versuch einer mathematische theorie der koagulationskinetic kolloider lösungen. *Z. Phys. Chem.*, **92**:129–168.

Spudich, J. L. and Koshland, D. E., Jr. (1976). Non-genetic individuality: chance in the single cell. *Nature*, **262**:467–471.

Tank, D. W., Wu, E.-S., Meers, P. R. and Webb, W. W. (1982). Lateral diffusion of gramicidin C in phospholipid multibilayers. *Biophys. J.*, **40**:129–135.

Torney, D. C. and McConnell, H. M. (1983). Diffusion-limited reaction rate theory for two-dimensional systems. *Proc. R. Soc. Lond. A*, 387:147–170.

Tranquillo, R. T. (1990). Theories and models of gradient perception. *In* J. P. Armitage and J. M. Lackie (Eds.), *Biology of the Chemotactic Response*, pp. 35–75. Cambridge: Cambridge University Press.

Tranquillo, R. T., Fisher, E. S., Farrell, B. E. and Lauffenburger, D. A. (1988). A stochastic model for chemosensory cell movement: application to neutrophil and macrophage persistence and orientation. *Math. Biosci.*, **90**:287–303.

Tranquillo, R. T. and Lauffenburger, D. A. (1987a). Analysis of leukocyte chemosensory movement. *Adv. Biosci.*, **66**:29–38.

Tranquillo, R. T. and Lauffenburger, D. A. (1987b). Stochastic model of leukocyte chemosensory movement. *J. Math. Biol.*, **25**:229–262.

Vaz, W. L. C., Goodsaid-Zalduondo, F. and Jacobson, K. (1984). Lateral diffusion of lipids and proteins in bilayer membranes. *Fed. Eur. Biochem. Soc.*, **174**:199–207.

Weisz, P. B. (1973). Diffusion and chemical transformation: an interdisciplinary excursion. *Science*, **179**:433–440.

White, J. M. and Littman, D. R. (1989). Viral receptors of the immunoglobulin superfamily. *Cell*, **56**:725–728.

Wickham, T. J., Granados, R. R., Wood, H. A., Hammer, D. A. and Shuler, M. L. (1990). General analysis of receptor-mediated viral attachment to cell surfaces. *Biophys. J.*, **58**:1501–1516.

Wiley, H. S. (1988). Anomalous binding of epidermal growth factor to A431 cells is due to the effect of high receptor densities and a saturable endocytic system. *J. Cell Biol.*, **107**: 801–810.

Wofsy, C. and Goldstein, B. (1987). The effect of cooperativity on the equilibrium binding of symmetric bivalent ligands to antibodies. *Mol. Immunol.*, **25**:151–161.

Wofsy, C., Goldstein, G. and Dembo, M. (1978). Theory of equilibrium binding of asymmetric bivalent haptens to cell surface antibody. *J. Immunol.*, **121**:593–601.

Young, S. H., McCloskey, M. and Poo, M.-M. (1982). *In* P. M. Conn (Ed.), *The Receptors*, pp. 511–539. New York: Academic Press.

Zigmond, S. H. (1977). Ability of polymorphonuclear leukocytes to orient in gradients of chemotactic factors. *J. Cell Biol.*, **75**:606–616.

Zwanzig, R. (1990). Diffusion-controlled ligand binding to spheres partially covered by receptors: an effective medium treatment. *Proc. Natl. Acad. Sci. USA*, **87**:5856–5857.

5

Signal Transduction

5.1 BACKGROUND

In Chapters 2–4, we examined models for receptor/ligand binding at the cell surface and for trafficking of receptors and ligands through the cell. In other words, we looked primarily at the state (free, bound, crosslinked) and location of the receptors. However, we did not examine the role of receptor state and location in the production of a cellular response. We turn now to this goal, as our ultimate aim is to predict cell behavioral responses that result from receptor/ligand binding.

Receptors, depending on their state and location, act to transmit biochemical signals. The simple binding of ligand to a cell surface receptor in itself is not sufficient to produce, for example, cell migration or secretion; the receptor must somehow convey to the cell interior the information that a binding event has occurred. In most cases, there is incomplete knowledge concerning the intracellular events transforming receptor-mediated signals into cell behavioral responses. Tremendous efforts are currently underway in many laboratories to identify and characterize the intracellular mediators and kinetic pathways that lie between receptor binding and cell responses.

As a result, a significant amount of information has been obtained during the past decade on the very *early signal generation processes* which occur immediately after receptor/ligand binding (see Chapters 8 and 9 of Gennis (1989) for a general background). These early signals are more closely connected to short time-scale responses, such as the secretion or release of certain cell products (*e.g.*, cyclic AMP, histamine, antibodies), than long time-scale behaviors of, for example, cell migration and proliferation. Consequently, there now is a substantial literature on mathematical models for early signal transduction processes as well as for short time-scale behavior.

In this chapter, we will examine some of the modeling approaches that have been used to describe the early events in signal transduction, with the hope of providing a foundation that may be extended to include events leading to long time-scale behaviors. Since the chief aim of this book is toward the quantitative understanding of long time-scale cell behavior (Chapter 6), we will do little more than briefly mention the application of these signal transduction models to short-time scale cell behavior.

Clearly it will be necessary to understand both the biochemistry and the mathematics of signal transduction in order to link the extent, timing, and nature of the cellular behavioral response with a given ligand dose. The mathematical work presented in this chapter is only a beginning, limited both by lack of knowledge of relevant biological details and by incomplete modeling efforts in this relatively new area. We focus first on the development of mathematical models capable of predicting the quantitative dependence of second messenger generation rates on receptor, ligand, and accessory component properties. We then discuss models which aim to account for experimentally-observed features such as intracellular second messenger oscillations, sensitivity of responses mediated by covalent modification, and adaptation of the cellular response.

5.1.1 Classes of Receptor Activity

As shown in Figure 5–1, there are three main classes of signal transduction models for cell surface receptors bound with ligand: receptors can act as *ion channels*; receptors can interact with G-proteins to produce other signaling molecules termed *second messengers*; and receptors can perform as *enzymes* (Gennis, 1989; Neubig and Thomsen, 1989). Interestingly, these three classes generally correspond to distinct *time-scales* for physiological responses, of fractions of seconds, seconds to minutes, and hours, respectively. Further, the typical *affinity* of ligand for receptors differs between each of the three modes: low affinity, intermediate affinity, and high affinity, respectively (Taylor, 1990).

In the first class of receptor activity, the receptor itself controls the passage of

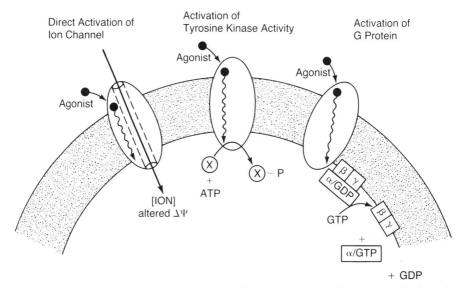

Figure 5–1 Three classes of receptor signal transduction activity. Receptors may act as ion channels, interact with G-proteins, or perform as enzymes (typically tyrosine kinases). Reprinted from Gennis (1989).

ions such as K^+ and Na^+ across the cell membrane. These ions typically regulate short time-scale behavioral phenomena, including the release of cell secretory vesicle contents and membrane electrical potential wave propagation. Receptors using this mode of signal transduction are mainly neurotransmitter receptors (*e.g.*, the nicotinic acetylcholine receptor, the glycine receptor, and the γ-aminobutyric acid (GABA) receptor), which regulate extremely rapid physiological responses such as muscle contraction. Because of our overriding interest in the long time-scale cell phenomena, and because the mathematical details of ion channels and electrophysiology can be found elsewhere (*e.g.*, Hille, 1991; Nossal and Lecar, 1991), we will neglect consideration of ion channel transduction processes in this text.

In the second class, receptors interact with and activate membrane-associated, GTP-binding G-proteins. Activated G-proteins may in turn act on other enzymes to induce the generation of small molecules ("second messengers") such as cAMP, cGMP, Ca^{2+}, and phospholipid metabolites within the cytoplasm (see Figure 5–2). These molecules are then capable of either directly stimulating short-term responses or modulating intracellular enzyme activities for indirect regulation of long-term responses. Receptors using this mode of transduction are many and diverse, but typically regulate physiological responses occurring on the time scale of minutes; examples include the chemotactic peptide receptor and the α and β-adrenergic receptors. More detailed information about second messengers and G-proteins is found below.

In the third class of receptor activity, the enzymatic activity of the receptor itself, typically the covalent modification activity of phosphorylation, is affected

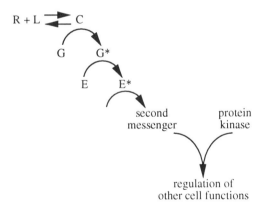

examples:

enzyme or effector	second messenger
adenylate cyclase	cAMP
phospholipase C	IP_3, DAG, calcium

Figure 5–2 Cascade of events following activation of G-proteins by receptor/ligand complexes. Activated G proteins can activate other enzymes, which in turn may induce the generation of second messengers. Many second messengers bind to protein kinases. G* and E* represent the active forms of G-protein and enzyme, respectively.

by ligand binding. The receptor then modulates activities of other intracellular enzymes for ultimate regulation of long-term behavioral responses. Most receptors known to utilize this mode of transduction are growth factor receptors acting as kinases and regulating physiological responses over a time scale of hours. Receptor kinases will be discussed in more detail shortly.

5.1.2 Second Messengers

The sequence of signal transduction events in many cells might be described as a series of messages. The ligand itself, a "first messenger", binds to cell surface receptors. As a result of this binding, various molecules inside the cell, as a group termed "second messengers", are generated or released from intracellular compartments. These second messengers then diffuse to appropriate targets within the cell to bring about a physiological response.

The existence of multiple messengers, or molecules working to accomplish the cellular response to a ligand, is now seen to be a common theme in signal transduction. As a first example, bound receptors may activate the enzyme *adenylate cyclase* and thus cause the production of the second messenger *cAMP* (cyclic AMP). A wide variety of receptors are now known to transduce ligand binding signals by means of the adenylate cyclase system, some by stimulating adenylate cyclase generation of cAMP and some by inhibiting this reaction. A few receptors, in fact, exist in two subtypes – one stimulatory and one inhibitory for adenylate cyclase. Evidently both positive and negative signals can be mediated via the adenylase cyclase system. A primary intracellular target of cAMP is the enzyme known as *protein kinase A* (pKA, also known as the cAMP-dependent protein kinase) (Edelman *et al.*, 1987). Binding of cAMP to the regulatory subunit of pKA causes dissociation of a catalytic subunit that is able to phosphorylate serine and threonine residues on protein substrates and thus modulate their activity. Among these substrates are the nicotinic acetylcholine and β-adrenergic receptors, troponin I, and protein phosphatase inhibitor I.

A second example of the involvement of second messengers in signal transduction is the case of receptors which activate the enzyme *phosphatidylinositol-specific phospholipase C* (PLC) (Berridge and Irvine, 1989). PLC acts to break down the membrane phospholipid phosphatidylinositol-4,5-bisphosphate (PIP_2) to form two second messengers: *inositol 1,4,5-trisphosphate* (IP_3) and *1,2-diacylglycerol* (DAG). IP_3 diffuses into the cytoplasm, where it causes release of *calcium* from various internal stores (Burgoyne and Cheek, 1991). Calcium is a seemingly ubiquitous regulator of a diverse range of cell functions and can bind to a variety of proteins, particularly calmodulin (Heizmann and Hunziker, 1991). Calcium–calmodulin complexes modulate the activity of many intracellular enzymes, including various protein kinases as well as glycogen synthase and microtubule-associated proteins. DAG diffuses within the membrane and activates the membrane-associated enzyme known as protein kinase C (pKC); this enzyme phosphorylates protein substrates on serine and threonine residues and thus modifies their activity. Nishizuka (1986) provides a listing of the numerous membrane and cytoplasmic proteins likely serving as pKC substrates, including a variety of cell surface receptors (*e.g.*, EGF, insulin, transferrin, interleukin-2, and

β-adrenergic receptors). Cytoplasmic proteins, including the cytoskeleton contractile protein myosin and the cytoskeleton/adhesion receptor-binding vinculin (both apparently involved in cell migration) and membrane transport proteins for glucose, Na^+, and H^+ (likely involved in cell metabolism for growth) also may serve as pKC substrates. Hence, the phosphoinositol system possesses two second messenger branches, both leading to covalent modification of cellular enzymatic activity and thus control over the cellular response.

A third class of second messenger system is the *cGMP* (cyclic GMP) system. The cellular concentration of cGMP can be increased by activation of the enzyme *guanylate cyclase* (Schulz et al., 1989) (which forms cGMP from GTP) or lowered by the activation of the enzyme *cGMP-specific phosphodiesterase* (Chabre et al., 1988), depending on the identity of the receptor bound by ligand. Here, then, positive and negative signals are permitted by the two distinct enzymes. Similarly to the cAMP system, cGMP exerts its effects largely through a cyclic-nucleotide-dependent enzyme, protein kinase G (pKG) (Edelman et al., 1987). Substrates for this enzyme have not been well-characterized.

This brief discussion of types of second messengers is not meant to be all inclusive. For example, we have omitted discussion of the production of the second messenger arachidonic acid via a phospholipase of the A_2-type (Murayama and Ui, 1985; Burch et al., 1986) and the possible role of other inositol lipid metabolites such as IP_4 in signal transduction (Berridge and Irvine, 1989; Irvine, 1991). Ongoing research in this area is likely to extend significantly what is now known about these and other second messengers and their modes of action.

Two critical mechanistic features are shared by these diverse second messenger systems. One is their means of effect. As has already become evident, a major means by which small second messengers (*e.g.*, cAMP, cGMP, DAG, Ca^{2+}) exert their regulatory influence is through *binding to protein kinases* (pKA, pKG, and pKC). Protein kinases covalently modify numerous cell membrane, cytoplasmic, and cytoskeletal proteins, modulating the activity of those target proteins, and thus have a role in metabolic pathways, transmembrane signal transduction, protein synthesis, cytoskeletal organization, and other cellular functions (Hunter, 1987). The second common feature (though exceptions exist, for example a membrane form of guanylate cyclase which acts as a receptor for atrial natriuretic peptide (Chinkers et al., 1989) is their means of generation, receptor coupling to membrane-associated G-proteins.

5.1.3   G-Proteins

In section 2.2.4.a, we discussed G-proteins in the context of their effect on receptor/ligand binding. We now focus on their role as functional intermediates between signal-initiating receptor/ligand complexes and resultant second messengers. A schematic of the interaction of G-proteins with receptors is shown in Figure 5–3. The inactive form of the G-protein consists of α, β, and γ subunits with a molecule of GDP bound to the α subunit. The interaction of this inactive G-protein with a bound receptor promotes the release of GDP from the α subunit and the binding of GTP at the same site. The G-protein is then released from the receptor and dissociated into separate $\beta\gamma$ and α-GTP entities; α-*GTP* is the active

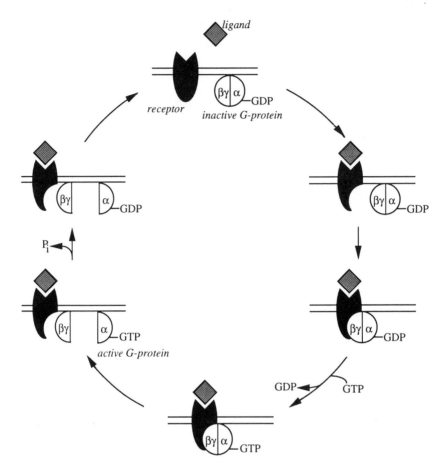

Figure 5–3 Activation of G-proteins by bound receptors. The inactive form of the G-protein couples to bound receptors. After GDP is replaced by GTP, the α subunit with attached GTP dissociates from the βγ subunit and acts on targets within the cell. Hydrolysis of GTP by the α subunit allows reformation of the inactive G-protein.

form of the G-protein. The bound receptors thus serve to catalyze the formation of active G-protein. Active G-proteins are returned to their inactive state upon hydrolysis of GTP by the GTPase activity found in the α subunit itself, and the α-GDP and βγ subunits can then recombine. We further note that the separate βγ subunit may also play an active role in signal transduction, though this role is not well-understood at present (Bourne, 1989).

The active form of the G-protein, α-GTP, can modulate the activity of the true signal generating enzyme or *effector*. For example, the activity of the effector adenylate cyclase (for the cAMP system discussed above) is stimulated by G-proteins termed G_s and inhibited by G-proteins denoted G_i.

Birnbaumer *et al.* (1990) list approximately 80 identified receptors that interact with 12 known G-proteins. In total, G-proteins are now known to influence the production of at least six second messengers: cAMP, cGMP, IP_3,

DAG, Ca^{2+}, and arachidonic acid. Further, G-proteins may have a role in stimulation of K^+ and Ca^{2+} channels (Yatani *et al.*, 1987a,b), which may blur the distinction made earlier in this chapter between signal transduction through second messenger production and through ion channels. Available evidence suggests that one receptor can couple to more than one G-protein and that one G-protein may be able to regulate more than one effector function (Birnbaumer *et al.*, 1990).

Considerable *amplification* of a receptor/ligand binding event is afforded by the G-protein system, because receptor/ligand complexes, active G-proteins, and effector enzymes all act catalytically. A single receptor/ligand complex may activate roughly five to ten different G-protein molecules before ligand dissociation. Each of these α-GTP complexes may then activate several effectors (*e.g.*, adenylate cyclase, PLC), although this number is difficult to estimate and clearly depends on the lifetime of the α-GTP complex itself. These effectors can then produce many second messengers. As a consequence of this cascade of catalytic events, for example, Sklar (1986) estimates that one bound formyl peptide receptor on the neutrophil may generate 100–1000 molecules of IP_3 and of the order of 10,000 molecules of calcium released in less than a minute.

When non-hydrolyzable GTP analog (such as GTPγS (guanosine-5'-γ-thiotriphosphate) or GppNHp (guanylyl imino-$\beta\gamma$-diphosphate)) are incorporated into the cell during ligand incubation and then bind to the α subunit of the G-protein, α apparently becomes stuck in the "on" position. (An analogous situation is thought to be responsible for the ability of the *ras* oncogene product, *Ras* protein, to cause malignant transformation of cells.) These non-hydrolyzable GTP analogs are widely used in experimental investigations of G-protein involvement in signal transduction. In addition, cholera toxin and pertussis toxin, which ADP-ribosylate the α subunits of the G-proteins G_s and G_i, respectively, are widely used to manipulate the activation of these G-proteins.

5.1.4 Receptors as Enzymes

We now turn to a distinct route for receptor-mediated signal transduction that has been discovered in the last decade; receptors can themselves be enzymes, typically protein kinases, and thus covalently modify intracellular proteins directly.

Excellent reviews of receptor kinases are provided by Carpenter (1987) and Yarden and Ullrich (1988). Hunter (1987) offers a provocative discussion of their importance in regulation of cell behavior. We will briefly summarize here the salient points. Primary among these is that receptor kinases found so far are exclusively *tyrosine kinases*; *i.e.*, these receptor/enzymes are capable of phosphorylating proteins on tyrosine residues. This is a striking observation, because other cellular kinases are predominantly serine/threonine kinases (some cellular kinases can apparently phosphorylate tyrosine, serine, and threonine residues (Lindberg *et al.*, 1992). To date there exist three major receptor tyrosine kinase families, all apparently composed of growth factor receptors: the EGF receptor family (including the EGF receptor and the *neu* and *erb*B proteins), the PDGF receptor family (including the PDGF and CSF-1 receptors), and the insulin receptor family (including the insulin and IGF-1 receptors). The general structure of these families

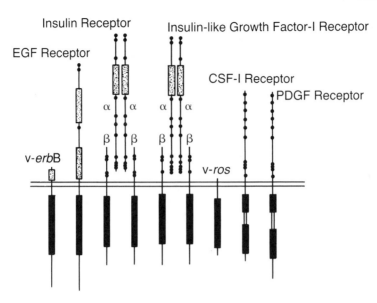

Figure 5–4 Schematic view of receptor tyrosine kinase structure. Filled boxes denote tyrosine kinase domains. Cysteine-rich domains and other cysteines are indicated by shaded boxes and small circles, respectively. Three families of receptors are shown: (1) EGF receptor and a related oncogene product v-*erb*B, (2) insulin receptor, IGF-1 receptor, and related oncogene product v-*ros*, and (3) CSF-1 receptor and PDGF receptor. Reprinted from Gennis (1989).

is illustrated in Figure 5–4. The receptors possess an extracellular ligand binding domain, a single transmembrane domain, and a large cytoplasmic domain. All families are believed to bind ligand monovalently.

The cytoplasmic domain contains the features necessary for kinase activity: an ATP-binding site (to provide the phosphate group) and a substrate binding site. In addition, there is typically at least one autophosphorylation site on the molecule. That receptor kinases always possess only a single transmembrane domain is thought to have some significance for their function; in contrast, receptors utilizing G-proteins for signal transduction typically contain seven distinct hydrophobic domains believed to span the membrane bilayer. The presence of a ligand-binding domain and a kinase domain within the same molecule allows these receptors to carry out their function of translating an extracellular stimulus into an intracellular signal directly, without requiring an accessory component such as a G-protein. At the same time, it is not clear how the ligand-binding event is transmitted to the enzymatic domain. Some investigators suggest that conformational change in the extracellular domain induced by ligand binding is somehow passed on to the cytoplasmic domain through the transmembrane region. Many investigators favor an intermolecular transfer mechanism, in which a receptor/ligand complex is able to induce stimulation of a second receptor by diffusion-limited encounter in the cell membrane (Ullrich and Schlessinger, 1990, Cadena and Gill, 1992). Analogy to systems in which receptor clusters are necessary for signal transduction, such as immunological activation of B-lymphocytes and mast cells by antigen (see section 4.3), may be relevant. In any case, the evidence for an intermolecular activation mechanism is not conclusive at present, so this remains an open question.

Discovery of the key cellular substrates for receptor kinases is an extremely important goal, which has until recently been elusive. Upon ligand stimulation, numerous cellular proteins exhibit a rapid increase in tyrosine phosphate content. Among these, of course, is the receptor itself, due to its autophosphorylation site(s). A few likely substrates centrally involved in resulting cell behavioral responses include phospholipase C, phosphatidylinositol 3-kinase, and GTPase-activating protein (Ullrich and Schlessinger, 1990). Recently, the presence of a common domain termed the SH2 domain in proteins that are targets for receptor tyrosine kinases has been found and is thought to govern the interaction of these target proteins with the autophosphorylated receptor (Koch et al., 1991).

Though there is growing appreciation of the central role of covalent protein modification in regulating many and varied cell behavioral processes, the special relationship between receptor kinases and control of cell proliferation in particular is not understood. No mathematical modeling efforts have yet been brought to bear on this issue. However, some general features of covalent modification have been analyzed theoretically by Goldbeter and Koshland (1981) and will be discussed later in this chapter.

Finally, we would be remiss to neglect the recent finding that some receptors may act as tyrosine phosphatases (Fischer et al., 1991). In particular, the T cell surface molecule CD45 is now known to be a tyrosine phosphatase, although the identity of its ligand is unknown. Many researchers speculate that a shift in research focus from tyrosine kinases to tyrosine phosphatases is in order. This information, in addition to the role of some receptors as guanylate cyclases mentioned earlier, indicates that receptors acting as enzymes are not solely tyrosine kinases.

5.1.5 Desensitization or Adaptation of the Cellular Response

The events of signal transduction following receptor/ligand binding may also act to desensitize the cell's ability to respond to that and other ligands. Desensitization of a receptor-mediated cell response refers to the termination or diminution of the response despite the continuing or repeated application of ligand. Desensitization may be *homologous* (ligand specific) or *heterologous* (ligand non-specific).

In the β-adrenergic receptor system, a combination of three mechanisms is thought to be responsible for the rapid desensitization of the receptor-stimulated adenylate cyclase response (Lohse et al., 1990). First, receptors may be actively *sequestered* from extracellular ligand by a process such as endocytosis. Thus there may be a clear link between the earlier analysis of endocytosis in section 3.2 and the production, or lack of production, of a cellular response, although other sequestration mechanisms can also be postulated. Second, phosphorylation of β-adrenergic receptors by protein kinase A may *uncouple* these receptors from their effector system; for example, coupling of the receptors with G-proteins may be impaired. This effect of protein kinase A is thought to be heterologous, for the activity of protein kinase A is effected by the binding of several different ligands. Third, phosphorylation of β-adrenergic receptors by the β-adrenergic receptor

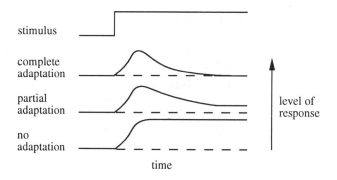

Figure 5–5 Adaptation following a step change in stimulus concentration. The magnitude of an unspecified cellular response as a function of time is sketched for the cases of complete, partial, and no adaptation.

kinase plays a role in homologous desensitization of this system. Lohse and coworkers find that these three mechanisms can together produce up to 70% desensitization of the signal transduction activity of the β-adrenergic receptor system in A431 cells. More generally, however, the mechanisms of desensitization are poorly understood for most receptor systems.

Desensitization may also be termed adaptation, particularly in reference to the readjustment of cell sensitivity to a constant stimulus level. Another well-studied example of adaptation or desensitization in eukaryotic cells is the synthesis of cAMP and cGMP by the cellular slime mold *Dictyostelium discoideum* in response to binding of extracellular cAMP to its cell surface receptor (Devreotes and Sherring, 1985). It has also been conjectured that the chemotactic response of neutrophils to peptide attractants, requiring a spatial comparison of extracellular ligand concentration, depends on adaptation of the receptor-mediated second messenger generation (Devreotes and Zigmond, 1988).

When a stimulus is introduced and held constant, a *completely-adapting system* will exhibit a response that increases to a maximum then decreases back to prestimulus behavior. Under these same conditions, the response of a *partially-adapting system* will also go through a maximum but then decrease to an intermediate level. These receptors are sketched in Figure 5–5; note that we are not concerned at this point with the shape of the transient but with the level of response reached at long times. Later in this chapter, we examine a model that offers an approach to predicting the signal-generating abilities of various receptor states in a completely-adapting system.

5.2 MODELS OF RECEPTOR/G-PROTEIN/EFFECTOR INTERACTIONS

As detailed earlier in this chapter, many receptors interact with G-proteins in order to transduce a signal. Modeling efforts to describe such interactions vary in their scope and system applicability. Some models aim to link receptor/ligand binding with the production of a second messenger; we will discuss below models describing the production of cAMP or the release of sequestered intracellular calcium following ligand binding at the cell surface. Other models aim to describe features in the second messenger response without explicitly incorporating the role of receptors and G-proteins. In this category fall several recently published models for intracellular calcium oscillations. Finally, a few models have focused only on the receptor/G-protein interaction alone, without incorporating the details of the ensuing second messenger production. We will also describe one of these efforts. In all cases, however, the ultimate goal is the quantitative description of the link between ligand binding and signal transduction with the future aim of connecting this to a cellular response such as secretion, growth, contraction, movement, or adhesion.

5.2.1   Activation of Adenylate Cyclase and Production of cAMP

Models of second messenger generation following receptor/ligand binding have been explored extensively in the pharmacological literature, particularly for application to the β-adrenergic receptor system. This receptor binds a class of hormone ligands known as catecholamines, *e.g.* epinephrine, resulting in intra-cellular cAMP production (as well as generation of other second messengers) and increased cell metabolic rate. Given the large number of molecular components in a G-protein system – ligand, receptor, three G-protein subunits, GTP/GDP, and effector enzyme (adenylate cyclase, in this instance) – many choices for basic model assumptions are possible. Not surprisingly, many models of receptor occupancy and cAMP production have been offered for this system. Excellent, comprehensive reviews of this vast modeling literature are available (Tolkovsky, 1983; Levitzki, 1984, 1986). Our focus here will be to analyze the most successful of these models, the collision coupling model.

5.2.1.a   Experimental Observations
Levitski, Tolkovsky, and coworkers (reviewed in Tolkovsky (1983)) performed a number of experiments designed to probe the kinetics of activation of adenylate cyclase and the accumulation of cAMP during periods of *steady state receptor/ligand binding* of L-epinephrine to the β-adrenergic receptor on turkey erythrocytes. In their experiments, they explicitly tested the dependence of these kinetics on ligand concentration and the number of receptors, G-proteins, and enzyme molecules. In each of the experiments described below, the presence of the non-hydrolyzable GTP analog GppNHp prevented the deactivation of activated G-proteins. We note, before moving on to more sophisticated measurements, that ligand binding in this system does not show cooperativity.

The investigators examined the activation of the enzyme adenylate cyclase as a function of time and ligand concentration with the quantitative findings shown in Figure 5–6. Figure 5–6a shows the intracellular accumulation of cAMP as a function of time and ligand concentration during continual incubation of cells with ligand in the presence of GppNHp. Note the apparent positive curvature of the curves, which can be interpreted as a lag period before linear accumulation of cAMP. Figure 5–6b gives the amount of active adenylate cyclase present as a function of time and ligand concentration. This measurement was accomplished

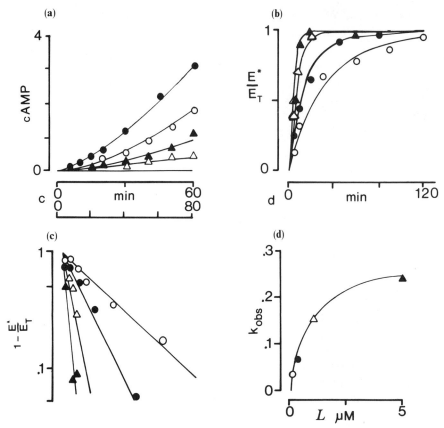

Figure 5–6 Activation of adenylate cyclase and generation of cAMP following stimulation of turkey erythrocytes with epinephrine in the presence of GppNHp. (a) Accumulation of cAMP is plotted as a function of time and ligand concentration. Ligand concentrations are denoted with open triangles (0.5 μm), filled triangles (1.0 μm), open circles (3.0 μm), and filled circles (15 μm). (b) Normalized amount of active adenylate cyclase (active adenylate cyclase at time t/total active adenylate cyclase, or $E^*(t)/E_T$) is plotted as a function of time and ligand concentration. Ligand concentrations are denoted with open circles (0.1 μm), filled circles (0.4 μm), open triangles (1.0 μm), and filled triangles (5.0 μm). (c) Most of the data shown in (b) are replotted as $\ln\{1 - [E^*(t)/E_T]\}$ versus time. (d) The calculated apparent first order rate constant k_{obs} is shown as a function of the ligand concentration. Redrawn from Tolkovsky (1983).

by transferring the cells from the agonist (activating) ligand epinephrine to the antagonist (non-activating) ligand propranolol at a sequence of times. The amount of active adenylate cyclase is then determined by measuring cAMP accumulation for a fixed time period following this transfer; under these conditions, *i.e.* no further stimulation but continued activation due to the presence of GppNHp, the second messenger will evolve linearly. The data from Figure 5–6b are replotted in semi-log fashion in Figure 5–6c; the slopes of the lines shown in that plot can be used to derive an overall or lumped *first order rate constant* k_{obs} for the activation of adenylase cyclase. Figure 5–6d shows that the value of k_{obs} is a saturable function of agonist concentration.

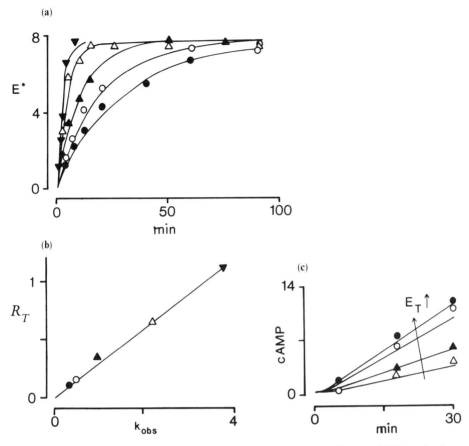

Figure 5–7 Influence of the number of receptor and enzyme molecules on cAMP production. (a) Amount of active adenylate cyclase as a function of time for different numbers of total receptors. (b) The calculated apparent first order rate constant k_{obs} is shown as a function of total receptor number. (c) Accumulation of cAMP as a function of time and total enzyme concentration. Redrawn from Tolkovsky (1983).

Figure 5–7 shows data on the influence of receptor and enzyme numbers, R_T and E_T, respectively, on activation in this system. Various concentrations of antagonists are used to effectively block a fraction of the cell surface receptors from binding agonist. Figure 5–7a shows the amount of active enzyme versus time for different numbers of available receptors and at saturating agonist concentration. Figure 5–7b demonstrates a resulting linear dependence of the enzyme activation rate constant k_{obs} on available total receptor number, R_T, where k_{obs} has been found as in Figure 5–6. In Figure 5–7c, the accumulation of cAMP is shown as a function of time and the total enzyme concentration E_T. One can derive from these data that the extent of activation of adenylate cyclase is a function of E_T but that the first order rate constant k_{obs} is independent of E_T.

Finally, Figure 5–8 illustrates the influence of G-protein level, accomplished by varying GppNHp concentration, on the activation of adenylase cyclase.

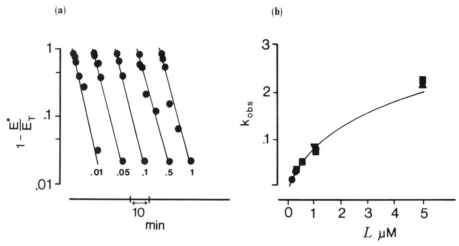

Figure 5–8 Influence of G-protein concentration on activation of adenylate cyclase. (a) $1 - [E^*(t)/E_T]$ is plotted as a function of time and GppNHp concentration. (b) The calculated apparent first order rate constant k_{obs} is shown as a function of agonist concentration. Different symbols represent different concentrations of GppNHp. Redrawn from Tolkovsky (1983).

Figures 5–8a and 5–8b demonstrate that the first order rate constant for enzyme activation is independent of GppNHp concentration and, therefore, effective G-protein level.

In summary, then, the major observations regarding adenylate cyclase activation that the mathematical model aims to describe can be summarized by the simple equation describing the evolution of the active enzyme E^* concentration

$$\frac{dE^*}{dt} = k_{obs}(E_T - E^*) \tag{5-1a}$$

or its solution

$$E^*(t) = E_T[1 - \exp(-k_{obs}t)] \tag{5-1b}$$

where k_{obs} is a saturable function of agonist concentration, linearly dependent on the available total receptor number, and independent of both the total enzyme concentration and the amount of G-protein. The models developed by Levitski, Tolkovsky, and coworkers focus on accounting for these qualitative observations.

5.2.1.b Collision Coupling Model

In proposing models to explain the type of data shown above, investigators need to incorporate the presumed roles of receptor, G-protein, and enzyme in producing the second messenger cAMP. In the context of the known biology, that is, that receptors activate G-proteins and that G-proteins activate enzymes which then produce the second messenger (Figure 5–2), there are a variety of alternative mathematical and biological pictures. One important distinction involves the difference between *precoupling* and *collision coupling*. In the models, precoupled molecules are assumed to be found together before stimulation with ligand. For example, Neubig *et al.* (1988) and Fay *et al.* (1991) have found that approximately one-third of the α_2-adrenergic receptors on platelets and the formyl peptide receptors on neutrophils, respectively, may be preassociated or precoupled with G-proteins, although the applicability of this tenet to other systems is not known. Alternatively, one could envision that G-proteins and enzymes are precoupled, or even that receptors, G-proteins, and enzymes are precoupled in a trimolecular unit of sorts. Note that when molecules are not precoupled but the coupling event is fast compared to the preceding or ensuing activation event, one can use the the mathematically simplifying assumption of precoupling as a valid approximation to simplify the model.

Collision coupled molecules are separate entities that may couple only when they "find" each other – what is envisioned here is that diffusion of, for example, receptors and G-proteins in the membrane will lead to collisions of the two molecules. If the receptor is occupied, it may then act to *catalyze* the formation of the active form of the G-protein. This is the picture offered in section 2.2.4.b, in the context of the mobile receptor hypothesis. One could also imagine that the G-protein/enzyme interaction involves collision coupling. In these cases, inclusion of the coupling event explicitly in a mathematical model implies that the coupling event is on the same time scale as the ensuing activation event.

Tolkovsky and Levitzki (1981) offer a successful model (in terms of explaining the data just shown) that includes both precoupled and collision coupling interactions. They term the model the *collision coupling receptor/effector model*, in reference to the mode of interaction they propose for receptors with effectors. They do not separate the roles of G-protein and enzyme; rather, these two molecules are considered precoupled and are treated as one entity. Thus the model can be shown schematically by:

$$L + R \underset{k_r}{\overset{k_f}{\rightleftharpoons}} C$$

$$E + C \underset{k_u}{\overset{k_c}{\rightleftharpoons}} E\text{-}C \overset{k_a}{\to} E^* + C \overset{k_d}{(\to)} E + C$$

Receptors (R) and ligands (L) bind to form complexes (C). Complexes then associate with effectors (E), here the precoupled G-protein/enzyme complex, to form an intermediate (E–C). The receptor/ligand complexes and G-protein/ enzyme complexes are presumably collision coupled with a rate constant that is a function of the diffusion coefficient. That diffusion limitations (section 4.2) play a measurable role is supported by the data of (Hanski *et al.*, 1979), who found that the overall rate constant for activation of adenylate cyclase via β-adrenergic receptors was influenced by the fluidity of the membrane (*i.e.* the diffusivity of receptors and/or G-proteins and/or adenylate cyclase in the membrane).

The intermediate complex E–C is then activated with a rate constant k_a, representing the activation of the enzyme adenylate cyclase via the exchange of GDP for GTP by the G-protein or some other rate-limiting step. E^* represents the active form of the enzyme. The enzyme can be deactivated with rate constant k_d; this pathway is shown in parentheses because the experiments to which we will compare the model results were all done in the presence of the non-hydrolyzable GTP analog GppNHp. Recall that G-proteins so activated are stuck in the "on" position, so that the activation of enzyme is essentially irreversible. Note that the net result of the proposed pathway is that effectors are activated by a transient encounter with a receptor/ligand complex and that a single receptor/ligand complex can catalyze the formation of many active effectors.

Before beginning the mathematical analysis of this model, we note the similarity between this model and the ternary complex (receptor/ligand/accessory component) model of section 2.2.4. In this earlier chapter, we introduced and investigated the ligand-binding properties of this model. We now look at the model in terms of its ability to account for the signal generation data of Figures 5–6 to 5–8.

The mathematical analysis of the collision coupling receptor/effector model proceeds as follows. Free ligand concentration, L, is assumed constant. Further, receptor/ligand binding is considered to be at quasi-equilibrium. All data under consideration here are obtained in the presence of non-hydrolyzable GTP analogs, so the effector deactivation step represented with rate constant k_d can be neglected. A key assumption in the analysis is that the amount of receptor/ligand/effector ternary complex (E–C) is small compared to the amounts of receptor/ligand complexes, free receptors, and free effectors. This latter assumption can also be stated as an assumption that k_a and k_u are large compared to k_c, so that the coupled form of receptor/ligand complex with effector enzyme is short-lived and present only at very low levels.

Because $E\text{–}C \ll C$, receptor/ligand binding at equilibrium is essentially unchanged from the simple case of homogeneous receptor/ligand binding, or

$$C = \frac{R_T L}{K_D + L} \tag{5–2}$$

where $K_D = k_r/k_f$. This equation is, of course, identical to the simple receptor/ligand binding relationship of Eqn. (2–13). Thus the prediction is for simple non-cooperative binding behavior, in agreement with experimental data on this system.

Now, the amount of active enzyme is governed by the kinetic equation

$$\frac{dE^*}{dt} = k_a E\text{–}C \tag{5–3}$$

If we are to determine the evolution of E^* with time, we will need to express $E\text{–}C$ in the above equation in terms of known quantities, and then integrate Eqn. (5–3) to get $E^*(t)$.

We can also write a differential equation describing the rate of change of $E\text{–}C$. If we assume that the amount of $E\text{–}C$ is small and does not change significantly during the experiment, we can evoke the pseudo-steady state assumption that $dE\text{–}C/dt$ is approximately equal to 0. Thus we have:

$$\frac{dE\text{–}C}{dt} = k_c EC - (k_u + k_a)E\text{–}C \approx 0 \tag{5–4}$$

Conservation of effectors, together with the assumption that $E–C$ is small compared with E and E^*, gives the relation $E_T \approx E + E^*$. Substituting $E = E_T - E^*$ into Eqn. (5–4) and solving for $E–C$ gives:

$$E–C = \frac{k_c C(E_T - E^*)}{k_u + k_a} \qquad (5–5)$$

Substituting this into Eqn. (5–3) yields a differential equation that, together with the initial condition that there is no active enzyme at time 0, can be integrated to obtain the amount of active enzyme as a function of time,

$$E^*(t) = E_T[1 - \exp(-k_{obs}t)] \qquad (5–6a)$$

with the overall first order rate constant k_{obs} defined by:

$$k_{obs} = \left(\frac{k_c k_a}{k_u + k_a}\right)\left(\frac{L}{K_D + L}\right) R_T \qquad (5–6b)$$

Note that Eqn. (5–6a) is identical to Eqn. (5–1b). According to Eqn. (5–6), then, the result of the collision coupling model is the prediction of a first-order adenylate cyclase activation rate. The dependence of the apparent activation rate constant k_{obs} on L and R_T is saturable and linear, respectively, in agreement with the experimental data in Figures 5–6d and 5–7b. Note also that the first order rate constant is not dependent on E_T, but that the total amount of active enzyme E'^* is dependent on E_T. Furthermore, since G-proteins and adenylate cyclase molecules are assumed coupled in this model, the dependence of activation on G-protein molecules is the same as for E_T. Hence, this model is consistent with all the epinephrine/β-adrenergic receptor system experimental data discussed earlier in this section. Some experimental observations on purified enzyme preparations support this model by showing copurification of adenylate cyclase with G-protein (Arad et al., 1984).

It is interesting to note here that the mathematical model is supported by the data without explicit identification of parameter values or fitting of the model to the data. Rather, identification of the type of behavior (linear, non-linear, etc.) played a major role. Several other models suggested were rejected at this point in their development by the same comparison (Tolkovsky and Levitzki, 1981). Clearly a next step in the development of the collision coupling model is to measure values for the model parameters and then examine the consistency of the data presented earlier with the model's predictions using these parameter values.

5.2.1.c Shuttle Model

We next turn to the details of the G-protein/effector enzyme interaction. This interaction was neglected in the collision coupling model discussed above, for the two molecules were assumed to be essentially one unit. We now examine the possibility that the G-protein can assume a role independent of the enzyme it activates.

There are many possible models to suggest; we will very briefly examine the model shown schematically as:

$$L + R \underset{k_r}{\overset{k_f}{\rightleftharpoons}} C$$

$$C + G \underset{k_u}{\overset{k_c}{\rightleftharpoons}} C\text{-}G \rightarrow C + G^* \rightarrow G^*\text{-}E^* + C$$
$$+ E$$

The first step is the binding of receptor (R) and ligand (L) to form a receptor/ligand complex (C). G-protein (G) then interacts with the receptor/ligand complex, dissociates from the complex as an activated form (G*), and then interacts with the enzyme (E) to produce the active enzyme E*. This model is termed the *G shuttle binary model* because G-protein shuttles between C and E and because G-protein is activated via a binary complex of C and G.

Under the condition that the coupling of C and G is rate limiting, *i.e.*, that the time scale of C and G coupling is slow compared to the time scale of G* and E coupling, the G shuttle binary model simply reduces to the collision coupling model and the amount of active enzyme G*–E* is governed by Eqn. (5–6). Thus in this limiting case, the G shuttle binary model is also consistent with all the experimental data presented earlier. An interesting test of this idea can then be suggested: if the concentration of available adenylate cyclase can be reduced, then the kinetics of G* and E coupling will begin to play a role in the prediction of adenylate cyclase activation (Tolkovsky, 1983). To the best of our knowledge, the results of such tests have not yet proved conclusive.

Under more general parameter conditions, the G shuttle binary model does not predict that the overall observed rate constant for adenylate cyclase activation, k_{obs}, is independent of the total amount of G-protein and is therefore not consistent with the experimental data. The full model analysis, along with other suggested but less successful models, is presented by Tolkovsky (1983) and Levitzki (1984) and will not be repeated here.

At present, then, the data on the activation of adenylate cyclase under steady-state binding of epinephrine to the β-adrenergic receptor on turkey erythrocytes is best explained by a model in which the G-protein and adenylate cyclase molecules are precoupled or are coupled quickly in comparison to other events, namely the activation of G-proteins via collision coupling with bound receptors.

5.2.2   Activation of Phospholipase C and the Production of Calcium

5.2.2.a   *Experimental Observations*
The last section focused on models for receptor/G-protein/adenylate cyclase interaction. We now turn to the activation of phospholipase C (PLC) by G-proteins. Recall that PLC acts to break down PIP_2 into IP_3 and DAG; IP_3 binds to a receptor on the surface of intracellular stores to release Ca^{2+}. As a result, the concentration of intracellular free Ca^{2+}, which we will denote as $[Ca^{2+}]_i$, increases. The activation of this pathway can be followed by assaying for the production of IP_3 in cells or by following calcium-dependent cell

physiological processes such as contraction or opening of ion channels. In addition to increases in $[Ca^{2+}]_i$ due to release of calcium from internal stores, in many cells it is believed that receptor occupation can lead to an increased flux of calcium across the plasma membrane and into the cell (Benham and Tsien, 1987; Pandol *et al.*, 1987). With the recent development of calcium-sensitive fluorescent probes such as quin-2, fura-2, and indo-1 (Tsien, 1989), it has become common to follow these changes in $[Ca^{2+}]_i$ in cell populations, or even single cells, as a function of time. These probes allow the measurement of the free calcium concentration only; the intracellular concentration of bound Ca^{2+} is generally several orders of magnitude greater and is typically not measured (an exception can be found in Chandra *et al.* (1989)).

The responses of several cells to ligand stimulation is shown in Figure 5–9. $[Ca^{2+}]_i$ typically increases from a baseline value of approximately 100 nM to a peak value of 300–1000 nM. There is great *variability* in the responses of individual cells, even when the responses of neighboring cells in the same field of view are compared. In two cell types tested, bovine aortic smooth muscle cells and BC$_3$H1 cells, the fraction of cells responding increases with ligand concentration (Figure 5–10a) but the peak height is only weakly dependent on ligand concentration (Mahoney *et al.*, 1992; Mahama and Linderman, 1993a). Furthermore, in these two cell types and others (*e.g.*, hepatocytes and A10 cells (Monck *et al.*, 1988)), it is found that the time between ligand addition and a significant $[Ca^{2+}]_i$ increase (latency) is a decreasing function of ligand dose (Figure 5–10b).

The data shown in Figure 5–9 show a single peak in $[Ca^{2+}]_i$ after cells are treated with a step change in ligand concentration. In contrast, in a wide range of electrically non-excitable cell types responding to a variety of stimuli, periodic

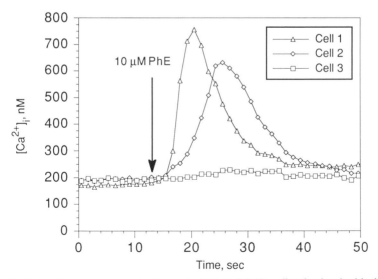

Figure 5–9 Cytosolic calcium concentration in individual BC$_3$H1 cells stimulated with the ligand phenylephrine. Three neighboring cells were stimulated with 10 μM phenylephrine at the time indicated. Plotted is the concentration of cytosolic calcium in each cell as a function of time.

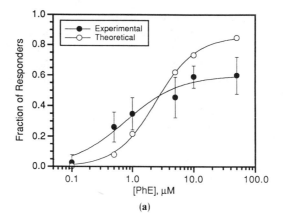

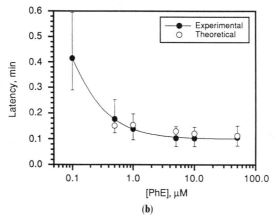

Figure 5–10 Experimental data and model predictions for the response of individual BC_3H1 cells stimulated with the ligand phenylephrine. In both figures, model predictions are the result of simulating the responses of 1000 cells. (a) The fraction of cells responding is plotted as a function of ligand concentration. A cell was scored as a responder if a calcium rise of at least 30% above basal level was observed within 2 min after ligand addition. (b) The time between ligand addition and the maximum rate of change in calcium concentration was defined as the latency between ligand addition and response. For responding cells only, the latency is plotted as a function of the ligand concentration. Model parameter values: $k_f = 1.0 \times 10^7 \text{ M}^{-1} \text{ s}^{-1}$, $k_r = 60 \text{ s}^{-1}$, $k_c = 2 \times 10^{-5} (\#/\text{cell})^{-1} \text{ s}^{-1}$, $k_{g1} = 0.01 \text{ s}^{-1}$, $k_{g2} = 2 \times 10^{-5} (\#/\text{cell})^{-1} \text{ s}^{-1}$, $k_{g3} = 1 \times 10^{-5} (\#/\text{cell})^{-1} \text{ s}^{-1}$, $k_6 = 2.0 \text{ s}^{-1}$, $k_{i1} = 1600 \text{ s}^{-1}$, $K_{i1} = 0.07 \text{ μM}$, $k_{i2} = 0.1 \text{ s}^{-1}$, $k_{ca1} = 0.1 \text{ s}^{-1}$, $K_{ca1} = 1.0 \times 10^7 \#/\text{cell}$, $k_{ca2} = 0.05 \text{ s}^{-1}$, $k_{ca3} = 0.5 \text{ μM s}^{-1}$, $K_{ca2} = 0.2 \text{ μM}$, $R_T = 1.9 \times 10^3 \#/\text{cell}$, $G_T = 1 \times 10^5 \#/\text{cell}$, $P_T = 2.0 \times 10^4 \#/\text{cell}$, $Ca_T = 2.0 \text{ μM}$, $V_s/V_{cy} = 0.16$, $n1 = 3$, $n2 = 2$. All parameter values except $n1$ and $n2$ were allowed to vary according to a Gaussian distribution with a standard deviation of 20% of the mean value. Taken from Mahama and Linderman (1993a).

oscillations in $[Ca^{2+}]_i$ have been observed. These are also sometimes referred to as repetitive spikes because of the sharp peaks characteristically found. For example, $[Ca^{2+}]_i$ oscillations in hepatocytes treated with three different concentrations of the peptide hormone vasopressin are shown in Figure 5–11. It is important to recognize that the ligand concentration is held constant during each of the three treatments, or that the oscillations continue with no change in stimulus.

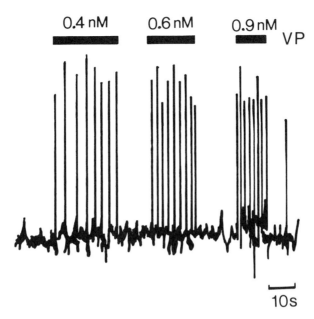

Figure 5–11 Oscillations in cytosolic calcium concentration in single hepatocytes exposed to the ligand vasopressin (VP). The period of the oscillations varies with VP concentration. Reprinted from *Nature* (Woods *et al.*, vol. 319, pp. 600–602). Copyright 1986.

Note also that the amplitude of the oscillations is not significantly influenced by the ligand concentration. Good reviews of the experimental data on this phenomenon are provided by Rink and Hallam (1989), Berridge (1990), and Meyer and Stryer (1991) and some examples are listed in Table 5–1. Notice that the period of the oscillations can vary from less than one second to a few minutes. The oscillations are generally asynchronous and variable from cell to cell in the same experiment. The cause of this variability is unknown; however, recent data show that there are cell-specific patterns that can be repeated, arguing for differences in parameter values between cells, rather than the effects of a stochastic process, as the cause of the variability (Prentki *et al.*, 1988; Byron and Villereal, 1989).

Table 5–1 Examples of cells exhibiting $[Ca^{2+}]_i$ oscillations

Cell	Stimulus	Period (s)
Rat myocyte	Caffeine	0.3–3
Rat hepatocyte	Vasopressin	18–240
Macrophage	Cell spreading	19–69
Smooth muscle	Phenylephrine or histamine	30–48
REF52 fibroblasts	Gramicidin + vasopressin	35–100
Endothelial cells	Histamine	40–125
B lymphocytes	Antigen	50–75
Mouse oocyte	Fertilization	600–1800

Abbreviated from Berridge (1989).

The characteristics of the oscillations can vary significantly with cell type and ligand. In many cases, the oscillation frequency increases with increasing ligand concentration (Figure 5–11), but this increase is not marked in all cell types (Mahoney *et al.*, 1992). In some systems, oscillations can occur and persist for a while despite the absence of calcium in the extracellular medium (Ambler *et al.*, 1988; Millard *et al.*, 1989), indicating a central role for release from intracellular stores. On the other hand, in some cell types the oscillation frequency is strongly influenced by the extracellular calcium concentration (Holl *et al.*, 1988; Mahoney *et al.*, 1992), suggesting that Ca^{2+} channels are involved in the entry of Ca^{2+} into the cytosol or in the refilling of intracellular stores.

Although increases in cytosolic calcium concentration and subsequent activation of calcium-binding proteins have been linked to a number of cellular responses (*e.g.*, secretion, T lymphocyte activation, contraction), it has been difficult thus far to make an explicit connection between the observed $[Ca^{2+}]_i$ oscillations and a physiological response (Tsien and Tsien, 1990). An exception is the recent report by Holl *et al.* (1988), who found that the amount of growth hormone released by individual spontaneously secreting pituitary cells correlated with the frequency and the amplitude of calcium oscillations in that same cell. On the other hand, there are suggestions that in some cell types, in fact, oscillations are only observed under non-physiological conditions (Gray, 1988; Harootunian *et al.*, 1988). Clearly, more work is needed in order to relate calcium oscillations in individual cells with physiologically significant responses of those cells.

Finally, several recent observations have been made of waves of calcium traveling through a single cell. In general the causes and significance of such waves are not known, and only recently have experimental reports (*e.g.*, Rooney *et al.*, 1990) and a simple mathematical model (Meyer, 1991) been offered. Our discussion in this text, however, will focus on models for the single calcium peak and for calcium oscillations, neglecting any spatial gradients within the cytoplasm of a single cell.

5.2.2.b *Single Calcium Rise*

To investigate the presumed roles of receptors, G-proteins, PLC, and IP$_3$ in eliciting the calcium increases that are linked with cellular responses, one can begin by formulating models similar to those proposed for the adenylate cyclase system. We start by describing the single (non-oscillatory) calcium rise seen in many cells (Figure 5–9). There are more steps to consider than in the case of generation of cAMP by adenylate cyclase: in the present case, the active enzyme that is generated (PLC) then acts to produce IP$_3$ from PIP$_2$, and IP$_3$ acts on receptors inside the cell to cause the release of sequestered calcium into the cytosol. Furthermore, the transient nature and timing of the single calcium rise data suggests that receptor/ ligand *binding kinetics*, and not simply equilibrium binding data, should be explicitly included in the mathematical model.

The model of Mahama and Linderman (1993a), developed for the stimulation of BC$_3$H1 cells by binding of phenylephrine to α_1-adrenergic receptors, describes the timing, magnitude, and cell-to-cell variability of a single calcium rise and the relationship of these quantities to receptor/ligand binding kinetics, number of G-protein molecules, and reaction kinetics. Equations describing the rate of change of the numbers of receptor/ligand complexes (C), activated alpha subunits of G-proteins (G^*-GTP), inactivated alpha subunits of G-proteins (G^*-GDP), activated phospholipase C(P-G^*-GTP), IP$_3$(I), and cytosolic calcium concentration (Ca) are given by:

$$\frac{dC}{dt} = k_f RL - k_r C \tag{5-7a}$$

$$\frac{d(G^*\text{-}GTP)}{dt} = (k_{g1} + k_c C)(G\text{-}GDP) - (k_6 + k_{g2}P)(G^*\text{-}GTP) \tag{5-7b}$$

$$\frac{d(G^*\text{-}GDP)}{dt} = k_6[(G^*\text{-}GTP) + (P\text{-}G^*\text{-}GTP)] - k_{g3}(G^*\text{-}GDP)(BG) \tag{5-7c}$$

$$\frac{d(P\text{-}G^*\text{-}GTP)}{dt} = k_{g2}P(G^*\text{-}GTP) - k_6(P\text{-}G^*\text{-}GTP) \tag{5-7d}$$

$$\frac{dI}{dt} = k_{i1}(P\text{-}G^*\text{-}GTP)\left(\frac{Ca}{Ca + K_{i1}}\right) - k_{i2}I \tag{5-7e}$$

$$\frac{dCa}{dt} = \left[k_{ca1}\left(\frac{I^{n1}}{I^{n1} + K_{ca1}^{n1}}\right) + k_{ca2}\right]\left(\frac{V_s}{V_{cy}}\right)(Ca_s - Ca) - k_{ca3}\left[\frac{Ca^{n2}}{Ca^{n2} + K_{ca2}^{n2}}\right] \tag{5-7f}$$

$G - GDP$, BG, and P represent the numbers of inactive G-proteins, $\beta\gamma$ subunits, and inactive PLC molecules, respectively. V_s and V_{cy} are the volumes of the calcium store and cytosol, and Ca_s is the concentration of mobile calcium in the store. Note that unlike the model described earlier for adenylate cyclase activation, this formulation can incorporate the dynamics of receptor binding following a change in ligand concentration. Also necessary are the *conservation relations* for receptors, G-proteins, PLC and calcium: $R_T = R + C$; $G_T = (G\text{-}GDP) + (G^*\text{-}GTP) + (G^*\text{-}GDP) + (P\text{-}G^*\text{-}GTP)$ (BG is assumed equal to $(G^*\text{-}GTP) + (G^*\text{-}GDP) + (P\text{-}G^*\text{-}GTP)$), $P_T = P + (P\text{-}G^*\text{-}GTP)$ and $Ca_T = Ca_s(V_s/V_{cy}) + Ca$, where R_T is the total number of cell surface receptors, G_T is the total number of G-proteins, P_T is the total number of PLC molecules, and Ca_T is the total concentration of releaseable calcium in the cell on the basis of the cytosol volume.

There are several other important assumptions that are part of this model. First, *collision coupling* of complexes to G-proteins is assumed to result in the production of activated G-proteins. The rate constant k_c for the production of active G-proteins from inactive G-proteins is assumed to be 2×10^{-5} (#/cell)$^{-1}$ s^{-1}, a diffusion-limited value calculated by simulations of membrane collision events (Mahama and Linderman, 1993b) and similar to a value-estimated from Eqn. (4–37b). Recall that this value represents a maximum for the rate constant, and that the value may be decreased if not all complexes and G-proteins that collide with each other result in the successful production of an active G-protein. The rate constants for the coupling of activated alpha subunits of G-proteins and PLC (k_{g2}) and the recombination of inactivated alpha subunits of G-proteins and $\beta\gamma$ subunits (k_{g3}) are also estimated from these simulations.

The rate constant k_{g1} describes the slow basal exchange of GTP for GDP (Taylor, 1990). The rate constant for the inactivation of alpha subunits of

G-proteins, k_6, is assumed equal to 2.0 s^{-1}, a value calculated from the model of Thomsen and Neubig (1989) described in section 5.2.3.

Eqn. (5–7e) describes the production of IP_3 following PLC activation. Formation of IP_3 depends on the amount of activated PLC and is a saturable function of the cytosolic calcium concentration; IP_3 is degraded with rate constant k_{i2}.

Changes in $[Ca^{2+}]_i$ are described by Eqn. (5–7f). Increases in $[Ca^{2+}]_i$ are assumed to result only from release of sequestered calcium into the cytosol and not from increased transport of calcium across the plasma membrane, a simplifying assumption based on the data of Ambler $et\ al.$ (1988) who found that the agonist-stimulated $[Ca^{2+}]_i$ increases in this cell line are not influenced significantly by changes in the concentration of calcium in the medium. IP_3 is assumed to act cooperatively (the Hill coefficient $n1$ is greater than one) to release calcium from intracellular stores (Champeil $et\ al.$, 1989). A small leak of calcium into the cytosol is assumed to be proportional to the calcium gradient and is included with the term k_{ca2} in Eqn. (5–7f). Calcium is resequestered in stores due to the action of a calcium ATPase.

Using reasonable parameter estimates, this deterministic model is able to describe the timing of the single calcium rise as a function of ligand dose (Figure 5–10b) and the magnitude of the rise. This formulation of the model does not, however, suggest an explanation for the observation that the fraction of cells responding is an increasing function of ligand dose. There are two possibilities: cellular responses may be due to such low numbers of molecules that stochastic effects play a role, or the variable responses may be due to physical $parameter$ $variation$ from cell to cell. The latter possibility is easily examined with the mathematical formulation of the model given above, by adding one additional feature. The parameters in the model ($e.g.$, rate constants, number of molecules) can be varied around a mean value according to a normal distribution, and the responses of many cells can be simulated. The results are shown in Figure 5–10a. For each ligand dose, the responses of 1000 cells were simulated and the fraction of cells responding is plotted. Although this simple explanation can explain the experimental data of Figure 5–10a, methods of confirming the variability of specific parameter values from cell to cell are difficult to identify.

In addition, the declining phase in the calcium response needs to be explained, for the present model simply predicts an increase in calcium concentration to a new steady state value. It may be that desensitization plays a role, or, alternatively, the mechanisms described in the next section as underlying calcium oscillations may be a key factor.

5.2.2.c Calcium Oscillations

The model detailed above describes only the timing, magnitude, and cell-to-cell variability of a single calcium rise. As mentioned earlier, however, the calcium levels in many non-excitable cells oscillate in the continual presence of a constant concentration of ligand. This has proved fertile ground for many mathematical modelers.

There are two questions that may be addressed by mathematical models of the calcium oscillations. First, what biochemical/biophysical mechanisms might be responsible for the oscillations? Recently, a number of models have been proposed which can account for some features of the oscillations in some types of cells (Meyer and Stryer, 1988; Goldbeter *et al.*, 1990; Swillens and Mercan, 1990; Cuthbertson and Chay, 1991; Somogyi and Stucki, 1991; Keizer and De Young, 1992). Second, what is the physiological significance or benefit of an oscillating second messenger? Only limited attempts to answer this question have been made (Goldbeter *et al.*, 1990).

Let us look first at models for the mechanisms underlying calcium oscillations. Any set of mathematical equations producing oscillations must have non-linear terms, although a system with these characteristics may not always exhibit oscillations. There are many potential models which may predict the $[Ca^{2+}]_i$ oscillations that have been observed. The goal of the mathematical modeling is to develop models that can be tested, that is, models that make predictions which suggest experiments to support or rule out the proposed mechanism.

Two general categories of mechanisms which may produce $[Ca^{2+}]_i$ oscillations in non-excitable cells can be defined (following Berridge, 1990): (1) *receptor-controlled oscillations*, in which periodic calcium release from intracellular stores is due to periodic generation of the calcium-mobilizing intermediate IP_3 resulting from feedback mechanisms operating at the plasma membrane; and, (2) *second messenger-controlled oscillations*, in which calcium itself drives the periodicity and the concentration of IP_3 does not oscillate. Alternatively, the models can be classified as to whether $[Ca^{2+}]_i$ acts to increase (*positive feedback*) or decrease (*negative feedback*) its own concentration. The proposed mathematical models for $[Ca^{2+}]_i$ oscillations all fit into these categories. These models and mechanisms are listed and classified in Table 5–2.

All of the proposed models for calcium oscillations lack the explicit detail found in the model of section 5.2.2.b concerning the link between receptor/ligand binding, G-protein activation, and IP_3 generation. Instead, these models concentrate on the non-linearities and feedback in the rates of production of IP_3 and $[Ca^{2+}]$ and their ability mathematically to generate oscillations in the simulated presence of a constant ligand concentration. Presumably an oscillation model will soon be linked with a model describing these earlier details as well as adaptation (section 5.4) for a complete description of the calcium transients.

The detailed mathematical descriptions of these oscillation models can be found in the original references; for brevity we will thoroughly describe here only one of the models. The calcium oscillation mechanism proposed by Meyer *et al.* (1988) is shown in Table 5–2 as model 1. Before turning to the mathematical equations describing this model, let us first describe the physical situation which is proposed to lead to oscillations. Binding of ligand to receptor causes an increase in IP_3 concentration via G-protein activation of PLC. This model includes the highly cooperative (and thus non-linear) release of Ca^{2+} from intracellular stores in response to IP_3, as demonstrated in permeabilized rat basophilic leukemia cells by Meyer *et al.* (1988). Further impetus for the steep initial rise in the calcium spikes is assumed to come from positive feedback, specifically that Ca^{2+} enhances the activity of PLC. Release of Ca^{2+} from intracellular stores continues until the stores are significantly depleted. The stores then refill via their Ca^{2+} pumps, IP_3 is hydrolyzed back to PIP_2, and the cycle can begin again. This model predicts

Table 5–2 Proposed mathematical models for $[Ca^{2+}]_i$ oscillations

Model no.	Schematic	Nature of calcium feedback	IP_3 oscillations?	Reference
1	LR → G → PLC → IP_3 → Ca release → exhaustion (−); DAG; (+) feedback	Positive feedback via PLC	Yes	Meyer and Stryer (1988)
2	LR → G → PLC → IP_3 → Ca release → exhaustion (−); DAG; CICR from separate pool (+)	Positive feedback via CICR	No	Goldbeter et al. (1990)
3	LR → G → PLC → IP_3 → Ca release; DAG → PKC; (+) and (−) feedback loops	Indirect positive feedback via PLC; negative feedback via PKC	Yes	Cuthbertson and Chay (1991)

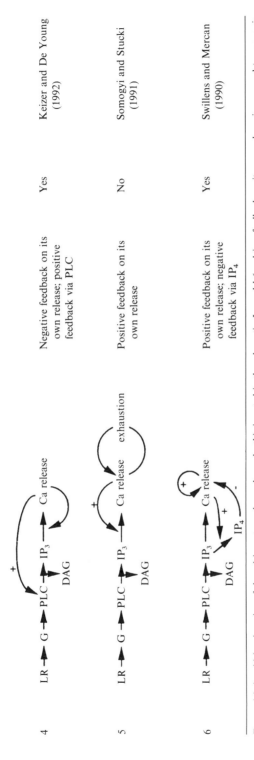

4		Negative feedback on its own release; positive feedback via PLC	Yes	Keizer and De Young (1992)
5		Positive feedback on its own release	No	Somogyi and Stucki (1991)
6		Positive feedback on its own release; negative feedback via IP_4	Yes	Swillens and Mercan (1990)

For models in which exhaustion of the calcium store plays a key role, this is noted in the schematic. In model 6, calcium feedback on its own release is proposed to occur via inhibition of calcium store efflux by the calcium concentration in the store. Lowering of store calcium removes this inhibition and thus acts as positive feedback. Adapted from Tsien and Tsien (1990).

that both IP$_3$ concentration and [Ca^{2+}]$_i$ will oscillate and is thus classified as a receptor-controlled mechanism.

With these foundations, kinetic species balance equations can be written for the three concentration variables, x = cytosolic Ca^{2+} concentration on a whole-cell basis, y = cytosolic IP$_3$ concentration on a whole-cell basis, and z = Ca^{2+} concentration in the stores on a whole-cell basis:

$$\frac{dx}{dt} = k_i \left(\frac{y^3}{K_1 + y^3} \right) z - k_p \left(\frac{x}{K_2 + x} \right)^2 + k_s z^2 + k_\theta \left[1 - \left(\frac{x}{K_4} \right)^{3.3} \right] \qquad (5-8a)$$

$$\frac{dy}{dt} = k_g \left(\frac{x}{K_3 + x} \right) F - k_h y \qquad (5-8b)$$

$$\frac{d(x + z)}{dt} = -k_\theta \left[1 - \left(\frac{x}{K_4} \right)^{3.3} \right] \qquad (5-8c)$$

The first term on the right-hand side of Eqn. (5–8a) describes the rate of IP$_3$-induced efflux of calcium from stores; its third-order form is based on the cooperativity found by Meyer et al. (1988). k_i is the rate constant for IP$_3$-induced efflux of calcium from stores and K_1 is a saturation constant. The second term represents the pumping of calcium back into stores due to a calcium ATPase believed to transport two Ca^{2+} ions per cycle (Carafoli, 1987), hence the second-order form (Inesi et al., 1980). k_p and K_2 are corresponding rate and saturation constants. The third term represents the movement of Ca^{2+} out of the stores at high loading levels, probably due to the reversal of the ATPase. The fourth term on the right-hand side of Eqn. (5–8a) represents removal of cytosolic calcium to other intracellular locations, buffering of calcium, or extrusion of calcium to the extracellular medium. Meyer and Stryer (1988) find that this term is required for the model to generate realistic oscillations given reasonable parameter values. In the original formulation of the model, Meyer and Stryer assumed that this term accounted for mitochondrial sequestration of calcium. The exponent of 3.3 is based on an empirical observation that the rate of influx to the mitochondria is proportional to the 3.3 power of cytosolic Ca^{2+} concentration whereas the rate of efflux from mitochondria is constant (Nicholls and Akerman, 1982). Recent data on the effects of mitochondrial inhibitors on calcium oscillations argue against a role for mitochondria in the production of oscillations (Millard et al., 1989). However, this term can simply be reinterpreted as another calcium-buffering mechanism which acts as an IP$_3$-insensitive calcium store without effect on the predictions of the model (Tsien and Tsien, 1990).

The first term on the right-hand side of Eqn. (5–8b) represents the rate of IP$_3$ generation and includes dependence on cytosolic calcium, providing positive feedback. In addition, this term is proportional to F, the degree of stimulation of IP$_3$ production from receptors acting through G-proteins. Note that the model does not explicitly include receptor/ligand binding. Rather, the assumption is made that during the continuous presence of ligand, the parameter F, which is a function of the ligand concentration and might most simply be interpreted as the fraction of receptors actively generating a signal, is constant. Thus this mathematical formulation of the model attempts to explain the steady generation of oscillations but not the initial transients in $[Ca^{2+}]_i$ which occur on a time scale similar to that for receptor/ligand binding in many systems. The second term on the right-hand side of Eqn. (5–8b) represents the rate of loss of IP_3 due to hydrolysis in an assumed first order process. Finally, Eqn. (5–8c) follows the change in the sum of cytosolic and store calcium; in other words, the sum of cytosolic, store, and mitochondrially-sequestered calcium is a constant.

Meyer and Stryer (1988) offer plausible estimates for the model parameters, largely based on their own experiments and some literature references: $k_i \sim 6.6 \text{ s}^{-1}$, $K_1 \sim 0.1 \text{ μM}$, $k_p \sim 5 \text{ μM s}^{-1}$, $K_2 \sim 0.15 \text{ μM}$, $k_s \sim 3 \times 10^{-5} \text{ μM}^{-1} \text{ s}^{-1}$, $k_g \sim 1 \text{ μM s}^{-1}$, $K_3 \sim 1 \text{ μM}$, $k_h = 2 \text{ s}^{-1}$, $k_\theta = 0.5 \text{ μM s}^{-1}$, and $K_4 = 0.6 \text{ μM}$. This set of values is used in the computations to be discussed below, along with a range for F between 0 and 1.

Repetitive calcium spiking with reasonable frequencies is predicted by this model, as shown in Figure 5–12, and the frequency of spiking increases as F (*i.e.*, ligand concentration) increases, Notice, however, that when the receptor activity reaches a certain level the oscillations are quickly damped to a persistently elevated level of cytosolic calcium (at the mitochondrial set-point of 0.6 μM). Elevated levels of calcium at high ligand concentrations have been observed in some cells (Jacob *et al.*, 1988), although the pattern of concentration changes is not consistent with Figure 5–12. Another limitation of the Meyer and Stryer model is the absence of a term involving extracellular calcium, which, as mentioned previously, is seen in some systems to influence oscillation frequency.

As seen from Table 5–2, other mathematical models to describe calcium oscillations have also been proposed. Briefly, the model of Goldbeter *et al.* (1990), model 2 in Table 5–2, postulates a different mechanism for the production of $[Ca^{2+}]_i$ oscillations. Its essential feature is the existence of two types of intracellular calcium stores, one sensitive to IP_3 levels and the other not. IP_3 produced by a receptor/G-protein-activated PLC modulates calcium release from the sensitive store into the cytosol. This store is assumed to remain a relatively constant source by continual replenishment of Ca^{2+} from the extracellular medium, according to a mechanism proposed by Putney (1986), although this has not been established by experimental data. The consequent rise in $[Ca^{2+}]_i$ is then hypothesized to induce release of calcium from the IP_3-insensitive store. This feature has led to the description of this proposed mechanism as *calcium-induced calcium release* (*CICR*). Positive feedback is achieved by the autocatalytic effect of calcium on this latter store, with a restoring flux back into both intracellular stores and to the extracellular medium.

Using appropriate parameter values, repetitive calcium spikes of approximately equal amplitude can indeed be produced by the CICR model (model 2). Three parameters play a particularly important role in the production of oscillations. The first parameter, β, gives a measure of the signaling activity of bound receptors. Although no details of the generation of this signaling activity (in terms of G-proteins, PLC, IP_3, etc.) are given, β is presumably an increasing function of the ligand concentration. The parameter v_1 gives the maximum flux of calcium from the IP_3-sensitive calcium store; thus the parameter combination βv_1 gives the actual stimulated flux from this store. A third parameter v_0 describes a small constant influx of calcium (presumably dependent on the extracellular calcium

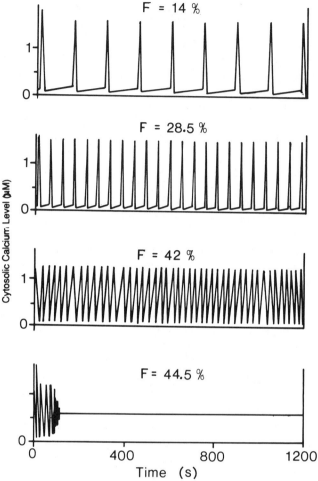

Figure 5–12 Calcium oscillations predicted by model 1 of Table 5–2. As the fraction of receptors bound increases, the parameter F increases. Model predictions for four different values of F are shown. Redrawn from Meyer and Stryer (1988).

concentration). Goldbeter *et al.* find that the sum of calcium fluxes $v_0 + \beta v_1$ is a governing parameter describing the existence and frequency of oscillations. For a small value of v_0, oscillations may begin or increase in frequency as the number of bound receptors, and hence β, increases. This result is similar to that of the model of Meyer and Stryer and the effect of their parameter F. The period decreases from a few seconds to a fraction of a second as β increases; qualitatively this behavior is similar to that found by Meyer and Stryer for the dependence of period on F. An additional feature of the Goldbeter *et al.* model is that extracellular calcium concentration has an effect on intracellular calcium concentration oscillation frequency through the parameter v_0. When oscillations exist, an increase in extracellular calcium concentration is predicted to lead to an increased oscillation frequency. An associated prediction is that oscillations may be induced without any receptor-mediated signaling at all, simply by a sufficiently large value of v_0. Such spontaneous oscillations have been observed in several cell types (Holl et al., 1988; Malgaroli et al., 1990).

The remaining four models of Table 5–2 are similar to the two models described above in that the explicit roles of receptors and G-proteins are not considered. Each of these models postulates that different key feedback mechanisms are primarily responsible for oscillations. Two of the models include roles for molecules not mentioned by the models 1 and 2: Cutherbertson and Chay (1991) include a role for protein kinase C and Swillens and Mercan (1990) model the proposed role for inositol 1,3,4,5-tetrakisphosphate (IP_4), produced from IP_3, in inhibiting calcium release. Most recently, Keizer and De Young (1992) have formulated a model based on key observations by Parker and Ivorra (1990) that calcium may actually inhibit (rather than induce, as in model 2) its own release.

As summarized most clearly by Tsien and Tsien (1990) and Harootunian et al. (1991a,b), the models shown schematically in Table 5–2 make *experimentally testable predictions*. For example, in several of the proposed models the IP_3 concentration is predicted to oscillate. This prediction is not consistent with observations in some cells in which injected IP_3 or a non-hydrolyzable analog of IP_3, assumed, perhaps in error, to insure a steady value of IP_3, elicited calcium oscillations (Wakui et al., 1989). Direct measurements of IP_3 in a cell population in order to test for IP_3 oscillations are generally not useful, because cells oscillate asynchronously. However, Harootunian et al. (1991b) have synchronized REF52 fibroblasts in order to follow the IP_3 concentration and their results suggest that IP_3 oscillations are likely, at least in these cells.

Other sophisticated tests of these models include: (1) microinjection or photorelease of calcium to test for positive or negative feedback of calcium on its own release; (2) pulsed release of IP_3 to test for perturbation of the oscillation phase; and (3) application of various pharmacological agents that are known to stimulate protein kinase C, block receptors on the calcium stores, chelate intracellular calcium, and so forth. By several of these experimental tests, Harootunian et al. (1991b) compare the predictions of models 1–4 of Table 5–2 and find that the Meyer and Stryer model (model 1) is most consistent with the oscillations they observe in REF52 fibroblasts. However, very recently Keizer and De Young (1992) have claimed that the model they propose (model 4) is also consistent with the REF52 data, and no definitive tests have yet been devised to distinguish between the two possibilities. In contrast, Wakui et al. (1990) find that model 2 is supported for mouse pancreatic acinar cells. It may be that oscillations are accomplished by different mechanisms in different cells. The very near future should bring experimental data bearing more criticallly on this issue.

5.2.2.d   Frequency-Encoded Signals

The functional significance of intracellular calcium oscillations in producing a cell behavioral response is unknown. Woods et al. (1987) and Berridge and Galione (1988) propose that some physiological processes may be under *frequency-dependent control* rather than amplitude-dependent control. For instance, fusion of intracellular secretion vesicles with the plasma membrane during stimulated release of vesicular contents may require a super-threshold intracellular calcium concentration. The overall secretion rate may then be proportional to the frequency of calcium spiking above the threshold concentration. In the context of calcium-modulated intracellular enzyme activity, as by means of the calcium-binding protein calmodulin, this modulation can also be envisioned to be frequency-dependent instead of amplitude-dependent. In fact, given the plethora of cell behavioral responses regulated in part by intracellular calcium, it may be easier to understand how these different responses could be differentially regulated if the frequency of concentration changes rather than absolute concentration is the key factor.

Goldbeter et al. (1990) present an extension of their oscillation model to analyze this possibility. Their premise is that the activity of a given intracellular enzyme may be covalently modulated, say by a kinase (adding a phosphate group) and by a phosphatase (removing a phosphate group). They postulate that the modulating kinase activity is stimulated by cytosolic Ca^{2+}, whereas the modulating phosphatase is independent of cytosolic Ca^{2+}. Denoting the fraction of the target enzyme in phosphorylated form by w^*, a kinetic balance on phosphorylated enzyme can be written:

$$\frac{dw^*}{dt} = \left(\frac{v_p}{W_T}\right)\left\{\left(\frac{v_k}{v_p}\right)\left[\frac{(1-w^*)}{K_k + (1-w^*)}\right] - \left[\frac{w^*}{K_p + w^*}\right]\right\} \tag{5-9a}$$

where W_T is the total amount of target enzyme in the cell, v_k and v_p are the rates of phosphorylation and dephosphorylation by the kinase and phosphatase, respectively, and K_k and K_p are saturation constants for these covalent modification reactions (Goldbeter and Koshland, 1981). The stimulation of kinase activity by cytosolic calcium is modeled by

$$v_k = V_{Mk}\left(\frac{x}{K_a + x}\right) \tag{5-9b}$$

where V_{Mk} is the maximal rate of modification and K_a is the saturation constant. x is the concentration of cytosolic calcium, predictable from mathematical models such as those described in Table 5–2.

Physiologically plausible values for the model parameters are used by Goldbeter et al. for numerical computations. Under conditions giving rise to repetitive spikes in x in their mathematical model of oscillations, the value of w^* is found to oscillate concomitantly. An example set of computations is shown in Figure 5–13a. Curves a and b in this figure correspond to low and high values of β, the receptor activity level mentioned in the previous section. Recall that increasing ligand concentration is presumed to increase β. For the higher β value, the frequency of oscillations is greater for both x and w^*. Most importantly, the mean value of the fraction of phosphorylated enzyme, denoted by $\langle w^* \rangle$, is also greater for higher β. The basis for this result resides in the kinetics of the associated enzyme phosphorylation and dephosphorylation cycles. For the smaller value of β and thus the lower frequency of Ca^{2+} oscillations, the target enzyme undergoes dephosphorylation to a greater extent during the Ca^{2+} concentration "valleys". A plot of this variation of $\langle w^* \rangle$ with Ca^{2+} oscillation frequency is shown in Figure 5–13b. Here, the two curves correspond to different sets of values for the

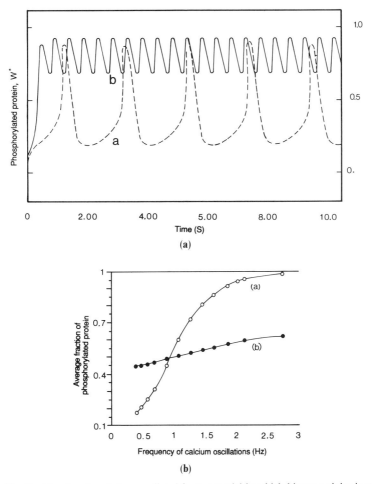

Figure 5–13 Protein phosphorylation predicted from a model in which kinase activity is stimulated by cytosolic calcium. (a) Predicted amount of phosphorylated protein w* as a function of time in a cell undergoing calcium oscillations. Curve a and curve b represent predictions for low and high ligand concentrations, respectively. (b) Average amount of phosphorylated protein, $\langle w^* \rangle$, as a function of the frequency of calcium oscillations. Recall that increasing ligand concentration is presumably reflected in an increasing frequency of oscillations. Curve a represents the case of $K_1 = K_2 = 0.01$. Curve b represents the case of $K_1 = K_2 = 10$. Redrawn from Goldbeter *et al.* (1990).

saturation constants (K_k, K_p) in the covalent modification rate expressions. An interesting observation is that the degree of target enzyme modification is more sensitive to oscillation frequency when K_k and K_p are small; that is, when the kinase and phosphatase are essentially saturated by their substrate – the target enzyme – and so operate in the "zero-order" kinetic regime with respect to the target enzyme. This result is consistent with the prediction by Goldbeter and Koshland (1981) that covalent modification is most effective in the zero-order regime. (We will discuss this point in detail in section 5.4.2.) A clear conclusion to be drawn here in summary is that oscillations in receptor-generated second messengers may indeed be capable of regulating cell behavioral responses in a frequency-dependent manner.

5.2.3 Details of G-protein Activation

The models described in sections 5.2.1 and 5.2.2 for the production of the second messengers cAMP and Ca^{2+} use rather simple descriptions of the mechanism of G-protein activation. For example, in the Levitzki and Tolkovsky (Tolkovsky, 1983; Levitzki, 1984) and Mahama and Linderman (1993a) models, the description of receptor/ligand complex and G-proteins coupling to produce active G-proteins is given simply by the term $k_c CG$ where k_c is an overall coupling rate constant. One might assume, for example, that k_c is equal to its diffusion limited value k_+. We have already discussed this transport step at length in section 4.2.3, where we calculated the rate constant for two membrane molecules, in the present case a receptor and a G-protein, to reach each other by diffusion. However, even this simple estimate is clouded if, for example, the local concentrations of molecules change with time (Mahama and Linderman, 1993b) or if the mobility of a molecule is a function of the membrane environment in which it is located. This may occur if the number of G-proteins available for activation changes significantly due to prior activation, if the number of bound receptors varies due to changes in the ligand concentration or deactivation of receptors, or if molecules diffuse more slowly in concentrated environments (Saxton, 1987). It is difficult to take these factors into account in our earlier estimate of the diffusion-limited value of the rate constant k_c, for our simple model assumed a uniform and constant distribution of molecules on a membrane and a constant value of the diffusion coefficient. Even more generally, the rate at which active G-proteins are produced may not at all be determined by the rate at which the two molecules find each other. Instead, another step, *e.g.* the exchange of GDP for GTP, might be rate limiting. This latter possibility will be examined here.

The goal, then, is to determine the *rate limiting step(s) in G-protein activation.* Thomsen *et al.* (1988) and Thomsen and Neubig (1989) investigate this question by examining the detailed kinetics of the steps required for G-protein activation in the adenylate cyclase system. In particular, they examine the ligand-mediated inhibition of forskolin-activated adenylate cyclase via the α_2-adrenergic receptor on platelet membranes. The G-protein known as G_i is responsible for the inhibition. Their model, shown in Figure 5–14, includes explicitly the steps of ligand binding, formation of the G-GDP/receptor/ligand complex, exchange of GDP for GTP, dissociation of the active form of the G-protein (α-GTP), and inactivation of α-GTP by hydrolysis. They assume that inhibition of adenylate cyclase is proportional to the concentration of α-GTP. The focus of their work was to find the individual rate constants of the steps in order to determine which step(s) are rate limiting in the overall process.

During receptor-mediated activation of G-proteins, the rate of change of concentrations of receptor/ligand complexes (C), inactive G-proteins (G-GDP), intermediate receptor/ligand/G-protein complexes with bound GDP (C-G-GDP), intermediate receptor/ligand/G-protein complexes with no bound nucleotide (C-G), intermediate receptor/ligand/protein complexes with bound GTP (C-G-GTP),

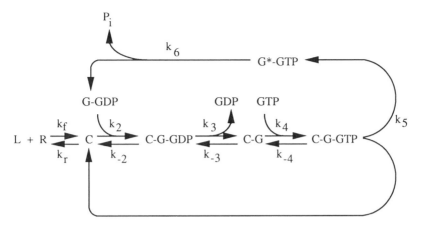

Figure 5–14   Model schematic of G-protein activation. Fixed parameters: $k_f = 5.0 \times 10^6 \text{ M}^{-1}\text{s}^{-1}$, $k_r = 0.5 \text{ s}^{-1}$, $k_2 G_T = 0.1 \text{ s}^{-1}$, $k_{-2} = 0.1 \text{ s}^{-1}$. Fit parameters: $k_3 = 5.0 \text{ s}^{-1}$, $k_{-3}\text{GDP} = 1.0 \times 10^{-4}\text{ s}^{-1}$, $k_4 = 5.0 \times 10^6 \text{ M}^{-1}\text{s}^{-1}$, $k_{-4} = 1.0 \text{ s}^{-1}$, $k_5 = 0.1 \text{ s}^{-1}$, $k_6 = 2.0 \text{ s}^{-1}$. Redrawn with permission from Thomsen and Neubig (1989). Copyright 1989 American Chemical Society.

and active G-proteins (G*-GTP) can be described by the differential equations

$$\frac{dC}{dt} = k_f RL + k_{-2}(C\text{-}G\text{-}GDP) - (k_r + k_2(G\text{-}GDP))(C)$$

$$+ k_5(C\text{-}G\text{-}GTP) \tag{5–10a}$$

$$\frac{d(C\text{-}G\text{-}GDP)}{dt} = k_2(G\text{-}GDP)(C) + k_{-3}(GDP)(C\text{-}G)$$

$$- (k_{-2} + k_3)(C\text{-}G\text{-}GDP) \tag{5–10b}$$

$$\frac{d(C\text{-}G)}{dt} = k_3(C\text{-}G\text{-}GDP) + k_{-4}(C\text{-}G\text{-}GTP)$$

$$- (k_{-3}GDP + k_4GTP)(C\text{-}G) \tag{5–10c}$$

$$\frac{d(C\text{-}G\text{-}GTP)}{dt} = k_4(GTP)(C\text{-}G) - (k_{-4} + k_5)(C\text{-}G\text{-}GTP) \tag{5–10d}$$

$$\frac{d(G^*\text{-}GTP)}{dt} = k_5(C\text{-}G\text{-}GTP) - k_6(G^*\text{-}GTP) \tag{5–10e}$$

The model does not include the steps of dissociation and recombination of the α and $\beta\gamma$ subunits of the G-protein. Furthermore, it is assumed that the number of receptors on the membrane R_T is constant, or that

$$R_T = R + C + (C\text{-}G\text{-}GDP) + (C\text{-}G) + (C\text{-}G\text{-}GTP) \tag{5–11}$$

It is also assumed that $(G\text{-}GDP) \gg R_T + (G^*\text{-}GTP)$, so that $(G\text{-}GDP)$ can be replaced by the total G-protein concentration G_T. Densensitization of receptors is assumed to be negligible.

For the agonist epinephrine, these investigators calculated values for all of the rate constants using a combination of methods (Thomsen and Neubig, 1989). First, analysis of the binding of the ligand agonist UK 14,304 to α_2-adrenergic receptor via a ternary complex model similar to that described in section 2.2.4 provided the rate constants k_f, k_r, k_2, and k_{-2} (Neubig et al., 1988). Because the binding properties of epinephrine are similar to UK 14,304, these values were assumed to hold for epinephrine as well. Second, a rapid-mix quench technique was used to measure the kinetics of cAMP production in the subsecond to second time frame. The concentrations of GTP and various ions were varied, and the percentage inhibition of the adenylate cyclase was determined. The remaining rate constants were then fitted to the data according to the form of the model, much as was done for the endocytosis model of section 3.2.1.b. The best fit values of the parameters (to within a factor 2) are given in the legend of Figure 5–14.

Two major conclusions can be drawn from this result. First, the rate constant describing the GTPase activity of the α subunit of the G-protein, k_6, was found to be substantially greater than the $2–4$ min^{-1} measured in reconstituted systems (Brandt and Ross, 1986; Tota et al., 1987). This reinforces the need for measurement of rates in systems as similar as possible to whole cells. Second, the data suggest that the steps of release of GDP and association of GTP are not rate-limiting in the activation of G-proteins, as had been suggested (Higashijima et al., 1987). When compared with these two steps, Thomsen and Neubig found that the release of active G-protein (α-GTP, or G*-GTP in the notation of the model being discussed here), is rate-limiting. Further, the kinetics of this step (rate constant k_5) are of the order of the kinetics of the association of inactive G-proteins with complexes, suggesting that a simplified mathematical model (ignoring the steps with rate constants k_3, k_{-3}, k_4, and k_{-4}) would also adequately model the data.

Depending on the system, then, one may need to include explicitly the kinetics of release of active G-protein (rate constant k_5 in the model above) in a model of second messenger production. Whether this step will be on the same time scale as complex association with inactive G-proteins will depend on among other things the diffusivity and relative concentrations of G-proteins and complexes. Finally, we note that it is not yet known whether this is a general result for other cases of receptor-mediated G-protein activation or whether it is specific to the system studied, that of adenylate cyclase inhibition via the action of the G-protein known as G_i.

5.3 MODEL FOR RECEPTORS AS ENZYMES

As discussed earlier in this chapter, the enzymatic activity of some receptors, particularly growth factor receptors, is altered by ligand binding. These receptors can then covalently modify a variety of cellular substrates. In fact, the covalent modification of cellular proteins, by receptors as well as various intracellular enzymes, appears to be intimately involved in control of many crucial cell processes. One estimate is that roughly 15% of all proteins in mammalian cells are regulated by covalent modification through phosphate groups and that the total amount of metabolic energy expended for covalent modification might be as great as 20% of the total energy available to the cell (Goldbeter and Koshland, 1987). Hunter (1987) guesses that up to one thousand different protein kinases will eventually be discovered to operate in eukaryotic cells, requiring devotion of at least 1% of the entire cell genome – and this does not include other modifying enzymes such as phosphatases, methylases, and demethylases. Taken together, this information strongly indicates that there must be tremendous benefit to the cell to regulate biochemical processes by this mechanism.

What might this tremendous advantage be? An attractive possibility can be found in the theoretical analysis of Goldbeter and Koshland (1981) on the stimulus–response sensitivity provided by a covalent modification mechanism. To give away the answer before going through the underlying mathematical model and analysis, consider the plot in Figure 5–15 of a generic cellular function as a function of stimulus concentration. This response may be a particular enzymatic activity, perhaps, or expression of a certain gene. The dotted curve represents a classical hyperbolic, or Michaelis–Menten, dependence of response on stimulus level. This dependence can arise when the response is proportional to the amount of an enzyme/stimulus complex – perhaps a receptor/ligand complex, for instance. From this curve it can be determined that a hundred-fold change in stimulus level is needed to cause a change in response from about 10% to 90% of maximal activity (Koshland, 1987). This is not a very sensitive response, hardly one that would qualify to serve as an "on/off" switch for an important cell function. The solid curve on the plot, denoting an *"ultrasensitive" response*, would be more useful for such a regulatory requirement.

The essential concept of Goldbeter and Koshland (1981) is that covalent modification of the enzyme activity generating the cell response in Figure 5–15 can provide the sharper, ultrasensitivity dependence. If this is the case, then protein kinases, such as a receptor kinase, can serve as efficient *"on/off" switches* for all functions needing fine control.

The Goldbeter–Koshland model is composed of three components: a protein W and two modifying enzymes E_1 and E_2. E_1 converts the substrate protein into its modified form, W^*, whereas E_2 accomplishes the reverse conversion, according to the scheme:

$$W + E_1 \underset{k_{u1}}{\overset{k_{c1}}{\rightleftharpoons}} U_1 \overset{k_{a1}}{\rightarrow} W^* + E_1$$

$$W^* + E_2 \underset{k_{u2}}{\overset{k_{c2}}{\rightleftharpoons}} U_2 \overset{k_{a2}}{\rightarrow} W + E_2$$

where U_1 and U_2 are the intermediate complexes of enzyme and substrate. k_{c1}, k_{c2}, k_{u1}, k_{u2}, k_{a1}, and k_{a2} are rate constants. It is assumed that other required species, such as ATP, ADP, H_2O, and inorganic phosphate, are present at constant, non-rate-limiting levels. Kinetic balance equations can then be written for the intracellular concentrations W, W^*, E_1, E_2, U_1, and U_2:

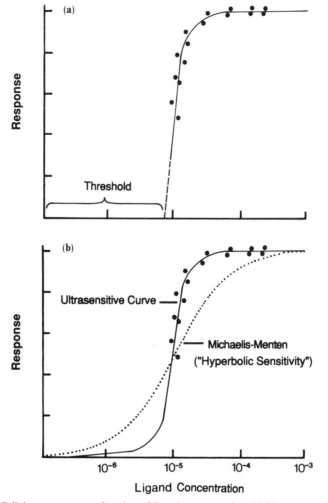

Figure 5–15   Cellular response as a function of ligand concentration. (a) No response is found until the ligand concentration has increased over a threshold value. (b) The data are fitted by a curve that is steeper than the classic Michaelis–Menten curve. Recall that a Michaelis–Menten curve can be described by a response that is proportional to $L/(K_M + L)$ where L is the ligand concentration and K_M is the Michaelis–Menten constant. Redrawn from Koshland (1987).

$$\frac{dW}{dt} = -k_{c1}WE_1 + k_{u1}U_1 + k_{a2}U_2 \qquad (5\text{–}12\text{a})$$

$$\frac{dU_1}{dt} = k_{c1}WE_1 - (k_{u1} + k_{a1})U_1 \qquad (5\text{–}12\text{b})$$

$$\frac{dW^*}{dt} = -k_{c2}W^*E_2 + k_{u2}U_2 + k_{a1}U_1 \qquad (5\text{–}12\text{c})$$

$$\frac{dU_2}{dt} = k_{c2}W^*E_2 - (k_{u2} + k_{a2})U_2 \qquad (5\text{–}12\text{d})$$

Additional restrictions are conservation equations for protein and enzymes: $W_T = (W + W^* + U_1 + U_2)$; $E_{1T} = (E_1 + U_1)$; $E_{2T} = (E_2 + U_2)$. In most of their analysis, Goldbeter and Koshland consider the protein to be present in great excess over the enzymes, so that U_1 and U_2 can be neglected compared to W_T, W, and W^* in the protein conservation equation.

We will examine the behavior of this system by calculating the change in the steady-state value of W^* following presentation of a stimulus. At steady-state, with all time derivatives in Eqns. (5–12a–d) set equal to zero, an expression for the fraction of protein in modified form, $w^* = W^*/W_T$, can be derived:

$$w^* = \left[2\left(\frac{V_1}{V_2} - 1\right)\right]^{-1} \left\{ \frac{V_1}{V_2} - 1 - K_{M2}\left(\frac{K_{M1}}{K_{M2}} + \frac{V_1}{V_2}\right) + \left[\left[\frac{V_1}{V_2} - 1 - K_{M2}\left(\frac{K_{M1}}{K_{M2}} + \frac{V_1}{V_2}\right)\right]^2 \right.\right.$$

$$\left.\left. + 4K_{M2}\left(\frac{V_1}{V_2} - 1\right)\left(\frac{V_1}{V_2}\right)\right]^{1/2} \right\} \tag{5–13}$$

where $V_1 = k_{a1}E_{1T}$, $V_2 = k_{a2}E_{2T}$, $K_{M1} = (k_{u1} + k_{a1})/k_{c1}W_T$, and $K_{M2} = (k_{u2} + k_{a2})/k_{c2}W_T$. V_1 and V_2 are the maximum conversion rates, and K_{M1} and K_{M2} are essentially Michaelis constants scaled to the total protein concentration. Note that V_1 and V_2, which combine conversion rate constants and conversion enzyme levels, can be thought of as representing the stimulatory and inhibitory activities of some control element further upstream.

For future reference, note also that by standard Michaelis–Menten kinetics, the rate of conversion of W into W^* is proportional to $W/(K_1 + W)$. These kinetics allow for two regimes to exist: when $K1 \ll W$ the kinetics are approximately zero order, and when $K1 \gg W$ the kinetics are approximately first order. Analogous statements can be made concerning kinetics of conversion of W^* into W.

Eqn. (5–13) can be rearranged to yield a convenient relation between the ratio of maximum conversion rates and the steady state fraction of modified protein:

$$\frac{V_1}{V_2} = \frac{w^*(1 - w^* + K_{M1})}{(1 - w^*)(w^* + K_{M2})} \tag{5–14}$$

with the scaled Michaelis constants as parameters. A plot of this relation is shown in Figure 5–16 for the simplifying case $K_{M1} = K_{M2}$. The fraction of modified protein, w^*, increases with (V_1/V_2) as expected. The sensitivity of w^* to (V_1/V_2) is weak when $K_{M1} = K_{M2} = 1$, but is extremely strong near (V_1/V_2) when $K_{M1} = K_{M2}$ is small – in this figure, equal to 10^{-2}. That is, there is a very sharp "on/off" switch for the modified protein when the conversion enzymes are nearly saturated with their substrate. In other words, an ultrasensitive dependence of response to stimulus is predicted when the covalent modification enzymes operate in a regime effectively *zero-order in substrate enzyme concentration*.

Let us consider a scenario to allow us to interpret this result in the context of a receptor kinase. E_1 can be the receptor itself, with its kinase activity stimulated by ligand binding. That is, the conversion rate constant k_{a1} would be effectively increased by ligand binding to the receptor. E_2 can be an intracellular, perhaps membrane-associated phosphatase, with an activity independent of ligand concentration. W is a key intracellular substrate for the receptor kinase and the

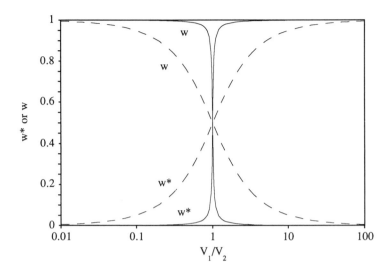

Figure 5–16 Fraction of protein in modified (w^*) or unmodified (w) form as a function of the ratio of maximum conversion rates, V_1/V_2. Solid curves represent $K_{M1} = K_{M2} = 0.01$, dashed curves represent $K_{M1} = K_{M2} = 1.0$. Note the sharp transition around $V_1/V_2 = 1$ for the lower values of K_{M1} and K_{M2}. Redrawn from Goldbeter and Koshland (1981).

phosphatase, with its modified form W* crucial perhaps in controlling expression of a certain gene. An increase in the number of receptor/ligand complexes might increase the conversion rate V_1 such that the ratio (V_1/V_2) consequently increases from below 1 to above 1. If K_{M1} and K_{M2} are small, then by Figure 5–16 this would yield a change in the modified protein fraction from close to 0 to near 1 – essentially, switching its activity from "off" to "on".

Goldbeter and Koshland quantify this sensitivity by defining the *response coefficient*, ρ_c, as the ratio of (V_1/V_2) required to give 90% of maximal w^* relative to the ratio required to give 10% of maximal w^*. Using Eqn. (5–14) with this definition yields:

$$\rho_c = 81 \frac{(K_{M1} + 0.1)(K_{M2} + 0.1)}{(K_{M1} + 0.9)(K_{M2} + 0.9)} \tag{5–15}$$

When K_{M1} and K_{M2} are very large, ρ_c approaches a value of 81. This is the same result that would be found for a pure, unmodified Michaelis–Menten system (see Figure 5–15). As K_{M1} and K_{M2} decrease, ρ_c asymptotically diminishes toward 1 – an infinitely steep transition. For $K_{M1} = K_{M2} = 10^{-2}$ as in Figure 5–16, $\rho_c \sim 1.2$. This means that only a 20% change in the conversion rate, due to receptor/ligand binding would be needed to move the substrate from 10% modified to 90% modified. An alternative interpretation of the above result can be stated in terms of an effective Hill coefficient by describing the response function as a Hill function:

$$w^* = \frac{(V_1/V_2)^{n_H}}{1 + (V_1/V_2)^{n_H}} \tag{5–16}$$

The greater the Hill coefficient, n_H, the greater the sensitivity of response to stimulus. The concentration of oxygen binding to its carrier protein hemoglobin for example, can be characterized by a Hill coefficient of about 2.9. This steepness is crucial to oxygen transport to tissues by erythrocytes. By comparison, for the case $K_{M1} = K_{M2} = 10^{-2}$ of Figure 5–16, $n_H \sim 13$. This provides an extremely sensitive dependence of modified protein level to the activating stimulus.

The framework of the Goldbeter and Koshland model has not yet been implemented in a full model for signal transduction via receptors acting as protein kinases. For the most part, missing information about later steps in the process, namely the precise functions of the phosphorylated proteins in terms of producing a final cellular response, are limiting mathematical development in this area. Later in this text (section 6.1), models for cell proliferation stimulated by growth factor binding will be presented; these models of necessity lack detail concerning the modifications and subsequent action of intracellular proteins.

5.4 MODELS FOR ADAPTATION

5.4.1 Simple Kinetic Models

The simplest approach to modeling adaptation, or an alteration in sensitivity of cells to a particular ligand, is simply to add a kinetic step removing cell receptors from an active (signaling) to an inactive (non-signaling) form. Thus the signal generation activity of the receptors will decrease with time. This is an appropriate approach if receptors are rendered non-signaling via, for example, covalent modification or sequestration, as has been suggested in a variety of systems (e.g., Sibley et al., 1987, Vaughan and Devreotes, 1988, Hargrave and McDowell, 1992). Non-signaling receptors may then be converted back into a signaling form; alternatively, the desensitization of receptors may be considered irreversible and resensitization of the cell may only occur with the expression of new receptors on the cell surface.

An example of this approach can be found in a model discussed earlier in this text (section 2.2.4.c). Sklar et al. (1989) and Sklar and Omann (1990) model the binding of formyl peptides to neutrophils with the scheme given by

$$
\begin{array}{ccccccc}
\text{G} & & & & & & \\
\diagdown & & & & & & \\
\text{L} + \text{R} & \rightleftharpoons & \text{LRG} & \longrightarrow & \text{LR} & \longrightarrow & \text{LR}_x & \longrightarrow & \text{LR}_{int} \\
& & & & \updownarrow & & \downarrow & & \\
& & & & \text{L} + \text{R} & & \text{L} + \text{R}_x & &
\end{array}
$$

where G is a G-protein, R_x is a slowly dissociating *desensitized receptor*, and R_{int} is the internalized receptor. We write LR rather than C here to refer to a receptor ligand complex, so that the state of the receptor (R or R_x) can be noted. The preferential loss of desensitized receptors from the surface due to internalization is included. No steps for new receptor synthesis, recycling of internalized receptors, or resensitization of desensitized receptors are included, for the time scale of the experiments to be described by this model was rapid compared to the expected time scale of these events. This model is based on a variety of binding data from both permeabilized and whole cells. The receptor involved in the ternary complex LRG is found to be slowly dissociating, the receptor in LR is rapidly dissociating, and the R_x form of the receptor is slowly dissociating and guanine-nucleotide insensitive.

This model accounts for desensitization by allowing for the conversion of actively signaling receptors into the R_x form which is apparently unable to signal. The time course of desensitization is determined by the values of the rate constants shown in the schematic. Sklar *et al.* find that the LRG forms and disappears very rapidly, but that conversion from LR to LR_x occurs with a half-time of about 10 s. This corresponds reasonably well with the time course of responses in these cells: a response is initiated within seconds of ligand addition, reaches a maximum after tens of seconds, and then decays slowly over the following minutes (Sklar and Omann, 1990). The precise mechanism of this desensitization is not well-understood, although it is believed that receptor occupancy is required for conversion into the R_x form.

For modeling of desensitization in the β-adrenergic receptor system, a similar approach has been taken. Su *et al.* (1980) and Levitzki (1984) describe the process of desensitization with the following scheme

$$ L + R \underset{}{\overset{K_D}{\rightleftharpoons}} LR \underset{k_{-x}}{\overset{k_x}{\rightleftharpoons}} LR_x \overset{k_e}{\longrightarrow} $$

R_x is a form of the receptor that cannot couple with adenylate cyclase. The rate constants k_x and k_{-x} characterize the forward and reverse process of desensitization. Note that the process of desensitization is considered reversible, in keeping both with the biological data and with the longer time scale of typical experiments with this system. Receptors of the form R_x may also be lost from the cell surface with rate constant k_e.

If k_e can be neglected compared with k_x and k_{-x}, the fraction of receptors in the form LR_x can be calculated, assuming a *pseudo-steady state*. Utilizing the receptor conservation relation $R_T = R + LR + LR_x$, the equilibrium dissociation constant $K_D = (L)(R)/LR$, and the equilibrium relationship $LR_x/LR = k_x/k_{-x}$, one can derive the relationship

$$ \frac{LR_x}{R_T} = \frac{\left(\dfrac{k_x}{k_{-x}}\right)L}{K_D + \left(\dfrac{k_x}{k_{-x}} + 1\right)L} \tag{5-17} $$

Note that the fraction of receptor in a desensitized form increases with both the increasing affinity of ligand for receptor and the ratio of the rates of conversion and reversion, k_x/k_{-x}. In the limiting case of saturating ligand concentration, $L \gg K_D$, Eqn. (5-17) reduces to

$$ \frac{LR_x}{R_T} = \frac{k_x}{k_x + k_{-x}} \tag{5-18} $$

Thus with this model, the steady state fraction of receptors in a desensitized or non-signaling form can be predicted from the rate constants and ligand concentration. This level of description should then be incorporated into a more complete model relating bound receptors to the levels of intracellular second messengers, for example, and ultimately the cellular response.

As more information about the mechanisms and rates of desensitization becomes available, these simple models may need to be extended. In particular, the data of Lohse *et al.* (1990) described earlier suggest that homologous and heterologous desensitization mediated by phosphorylation of receptors may be the result of the action of two different enzymes and that internalization may need to be taken into account separately. As discussed in Chapter 2 for the related problem of distinguishing cell surface events from receptor trafficking processes, approaches are likely to involve the experimental isolation of specific events by the use of pharmacologic agents, reconstituted systems, or careful choice of time scales and measurement procedures as well as the theoretical estimation of several model rate constants simultaneously from a variety of experimental data.

5.4.2   Consequences of Complete Adaptation

In describing earlier the relationship between a signal (the presence of ligand) and a cellular response, we discussed the distinction between completely and partially-adapting systems. For the sample responses shown in Figure 5–5, for which the signal of ligand concentration remains constant, adaptation must arise from some modification of the signal generation and/or response mechanism. For example, a molecular species necessary for signal generation could be depleted or altered in some way as to interfere with signal generation.

Experimental data for one adapting system are shown in Figure 5–17a. The secretion of cAMP by the cellular slime mold *Dictyostelium discoideum* in response to binding of extracellular cAMP to its cell surface receptor is shown (Knox *et al.*, 1986). After a step change in extracellular cAMP concentration, there is an initial rise in secretion rate to a maximum, followed by a gradual decrease to the original baseline level. This is a completely adapting system.

An attractive model for adaptation based on three key ideas has been formulated by Segel and coworkers (Segel *et al.*, 1986; Knox *et al.*, 1986). First, for a variety of receptors, ligand binding induces covalent modification (*e.g.*, phosphorylation, methylation) of the receptor cytoplasmic tail domains. Second, the kinetics and ligand-concentration dependence of this covalent modification correlate closely with adaptation of the receptor-mediated response. In particular, cAMP receptors on *Dictyostelium* have been shown to be phosphorylated with a time course similar to that of adaptation (Figure 5–17b) (Vaughan and Devreotes, 1988). Third, there may be a slight change in receptor/ligand binding affinity correlated with these modification processes (Devreotes and Sherring, 1985).

Segel *et al.* (1986) postulate a scheme in which the receptor and corresponding receptor/ligand complex can each exist in *two different states*, R_1 and R_2, and C_1

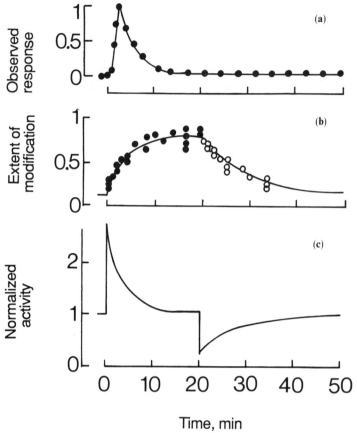

Figure 5–17 Adaptation in the secretion of cAMP by the slime mold *Dictyostelium discoideum* in response to binding cAMP to surface receptors. At time 0, the ligand concentration is increased from zero. (a) The normalized response is plotted as a function of time. The solid line represents model predictions. (b) The extent of receptor modification, $(R_2 + C_2)/R_T$, is plotted as a function of time. The solid line represents a fit of the model to the data. (c) The normalized activity $A(L)/A_0$ predicted by the model is plotted. Interconversion rate constants and weights a_i were determined as described in the text. Redrawn from Knox *et al.* (1986).

and C_2, respectively:

$$R_1 + L \underset{k_{r1}}{\overset{k_{f1}}{\rightleftharpoons}} C_1$$

$$k_{12} \downarrow\uparrow k_{21} \qquad\qquad k'_{12} \downarrow\uparrow k'_{21}$$

$$R_2 + L \underset{k_{r2}}{\overset{k_{f2}}{\rightleftharpoons}} C_2$$

This model was examined in section 2.2.3 when we discussed the implication of multiple receptor states on receptor/ligand binding (it is model C). The new feature here is a link between receptor state and signal generation. Segel *et al.* hypothesize that in the most general case, all receptor states are capable of signal generation. In addition, they make the simplifying assumption for this system that the *total signal transduction activity* of all receptor forms, A, is given by a weighted sum of the four forms of the receptor:

$$A(t) = a_1 R_1(t) + a_2 C_1(t) + a_3 C_2(t) + a_4 R_2(t) \qquad (5\text{--}19)$$

The weights a_i, $i = 1\text{--}4$, are constants representing the contribution of each receptor state to the total signal transduction activity. In the present example, cAMP production by *Dictyostelium* upon stimulation of the cAMP receptor by ligand, this activity A is due to coupling of receptor to G-proteins. Therefore, the assumption is that all receptor forms are capable of coupling with effectors but with different affinities; the weights a_i may be considered proportional to the coupling affinities if the binding of receptor to effector is sufficiently fast. In other systems, such as those involving ion channel or receptor kinase mechanisms, alternative biochemical interpretation of these weights is required.

Note that the relationship between activity A and the response of the system is not specified, but it is assumed that given an activity A only one level of response can be achieved. If this simple model is to account for complete adaptation, it must be that the value of A prior to ligand addition must be reached again after some time with ligand present. The model assumes that all adaptation occurs at the level of the receptor in the process of shifting receptors from one state to another. Because the model does not indicate the precise differences between the "1" and "2" receptor states, it is consistent with the idea of covalent modification of receptors as well as internalization of receptors. In the latter case, k_{12} and k'_{12} would represent internalization rate constants; k_{21} and k'_{21} would represent recycling rate constants. In the *Dictyostelium* system, as mentioned briefly above, the data are consistent with covalent modification of receptors as the mechanism for desensitization.

According to Eqn. (5–19), the theoretical prediction of signal generation activity following ligand addition or removal reduces to calculation of receptor numbers in the various states. Based on experimental data, Segel *et al.* assume that receptor/ligand association and dissociation reactions are rapid compared to the receptor state interconversion processes, so that binding can be considered to be at equilibrium for any given extracellular ligand concentration. By methods similar to those used in Chapter 2, equations can be written to describe the rate of change of the numbers of receptors in each of the four states. In addition, the total receptor number is assumed constant and equal to $R_T = R_1 + C_1 + R_2 + C_2$. With these equations and the calculated parameter values found by Devreotes and Sherring (1985) for the phosphorylation and dephosphorylation of the cAMP receptor on *Dictyostelium*, $k_{12} = 0.012 \text{ min}^{-1}$, $k_{21} = 0.104 \text{ min}^{-1}$, $k'_{12} = 0.22 \text{ min}^{-1}$, and $k'_{21} = 0.055 \text{ min}^{-1}$, the number of receptors in each of the four states as a function of time can be calculated analytically or numerically (see section 2.2.3). A good fit to the experimental data is found, as can be seen in Figure 5–17b.

It remains to determine the signal generation weights, a_i. This is accomplished by setting restrictive specifications corresponding to experimental observations. The crucial specification is the level of adaptation. For the *Dictyostelium* cAMP response, adaptation is complete; that is, cAMP secretion returns to the baseline level following a step change in ligand concentration (Figure 5–17a). Hence, we specify that at *steady-state* the signal generation activity A is independent of ligand concentration L. In other words, we require that the value of A when no ligand is present must be identical to the steady-state value of A for any ligand concentration.

For $L = 0$, the *basal activity* A_0 is:

$$A_0 = a_1 R_{10} + a_2 C_{10} + a_3 C_{20} + a_4 R_{20}$$

$$= \left(\frac{R_T}{1 + K_{21}}\right)(a_1 K_{21} + a_4) \tag{5-20}$$

where $K_{21} = k_{21}/k_{12}$ and the subscript "0" denotes the basal levels of the various receptor states. The simple final result arises from recognizing that for $L = 0$, $C_{10} = C_{20} = 0$. For any $L \neq 0$, the *steady-state activity* $A(L)$ is:

$$A(L) = a_1 R_{1ss} + a_2 C_{1ss} + a_3 C_{2ss} + a_4 R_{2ss}$$

$$= a_1 \left\{\frac{R_T}{Q[1 + (L/K_{D1})]}\right\} + a_2 \left\{\frac{R_T(L/K_{D1})}{Q[1 + (L/K_{D1})]}\right\}$$

$$+ a_3 \left\{\frac{R_T(L/K_{D2})(Q - 1)}{Q[1 + (L/K_{D2})]}\right\} + a_4 \left\{\frac{R_T(Q - 1)}{Q[1 + (L/K_{D2})]}\right\} \tag{5-21}$$

where

$$Q = 1 + \left(\frac{k_{12} + k'_{12}[L/K_{D1}]}{k_{21} + k'_{21}[L/K_{D2}]}\right)\left(\frac{1 + [L/K_{D2}]}{1 + [L/K_{D1}]}\right) \tag{5-22}$$

and $K_{D1} = k_{r1}/k_{f1}$ and $K_{D2} = k_{r2}/k_{f2}$.

To describe complete adaptation with this model, we next set $A_0 = A(L)$ by equating Eqns. (5-20) and (5-21). We then solve for any requirements on the coefficients a_i such that this equality holds, remembering that no restrictions can be placed on the value of L as exact adaptation occurs for any value of L.

The requirements on the coefficients a_i can be found corresponding to either the presence or absence of detailed balance in the cyclic reaction scheme (*e.g.*, Wyman, 1975). In the case of detailed balance, only a single thermodynamic cycle is present and it must be that $k'_{12}k_{21}K_{D2} = k_{12}k'_{21}K_{D1}$. However, when the receptor interconversion steps involve covalent modification and thus the expenditure of energy, additional thermodynamic cycles arise in this reaction scheme (see Appendix 2 of Segel *et al.* (1986)). In this situation detailed balance is not imposed.

Because there is experimental evidence for covalent modification of the cAMP receptor on *Dictyostelium*, we will focus on the case in which microscopic reversibility is not relevant. In this situation, the *requirements on the coefficients* a_i can be written as:

$$\frac{a_1 - (A_0/R_T)}{k_{12}} = \frac{a_2 - (A_0/R_T)}{k'_{12}} = -\frac{a_3 - (A_0/R_T)}{k'_{21}} = -\frac{a_4 - (A_0/R_T)}{k_{21}} \tag{5-23}$$

The final equality in this equation is an identity derived from Eqn. (5-20); it is included for physical insight but in actuality does not represent any additional requirement. Thus Eqn. (5-23) represents two equations and thus two restrictions on the coefficients. This implies that two of the weights a_i can be chosen arbitrarily by the cell but the remaining two must take on the values required by Eqn. (5–23) in order to produce exact adaptation according to this model.

An interesting interpretation of Eqn. (5–23) is that complete adaptation occurs when each of the weights is inversely proportional to the mean lifetime of the corresponding receptor state (Knox *et al.*, 1986). In other words, the system adapts completely when the shortest-lived receptor states are the most active in signal generation. Sample calculations based on this situation are plotted in Figure 5–18 for illustrative parameter values $k_{12} = k'_{21} = 0.01$ min^{-1}, $k_{21} = k'_{12} = 0.1$ min^{-1}, $K_{D1} = K_{D2}$, and $a_1 = 0.11$, $a_2 = 0.20$, $a_3 = 0.09$, $a_4 = 0$. In response to a step change in ligand concentration, the fraction of receptors in the "2"-state (R_2 and C_2) increases asymptotically. At the same time, the signal generation activity spikes briefly before exhibiting exact adaptation to the basal activity.

Comparison to experimental data on cAMP secretion in response to cAMP

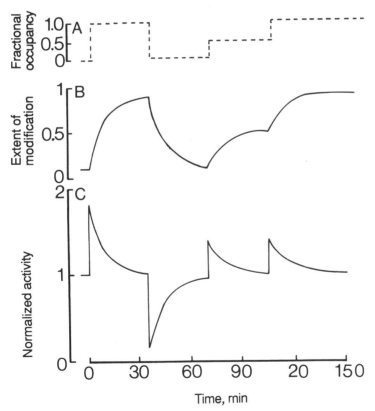

Figure 5–18 Sample predictions of the adaptation model. (a) Step changes in the ligand concentration produce step changes in receptor occupancy, due to the assumption of rapid binding kinetics. (b) For the range constants $k_{12} = k'_{21} = 0.01$ min^{-1}, $k_{21} = k'_{12} = 0.1$ min^{-1}, $K_{D1} = K_{D2}$, and $a_1 = 0.11$, $a_2 = 0.20$, $a_3 = 0.09$, $a_4 = 0$, the extent of receptor modification (fraction of receptors in the "2" form) is plotted. (c) The predicted normalized activity $A(L)/A_0$ is plotted as a function of time. Redrawn from Knox *et al.* (1986).

binding to *Dictyostelium* membrane receptors is provided in Figure 5–17a,b. The rate constants k_{12}, k_{21}, k'_{12}, and k'_{21} were set at the values determined previously by a fit to experimental data. The coefficient a_4 was arbitrarily set at 0 and $(A_0 R_T)$ was set equal to k_{21}, and the remaining three coefficients were found from Eqn. (5–23). The theoretically-predicted signal generation activity is plotted in Figure 5–17c and the predicted cAMP secretion rate is plotted in Figure 5–17a. The brief transient response predicted by the model is in excellent agreement with the secretion data; presumably this physiological response is linearly proportional to the signaling activity, at least within the range relevant here (see Segel *et al.* (1986) for further discussion of this point).

Finally, the adaptation model presented here suggests some approximate methods of dealing with signal transduction itself, particularly when quantitative understanding of the biochemical pathways between receptor occupancy and a cellular response is lacking. It may well be that useful models can be developed by using the idea that the response is proportional to the numbers of receptors in certain states within the cell or on the cell surface, much as was done here. The weighting coefficients a_i might be set to enable a particular hypothesis to be tested. For example, one might postulate that the response of a cell to receptors crosslinked by ligand increases with the size of the cluster, and that the weighting coefficients a_i could be modeled as increasing linearly according to the cluster size in which the receptor is found. Similarly, the responses of some cells may depend on the occupancy of several distinct and non-interconverting receptor species. One might test a model in which the response was assumed to be proportional to a weighted sum of the numbers of bound receptors of each type.

NOMENCLATURE

Symbol	Definition	Typical Units
a_i	weighting coefficient ($i = 1$–4)	$(\#/\text{cell})^{-1}$
A	total signal transduction activity	
A_0	basal level of total signal transduction activity	
BG	number of $\beta\gamma$ subunits of G-proteins	$\#/\text{cell}$
C	receptor/ligand complex number	$\#/\text{cell}$
C_1	receptor/ligand complex number (state 1)	$\#/\text{cell}$
C_2	receptor/ligand complex number (state 2)	$\#/\text{cell}$
Ca	cytosol calcium concentration	M
Ca_s	store calcium concentration	M
Ca_T	total concentration of calcium on cytosol volume basis	M
$C\text{-}G$	receptor/ligand-G protein complex number	$\#/\text{cell}$
$C\text{-}G\text{-}GDP$	receptor/ligand-G protein-GDP complex number	$\#/\text{cell}$
$C\text{-}G\text{-}GTP$	receptor/ligand-G protein-GTP complex number	$\#/\text{cell}$
E	unactivated enzyme number	$\#/\text{cell}$
E_T	total number of enzyme molecules	$\#/\text{cell}$
E_1	modifying enzyme concentration	M
E_{1T}	total concentration of enzyme E_1	M
E_2	modifying enzyme concentration	M
E_{2T}	total concentration of enzyme E_2	M
E^*	activated enzyme number	$\#/\text{cell}$
$E\text{-}C$	number of enzyme-receptor/ligand complexes	$\#/\text{cell}$

F	degree of stimulation of IP_3 production due to receptor occupation	
G_T	total G protein number	#/cell
GDP	GDP concentration	M
GTP	GTP concentration	M
$G\text{-}GDP$	G protein-GDP complex number	#/cell
$G^*\text{-}GDP$	inactivated G-protein alpha subunits	#/cell
$G^*\text{-}GTP$	activated G protein number	#/cell
I	number of IP_3 molecules	#/cell
k_a	rate constant for enzyme activation	min^{-1}
k_{a1}	rate constant for action of enzyme E_1	min^{-1}
k_{a2}	rate constant for action of enzyme E_2	min^{-1}
k_c	rate constant for coupling	$(\#/cell)^{-1}\,min^{-1}$
k_{c1}	rate constant for coupling of W and E_1	$M^{-1}\,min^{-1}$
k_{c2}	rate constant for coupling of W* and E_2	$M^{-1}\,min^{-1}$
k_{ca1}	rate constant for calcium release from stores	min^{-1}
k_{ca2}	rate constant for calcium leak from stores into cytosol	min^{-1}
k_{ca3}	rate constant for removal of calcium	M/min
k_d	rate constant for enzyme deactivation	min^{-1}
k_e	rate constant for loss of receptors from cell surface	min^{-1}
k_f	rate constant for association of receptor/ligand complexes	$M^{-1}\,min^{-1}$
k_{f1}	rate constant for association of receptor/ligand complexes	$M^{-1}\,min^{-1}$
k_{f2}	rate constant for association of receptor/ligand complexes	$M^{-1}\,min^{-1}$
k_g	rate constant for generation of IP_3	M/min
k_{g1}	rate constant for unstimulated exchange of GTP for GDP	min^{-1}
k_{g2}	rate constant for coupling of G*-GTP and PLC	$(\#/cell)^{-1}\,min^{-1}$
k_{g3}	rate constant for recombination of BG and G*-GDP	$(\#/cell)^{-1}\,min^{-1}$
k_h	rate constant for hydrolysis of IP_3	min^{-1}
k_i	rate constant for IP_3-induced calcium efflux from stores	min^{-1}
k_{i1}	rate constant for IP_3 production	min^{-1}
k_{i2}	rate constant for IP_3 degradation	min^{-1}
k_{obs}	overall observed rate constant for enzyme activation	min^{-1}
k_p	rate constant for pumping of calcium into stores	M/min
k_r	rate constant for dissociation of receptor/ligand complexes	min^{-1}
k_{r1}	rate constant for dissociation of receptor/ligand complexes	min^{-1}
k_{r2}	rate constant for dissociation of receptor/ligand complexes	min^{-1}
k_s	rate constant for calcium efflux from stores at high loading	$M^{-1}\,min^{-1}$
k_u	rate constant for uncoupling	min^{-1}
k_{u1}	rate constant for uncoupling of W and E_1	min^{-1}
k_{u2}	rate constant for uncoupling of W* and E_2	min^{-1}
k_x	rate constant for receptor desensitization	min^{-1}
k_{-x}	rate constant for receptor resensitization	min^{-1}
k_2	rate constant for coupling of C and G-GDP	$(\#/cell)^{-1}\,min^{-1}$
k_{-2}	rate constant for uncoupling of C and G-GDP	min^{-1}
k_3	rate constant for release of GDP	min^{-1}
k_{-3}	rate constant for binding of GDP	$M^{-1}\,min^{-1}$
k_4	rate constant for binding of GTP	$M^{-1}\,min^{-1}$
k_{-4}	rate constant for release of GTP	min^{-1}
k_5	rate constant for release of alpha subunit of G protein	min^{-1}
k_6	rate constant for G protein inactivation	min^{-1}
k_{12}	receptor conversion rate constant	min^{-1}
k_{21}	receptor conversion rate constant	min^{-1}
k'_{12}	receptor conversion rate constant	min^{-1}
k'_{21}	receptor conversion rate constant	min^{-1}
k_θ	rate constant for removal of calcium	M/min
K_a	saturation constant for stimulation of kinase activity	M

K_{ca1}	saturation constant for IP_3-induced calcium release	#/cell
K_{ca2}	saturation constant for calcium ATPase	M
K_D	equilibrium dissociation constant	M
K_{D1}	equilibrium dissociation constant	M
K_{D2}	equilibrium dissociation constant	M
K_{i1}	saturation constant for IP_3 production	M
K_k	scaled saturation constant for phosphorylation	
K_{M1}	scaled Michaelis constant for protein modification	M
K_{M2}	scaled Michaelis constant for protein modification	M
K_p	scaled saturation constant for dephosphorylation	
K_1	saturation constant for IP_3-induced calcium efflux from stores	M
K_2	saturation constant for pumping of calcium into stores	M
K_3	saturation constant for calcium feedback on generation of IP_3	M
K_4	set-point for calcium removal	M
K_{21}	ratio of conversion rate constants	
L	ligand concentration	M
n_H	Hill coefficient	
$n1$	exponent	
$n2$	exponent	
P_T	total number of PLC molecules	#/cell
P	total number of inactivated PLC molecules	#/cell
$P\text{-}G^*\text{-}GTP$	activated phospholipase C molecules	#/cell
R	free receptor number	#/cell
R_1	number of free receptors in state "1"	#/cell
R_2	number of free receptors in state "2"	#/cell
R_T	total surface receptor number	#/cell
R_x	desensitized receptor number	#/cell
t	time	min
U_1	number of intermediate complexes of W and E_1	#/cell
U_2^*	number of intermediate complexes of W* and E_2	#/cell
v_k	rate of phosphorylation of target enzyme	M/min
v_p	rate of dephosphorylation of target enzyme	M/min
V_{Mk}	maximal rate of kinase activity	M/min
V_{cy}	cytosol volume	μm^3
V_s	calcium store volume	μm^3
V_1	maximum conversion rate	(#/cell)/min
V_2	maximum conversion rate	(#/cell)/min
w^*	fraction of protein in modified form	
W	unmodified protein concentration	M
W_T	total protein W concentration	M
W^*	modified protein concentration	M
x	cytosolic calcium concentration on a whole cell basis	M
y	cytosolic IP_3 concentration on a whole cell basis	M
z	calcium concentration in stores on a whole cell basis	M
β	signaling activity of bound receptors	
ρ_c	response coefficient	
$\langle w^* \rangle$	mean value of w^*	

REFERENCES

Ambler, S. K., M., P., Tsien, R. Y. and Taylor, P. (1988). Agonist-stimulated oscillations and cycling of intracellular free calcium in individual cultured muscle cells. *J. Biol. Chem.*, **263**:1952–1959.

Arad, H., Rosenbusch, J. P. and Levitzki, A. (1984). Stimulatory GTP regulatory unit Ns and the catalytic unit of adenylate cyclase are tightly associated: mechanistic consequences. *Proc. Natl. Acad. Sci. USA*, **81**:6579–6583.

Benham, C. D. and Tsien, R. W. (1987). A novel receptor-operated Ca^{2+}-permeable channel activated by ATP in smooth muscle. *Nature*, **328**:275–278.

Berridge, M. J. (1990). Calcium oscillations. *J. Biol. Chem.*, **265**:9583–9586.

Berridge, M. J. and Galione, A. (1988). Cytosolic calcium oscillators. *FASEB J.*, **2**:3074–3082.

Berridge, M. J. and Irvine, R. F. (1989). Inositol phosphates and cell signalling. *Nature*, **341**:197–205.

Berridge, M. J. (1989). Cell signalling through cytoplasmic calcium oscillations. *In* A. Goldbeter (Ed.), *Cell to Cell Signalling: From Experiments to Theoretical Models*, pp. 449–459.

Birnbaumer, L., Abramowitz, J. and Brown, A. M. (1990). Receptor–effector coupling by G proteins. *Biochim. Biophys. Acta*, **1031**:163–224.

Bourne, H. R. (1989). Who carries what message? *Nature*, **337**:504–505.

Brandt, D. R. and Ross, E. M. (1986). Catecholamine-stimulated GTPase cycle. Multiple sites of regulation by beta-adrenergic receptor and Mg^{2+} studied in reconstituted receptor-G_s vesicles. *J. Biol. Chem.*, **261**:1656–1664.

Burch, R. M., Luini, A. and Axelrod, J. (1986). Phospholipase A2 and phospholipase C are activated by distinct GTP-binding proteins in response to alpha$_1$-adrenergic stimulation in FRTL5 thyroid cells. *Proc. Natl. Acad. Sci. USA*, **83**:7201–7205.

Burgoyne, R. D. and Cheek, T. R. (1991). Locating intracellular calcium stores. *Trends Biol. Sci.*, **16**:319–320.

Byron, K. L. and Villereal, M. L. (1989). Mitogen-induced $[Ca^{2+}]_i$ changes in individual human fibroblasts. Image analysis reveals asynchronous responses which are characteristic for different mitogens. *J. Biol. Chem.*, **264**:18234–18239.

Cadena, D. L. and Gill, G. N. (1992). Receptor tyrosine kinases. *FASEB J.*, **6**:2332–2337.

Carafoli, E. (1987). Intracellular calcium homeostasis. *Annu. Rev. Biochem.*, **56**:395–433.

Carpenter, G. (1987). Receptors for epidermal growth factor and other polypeptide mitogens. *Annu. Rev. Biochem.*, **56**:881–914.

Chabre, M., Bigay, J., Bruckert, F., Bornancin, F., Deterre, P., Pfister, C. and Vuong, T. M. (1988). Visual signal transduction: the cycle of transducin shuttling between rhodopsin and cGMP phosphodiesterase. *Cold Spring Harbor Symp. Quant. Biol.*, **53**:313–324.

Champeil, P., Combettes, L., Berthon, B., Doucet, E., Orlowski, S. and Claret, M. (1989). Fast kinetics of calcium release induced by myo-inositol trisphosphate in permeabilized rat hepatocytes. *J. Biol. Chem.*, **264**:17665–17673.

Chandra, S., Gross, D., Ling, Y. C. and Morrison, G. H. (1989). Quantitative imaging of free and total intracellular calcium in cultured cells. *Proc. Natl. Acad. Sci. USA*, **86**:1870–1874.

Chinkers, M., Garbers, D. L., Chang, M., Lowe, D. G., Chin, H., Goeddel, D. V. and Schulz, S. (1989). A membrane form of guanylate cyclase is an atrial natriuretic peptide receptor. *Nature*, **338**:78–83.

Cuthbertson, K. S. R. and Chay, T. R. (1991). Modelling receptor-controlled intracellular calcium oscillators. *Cell Calcium*, **12**:97–109.

Devreotes, P. N. and Sherring, J. A. (1985). Kinetics and concentration dependence of reversible cAMP-induced modification of the surface cAMP receptor in dictyostelium. *J. Biol. Chem.*, **260**:6378–6384.

Devreotes, P. N. and Zigmond, S. H. (1988). Chemotaxis in eukaryotic cells: a focus on leukocytes and *Dictyostelium*. *Annu. Rev. Cell Biol.*, **4**:649–686.

Edelman, A. M., Blumenthal, D. K. and Krebs, E. G. (1987). Protein serine/threonine kinases. *Annu. Rev. Biochem.*, **56**:567–613.

Fay, S. P., Posner, R. G., Swann, W. N. and Sklar, L. A. (1991). Real-time analysis of the assembly of ligand, receptor, and G protein by quantitative fluorescence flow cytometry. *Biochemistry*, **30**:5066–5075.

Fischer, E. H., Charbonneau, H. and Tonks, N. K. (1991). Protein tyrosine phosphatases: a diverse family of intracellular and transmembrane enzymes. *Science*, **253**:401–406.

Gennis, R. B. (1989). *Biomembranes: Molecular Structure and Function*. New York: Springer-Verlag.

Goldbeter, A., Dupont, G. and Berridge, J. (1990). Minimal model for signal-induced Ca^{2+} oscillations and for their frequency encoding through protein phosphorylation. *Proc. Natl. Acad. Sci. USA*, **87**:1461–1465.

Goldbeter, A. and Koshland, D. E., Jr (1981). An amplified sensitivity arising from covalent modification in biological systems. *Proc. Natl. Acad. Sci. USA*, **78**:6840–6844.

Goldbeter, A. and Koshland, D. E. J. (1987). Energy expenditure in the control of biochemical systems by covalent modification. *J. Biol. Chem.*, **262**:4460–4471.

Gray, P. T. A. (1988). Oscillations of free cytosolic calcium evoked by cholinergic and catecholaminergic agonists in rat parotid acinar cells. *J. Physiol.*, **406**:35–53.

Hanski, E., Rimon, G. and Levitzki, A. (1979). Adenylate cyclase activation by the beta-adrenergic receptors as a diffusion-controlled process. *Biochemistry*, **18**:846–853.

Hargrave, P. A. and McDowell, J. H. (1992). Rhodopsin and phototransduction: a model system for G protein-linked receptors. *FASEB J.*, **6**:2323–2331.

Harootunian, A. T., Kao, J. P. Y., Paranjape, S., Adams, S. R., Potter, B. V. L. and Tsien, R. Y. (1991a). Cytosolic Ca^{2+} oscillations in REF52 fibroblasts: Ca^{2+}-stimulated IP_3 production or voltage-dependent Ca^{2+} channels as key positive feedback elements. *Cell Calcium*, **12**:153–164.

Harootunian, A. T., Kao, J. P. Y., Paranjape, S. and Tsien, R. Y. (1991b). Generation of calcium oscillations in fibroblasts by positive feedback between calcium and IP_3. *Science*, **251**:75–78.

Harootunian, A. T., Kao, J. P. Y. and Tsien, R. Y. (1988). Agonist-induced calcium oscillations in depolarized fibroblasts and their manipulation by photoreleased $Ins(1,4,5)P_3$, Ca^{++}, and Ca^{++} buffer. *Cold Spring Harbor Symp. Quant. Biol.*, **53**:935–943.

Heizmann, C. W. and Hunziker, W. (1991). Intracellular calcium-binding proteins: more sites than insights. *Trends Biol. Sci.*, **16**:98–103.

Higashijima, T., Ferguson, K. M., Smigel, M. D. and Gilman, A. G. (1987). The effect of GTP and Mg^{2+} on the GTPase activity and the fluorescent properties of G0*. *J. Biol. Chem.*, **262**:757–761.

Hille, B. (1991). *Ionic Channels of Excitable Membranes*. Second edition. Sunderland, MA: Sinauer.

Holl, R. W., Thorner, M. O., Mandell, G. L., Sullivan, J. A., Sinha, Y. N. and Leong, D. A. (1988). Spontaneous oscillations of intracellular calcium and growth hormone secretion. *J. Biol. Chem.*, **263**:9682–9685.

Hunter, T. (1987). A thousand and one protein kinases. *Cell*, **50**:823–829.

Inesi, G., Kurzmack, M., Coan, C. and Lewis, D. E. (1980). Cooperative calcium binding and ATPase activation in sarcoplasmic reticulum vesicles. *J. Biol. Chem.*, **255**:3025–3031.

Irvine, R. F. (1991). Inositol tetrakisphosphate as a second messenger: confusions, contradictions, and a potential resolution. *BioEssays*, **13**:419–427.

Jacob, R., Merritt, J. E., Hallam, T. J. and Rink, T. J. (1988). Repetitive spikes in cytoplasmic calcium evoked by histamine in human endothelial cells. *Nature*, **335**:40–45.

Keizer, J. and De Young, G. W. (1992). Two roles for Ca^{2+} in agonist stimulated Ca^{2+} oscillations. *Biophys. J.*, **61**:649–660.

Knox, B. E., Devreotes, P. N., Goldbeter, A. and Segel, L. A. (1986). A molecular mechanism for sensory adaptation based on ligand-induced receptor modification. *Proc. Natl. Acad. Sci. USA*, **83**:2345–2349.

Koch, C. A., Anderson, D., Moran, M. F., Ellis, C. and Pawson, T. (1991). SH2 and SH3 domains: elements that control interactions of cytoplasmic signaling proteins. *Science*, **252**:668–674.

Koshland, D. E., Jr (1987). Switches, thresholds, and ultrasensitivity. *Trends Biochem. Sci.*, **12**:225–229.

Levitzki, A. (1984). Receptors: A Quantitative Approach. Menlo Park, CA: Benjamin/Cummings.

Levitzki, A. (1986). Beta-adrenergic receptors and their mode of coupling to adenylate cyclase. *Physiol. Rev.*, **66**:819–854.

Lindberg, R. A., Quinn, A. M. and Hunter, T. (1992). Dual-specificity protein kinases: will any hydroxyl do? *Trends Biochem. Sci.* **17**:114–117.

Lohse, M. J., Benovic, J. L., Caron, M. G. and Lefkowitz, R. J. (1990). Multiple pathways of rapid beta$_2$-adrenergic receptor desensitization. Delineation with specific inhibitors. *J. Biol. Chem.*, **265**:3202–3209.

Mahama, P. A. and Linderman, J. J. (1993a). Investigation of calcium signaling in BC3H1 cells using digital fluorescence imaging and mathematical modeling. Submitted for publication.

Mahama, P. A. and Linderman, J. J. (1993b). Simulations of membrane signal transduction events: Effect of receptor blockers on G-protein activation. Submitted for publication.

Mahoney, M. G., Randall, C. J., Linderman, J. J., Gross, D. J. and Slakey, L. L. (1992). Independent pathways regulate the cytosolic $[Ca^{2+}]$ initial transient and subsequent oscillations in individual cultured arterial smooth muscle cells responding to extracellular ATP. *Molec. Biol. Cell*, **3**:493–505.

Malgaroli, A., Fesce, R. and Meldolesi, J. (1990). Spontaneous $[Ca^{2+}]_i$ fluctuations in rat chromaffin cells do not require inositol 1,4,5-trisphosphate elevations but are generated by a caffeine- and ryanodine-sensitive intracellular Ca^{2+} store. *J. Biol. Chem.*, **265**:3005–3008.

Meyer, T. (1991). Cell signaling by second messenger waves. *Cell*, **64**:675–678.

Meyer, T., Holowka, D. and Stryer, L. (1988). Highly cooperative opening of calcium channels by inositol 1,4,5-trisphosphate. *Science*, **240**:653–656.

Meyer, T. and Stryer, L. (1988). Molecular model for receptor-stimulated calcium spiking. *Proc. Natl. Acad. Sci. USA*, **85**:5051–5055.

Meyer, T. and Stryer, L. (1991). Calcium spiking. *Annu. Rev. Biophys. Biophys. Commun.*, **20**:153–174.

Millard, P. J., Ryan, T. A., Webb, W. W. and Fewtrell, C. (1989). Immunoglobulin E receptor cross-linking induces oscillations in intracellular free ionized calcium in individual tumor mast cells. *J. Biol. Chem.*, **264**:19730–19739.

Monck, J. R., Reynolds, E. E., Thomas, A. P. and Williamson, J. R. (1988). Novel kinetics of single cell Ca^{2+} transients in stimulated hepatocytes and A10 cells measured using fura-2 and fluorescent videomicroscopy. *J. Biol. Chem.*, **263**:4569–4575.

Murayama, T. and Ui, M. (1985). Receptor-mediated inhibition of adenylate cyclase and stimulation of arachidonic acid release in 3T3 fibroblasts. Selective susceptibility to islet-activating protein, pertussis toxin. *J. Biol. Chem.*, **260**:7226–7233.

Neubig, R. R., Gantzos, R. D. and Thomsen, W. J. (1988). Mechanism of agonist and antagonist binding to alpha$_2$ adrenergic receptors: evidence for a precoupled receptor–guanine nucleotide protein complex. *Biochemistry*, **27**:2374–2384.

Neubig, R. R. and Thomsen, W. J. (1989). How does a key fit a flexible lock? Structure and dynamics in receptor function. *BioEssays*, **11**:136–141.

Nicholls, D. and Akerman, K. (1982). Mitochondrial calcium transport. *Biochim. Biophys. Acta*, **683**:57–88.

Nishizuka, Y. (1986). Studies and perspectives of protein kinase C. *Science*, **233**:305–312.

Nossal, R. and Lecar, H. (1991). *Molecular and Cell Biophysics*. Redwood City, CA: Addison-Wesley.

Pandol, S. J., Schoeffield, M. S., Fimmel, C. J. and Muallem, S. (1987). The agonist-sensitive calcium pool in the pancreatic acinar cell. Activation of plasma membrane Ca^{2+} influx mechanism. *J. Biol. Chem.*, **262**:16963–16968.

Parker, I. and Ivorra, I. (1990). Inhibition by Ca^{2+} of inositol trisphosphate-mediated Ca^{2+} liberation: a possible mechanism for oscillatory release of Ca^{2+}. *Proc. Natl. Acad. Sci. USA*, **87**:260–264.

Prentki, M., Glennon, M. C., Thomas, A. P., Morris, R. L., Matschinsky, F. M. and Corkey, B. E. (1988). Cell-specific patterns of oscillation free Ca^{2+} in carbamylcholine-stimulated insulinoma cells. *J. Biol. Chem.*, **263**:11044–11047.

Putney, J. W. (1986). A model for receptor-regulated calcium entry. *Cell Calcium*, **7**:1–12.

Rink, T. J. and Hallam, T. J. (1989). Calcium signalling in non-excitable cells: notes on oscillations and store refilling. *Cell Calcium*, **10**:385–395.

Rooney, T. A., Sass, E. J. and Thomas, A. P. (1990). Agonist-induced cytosolic calcium oscillations originate from a specific locus in single hepatocytes. *J. Biol. Chem.*, **265**:10792–10796.

Saxton, M. J. (1987). Lateral diffusion in an archipelago: the effect of mobile obstacles. *Biophys. J.*, **52**:989–997.

Schulz, S., Chinkers, M. and Garbers, D. L. (1989). The guanylate cyclase/receptor family of proteins. *FASEB J.*, **3**:2026–2035.

Segel, L. A., Goldbeter, A., Devreotes, P. N. and Knox, B. E. (1986). A mechanism for exact sensory adaptation based on receptor modification. *J. Theor. Biol.*, **120**:151–179.

Sibley, D. R., Benovic, J. L., Caron, M. G. and Lefkowitz, R. J. (1987). Regulation of transmembrane signaling by receptor phosphorylation. *Cell*, **48**:913–922.

Sklar, L. A. (1986). Ligand–receptor dynamics and signal amplification in the neutrophil. *Adv. Immunol.*, **39**:95–143.

Sklar, L. A., Mueller, H., Omann, G. and Oades, Z. (1989). Three states for the formyl peptide receptor on intact cells. *J. Biol. Chem.*, **264**:8483–8486.

Sklar, L. A. and Omann, G. M. (1990). Kinetics and amplification in neutrophil activation and adaptation. *Semin. Cell Biol.*, **1**:115–123.

Somogyi, R. and Stucki, J. W. (1991). Hormone-induced calcium oscillations in liver cells can be explained by a simple one pool model. *J. Biol. Chem.*, **266**:11068–11077.

Su, Y., Harden, T. K. and Perkins, J. P. (1980). Catecholamine-specific desensitization of adenylate cyclase. Evidence for a multistep process. *J. Biol. Chem.*, **255**:7410–7419.

Swillens, S. and Mercan, D. (1990). Computer simulation of a cytosolic calcium oscillator. *Biochem. J.*, **271**:835–838.

Taylor, C. W. (1990). The role of G proteins in transmembrane signalling. *Biochem. J.*, **272**:1–13.

Thomsen, W. J., Jacquez, J. and Neubig, R. R. (1988). Inhibition of adenylate cyclase mediated by the high affinity conformation of the alpha$_2$-adrenergic receptor. *Mol. Pharmacol.*, **34**:814–822.

Thomsen, W. J. and Neubig, R. R. (1989). Rapid kinetics of alpha$_2$-adrenergic inhibition of adenylate cyclase. Evidence for a distal rate-limiting step. *Biochemistry*, **28**:8778–8786.

Tolkovsky, A. M. (1983). The elucidation of some aspects of receptor function by the use of a kinetic approach. *Curr. Top. Membr. Trans.*, **18**:11–44.

Tolkovsky, A. M. and Levitzki, A. (1981). Theories and predictions of models describing sequential interactions between the receptor, the GTP regulatory unit, and the catalytic unit of hormone dependent adenylate cyclases. *J. Cyclic Nucleotide Res.*, **7**:139–150.

Tota, M. R., Kahler, K. R. and Schimerlik, M. I. (1987). Reconstitution of the purified porcine atrial muscarinic acetylcholine receptor with purified porcine atrial inhibitory guanine nucleotide binding protein. *Biochemistry*, **26**:8175–8182.

Tsien, R. W. and Tsien, R. Y. (1990). Calcium channels, stores, and oscillations. *Annu. Rev. Cell Biol.*, **6**:715–760.

Tsien, R. Y. (1989). Fluorescent indicators of ion concentrations. *Methods Cell Biol.*, **30**:127–156.

Ullrich, A. and Schlessinger, J. (1990). Signal transduction by receptors with tyrosine kinase activity. *Cell*, **61**:203–212.

Vaughan, R. A. and Devreotes, P. N. (1988). Ligand-induced phosphorylation of the cAMP receptor from *Dictyostelium discoideum. J. Biol. Chem.*, **263**:14538–14543.

Wakui, M., Osipchuk, Y. V. and Petersen, O. H. (1990). Receptor-activated cytoplasmic Ca^{2+} spiking mediated by inositol trisphosphate is due to Ca^{2+}-induced Ca^{2+} release. *Cell*, **63**:1025–1032.

Wakui, M., Potter, B. V. L. and Petersen, O. H. (1989). Pulsatile intracellular calcium release does not depend on fluctuations in inositol trisphosphate concentration. *Nature*, **339**:317–320.

Woods, N. M., Cuthbertson, K. S. R. and Cobbold, P. H. (1986). Repetitive transient rises in cytoplasmic free calcium in hormone-stimulated hepatocytes. *Nature*, **319**:600–602.

Woods, N. M., Cuthbertson, K. S. R. and Cobbold, P. H. (1987). Agonist-induced oscillations in cytoplasmic free calcium concentration in single rat hepatocytes. *Cell Calcium*, **8**:79–100.

Wyman, J. (1975). The turning wheel: a study in steady states. *Proc. Natl. Acad. Sci. USA*, **72**:3983–3987.

Yarden, Y. and Ullrich, A. (1988). Growth factor receptor tyrosine kinases. *Annu. Rev. Biochem.*, **57**:443–478.

Yatani, A., Codina, J., Brown, A. M. and Brinbaumer, L. (1987a). Direct activation of mammalian atrial muscarinic potassium channels by GTP regulatory protein Gk. *Science*, **235**:207–211.

Yatani, A., Codina, J., Imoto, Y., Reeves, J. P., Birnbaumer, L. and Brown, A. M. (1987b). A G protein directly regulates mammalian cardiac calcium channels. *Science*, **238**:1288–1291.

6

Receptor-Mediated Cell Behavioral Responses

This chapter provides a discussion of efforts directed toward quantitative understanding of the effects of receptor and ligand properties on cell behavioral responses. As outlined in Chapter 1, in most basic form we wish to have models that predict a quantitative measure of cellular behavior (*e.g.*, proliferation rate, adhesion strength, adhesion probability in fluid flow, migration speed, migration direction) in terms of biochemical and biophysical parameters that characterize interactions between receptors and ligands or of these molecules with other cell components. We believe that the time is truly at hand for pursuing this sort of work, because of the advances in molecular biology and genetic engineering. Mathematical models for the effects of receptor and ligand parameters on cell function can now be tested by deliberate alteration of relevant molecular properties. Thus, we predict an important role for this new approach, combining mathematical modeling at the cellular level with genetic engineering directed toward model validation.

Chapters 2–5 were primarily concerned with an ability to elucidate and measure parameters representing receptor and ligand processes. We now consider the use of those parameters in models for cell function regulated by receptor/ligand interactions. The following sections will address the state of modeling efforts in three major areas: cell proliferation, cell adhesion, and cell migration. Although a great number of models analyzing various aspects of these important cell behavioral phenomena have appeared in the literature, we restrict our attention to a small subset. We will largely neglect *phenomenological models, i.e.*, those that describe a cell function quantitatively but make no attempt at underlying mechanistic detail, but will instead concentrate on *mechanistic models*, which do aim to connect molecular properties to cell function. Further, we will narrow our discussion to mechanistic models that focus on *receptor/ligand regulation of cell function*, because of its centrality to understanding cell behavior.

6.1   CELL PROLIFERATION

6.1.1   Background

The importance of cell proliferation for biotechnological and medical applications is easy to recognize. Blood and tissue cells of higher organisms are increasingly touted for use in the bioprocess industry, primarily for production of therapeutic glycoproteins, because of their natural ability to synthesize active, properly folded and modified products of this class. Recovery of substantial quantities of product usually depends on generating high cell densities, typically requiring growth of the cell population from a smaller initial density. Some paramount medical problems revolve largely around regulation of cell proliferation: wound healing, immune host defense, and cancer. In the first two areas, the question is essentially one of how to stimulate multiplication of tissue and blood cells of appropriate types in a selective manner and to effective extents. Conversely, how to prevent cell growth is the issue in cancer. Finally, the new field of tissue engineering is based on an ability to generate viable cell populations of appropriate types organized into effectively functioning tissue, either *in vitro* or *in vivo*. There is little doubt that quantitative manipulation of receptor/ligand interactions that regulate cell proliferation will play a leading role in all these problems.

Most commonly, phenomenological models are used currently for quantitative description of cell proliferation in culture; in general, such models characterize a phenomenon without offering details concerning its operation. For cell proliferation, this class of models is primarily based on the kinetic framework originally developed by Monod (1949) for microbial population growth, as discussed in common texts (*e.g.*, Bailey and Ollis, 1986). In the *Monod model*, the *specific cell proliferation rate constant* k_g (time^{-1}) is related to a "growth-limiting" substrate concentration S by the expression

$$k_g = \frac{k_{gmax}S}{K_M + S} \tag{6-1}$$

where k_g can be defined in terms of the change in total cell number, N, versus time when no other processes are occurring:

$$k_g = \frac{1}{N}\frac{dN}{dt} = \frac{d(\ln N)}{dt} \tag{6-2}$$

k_{gmax} (time^{-1}) is the maximum specific cell growth rate constant, presumed to be observed when the substrate concentration is sufficiently great to not be limiting for cell growth. K_M (moles/volume) is a saturation constant, the value of S which provides for a half-maximal specific growth rate constant. The relevant substrate is typically a nutrient, such as glucose, that is needed for basic cell metabolism. Variations on Eqn. (6–1) exist (Bailey and Ollis, 1986), but for all these the approach remains entirely phenomenological.

From our perspective, a more fundamental and mechanistic approach is needed for two major reasons. First, in higher organisms, cell proliferation is a highly-regulated process. Unlike for prokaryotic cells and cells of lower eukaryotes, simple nutrients (carbon, nitrogen, and energy sources) are not the primary controlling factors for blood and tissue cells of plants and animals. Rather, it is a class of receptor-binding ligands known as *growth factors*, generally peptides of fairly low to intermediate molecular weight found in blood serum, that serve as the key regulatory molecules (Cross and Dexter, 1991). Subversion of control normally exerted by these factors is intimately linked to the development of transformed cells in cancer (Aaronson, 1991).

Because the mechanism by which growth factors stimulate cell growth is significantly different from that for simple nutrients, it is not clear that Monod kinetics will describe the case when S, the substrate, is a growth factor. Indeed, McKeehan and McKeehan (1981) found that Monod kinetics could satisfactorily describe the dependence of human fibroblast proliferation on the concentration of simple nutrients, but not of growth factors. At the same time, some investigators (*e.g.*, Glacken *et al.*, 1988) have demonstrated success in using this sort of approach for effects of serum on cell growth in culture.

More importantly, though, the phenomenological approach does not offer any possibility of relating molecular properties to proliferation responses. The parameters k_{gmax} and K_M, because they are quantities merely defined by the fit of the equation to data, do not indicate any underlying mechanisms by which growth factor binding induces cell proliferation. Nor do they suggest which properties of the key regulatory molecules are important. None of the details of growth factor binding, trafficking, or signal transduction, as have been described in the mathematical models developed earlier in this text (Chapters 2–5), are included in models of this type, yet it is clear that receptor and ligand properties affecting these processes may consequently influence a mitogenic response.

Hence, we offer in this section a small number of initial attempts at capturing some relationships between receptor/ligand interactions and cell proliferation. We will focus on several questions that can be raised in light of the mathematical models described in previous chapters. For instance: what is the relationship between growth factor/receptor complex formation and the later event of cell growth and division? Do rates of receptor and ligand trafficking events affect the proliferation response? What possibilities might exist for attempts to control cell growth by manipulation of growth factor or receptor molecular properties?

In order to build a foundation for the models to be described, a brief overview of control of cell proliferation by growth factors may be useful.

6.1.1.a Growth Factors and Their Receptors

At this point in time, well-defined *peptide growth factors* number at least two dozen, with continued discovery of many more to be anticipated (Cross and Dexter, 1991). Fairly comprehensive references providing information about properties of growth factors and their receptors can be found in Sporn and Roberts (1990), and the summary here is based on these. Among the more widely-studied examples are epidermal growth factor (EGF), interleukin-2 (IL-2), platelet-derived growth factor (PDGF), acidic and basic fibroblast growth factors (aFGF, bFGF), and insulin-like growth factors-I,-II (IGF-I, IGF-II). Growth factors can be functionally defined as peptide compounds required by certain cells for their proliferation in culture. Their molecular weights range from about 6–8 kDa for EGF and IGF-I,II to 35 kDa for PDGF, with some intermediate at approximately

15–18 kDa (IL-2, aFGF, bFGF). The unusually large size of PDGF is explained by its structure. It is a dimer of two peptide chains designated A and B, allowing for three distinct forms – AA, AB, and BB – possessing different receptor binding properties.

We note that transferrin (Tf), which satisfies our functional definition, is generally considered not a true growth factor, but rather is basically a transport protein for the required metabolite iron. Insulin may also be placed in a separate category because it regulates purely metabolic cell responses in addition to playing a role in proliferation.

Growth factor origins can be diverse. They are produced by cells in various tissues and organs, and the resulting chemical communication between source and target cells may fall into the classes of endocrine, paracrine, autocrine, and juxtacrine factors. *Endocrine factors* or *hormones* are those produced by a specific organ and then transported through the bloodstream before acting on their target cells. One important example of an endocrine factor is insulin, which is synthesized by beta cells of the pancreatic islets of Langerhans and acts on a wide range of cells in various tissues. Most growth factors appear to be *paracrine factors*, secreted by particular cell types in a given tissue and acting on target cells of a different type in that tissue. EGF and IGF-I,II are prominent examples. PDGF, produced by blood platelets activated at tissue sites during blood clotting, also usually acts as a paracrine growth factor. These factors, among others, are therefore found in the blood serum commonly added to cell culture media to facilitate cell proliferation. *Autocrine factors* are produced by the same cells as they act upon. IL-2, secreted by and then stimulating T-lymphocytes after exposure to antigen, is a well-known physiological example. Sometimes, cancer cells can arise by expression of oncogenes coding for autocrine growth factors (Sporn and Roberts, 1990). Transforming growth factor alpha (TGFα) is secreted by A431 epidermoid carcinoma cells and stimulates proliferation of these same cells via binding to the EGF receptor (Van de Vijver *et al.*, 1991). The most recently discovered mode of growth factor action is through *juxtacrine factors*. It has been found that a cell membrane-anchored form of TGF-α can stimulate proliferation of neighboring adherent cells (Brachmann *et al.*, 1989). The extent of generality of this mechanism is currently unclear.

Corresponding cell surface *membrane receptors for growth factors* have been identified, and many have been isolated and sequenced. These receptors are typically large glycoproteins, with molecular weights in the range 100–200 kDa. Excellent reviews of their properties and function are available (Carpenter, 1987; Gill, 1989; Czech *et al.*, 1990). The EGF receptor (EGF-R) is illustrated schematically in Figure 6–1, in which the key features, general to most growth factor receptors, are: a single transmembrane domain, a glycosylated ligand-binding extracellular domain, and a cytoplasmic regulatory domain. Some growth factor receptors are monomeric (EGF-R, IGF-II-R) and some dimeric (Insulin-R, IGF-I-R). The insulin and IGF-I receptors, in fact, are each composed of two dimers of non-identical α and β subunits. The PDGF and IL-2 receptors can both exist in three forms, all capable of ligand binding: a heterodimer form and the two individual monomer subunit forms. A new twist has been found recently in bFGF receptors: ligand binding involves a cell surface proteoglycan molecule in addition to a more traditional glycoprotein (Klagsbrun and Baird, 1991).

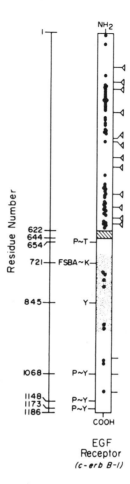

Residue Number

622
644
654

721 — FSBA~K

845 — Y

1068 — P~Y

1148
1173 — P~Y
1186 — P~Y

P~T

NH₂

COOH

EGF
Receptor
(c-erb B-l)

Figure 6-1 Schematic drawing of the EGF receptor, for illustration of key growth factor receptor features. The crosshatched region is the transmembrane domain. Upper region is the extracellular ligand binding domain; triangles denote probable glycosylation sites, and solid circles represent cysteine residues that typically participate in stabilizing three-dimensional conformation. Lower region is the cytoplasmic tail, with the tyrosine kinase signaling domain shaded. Ys indicate positions of tyrosine residues accessible for autophosphorylation by the receptor kinase, involved in regulation of receptor signaling and trafficking functions. Redrawn from Carpenter (1987). Reproduced, with permission, from the Annual Review of Biochemistry, Vol. 56, © 1987 by Annual Reviews Inc.

Growth factor binding leads to activation of the receptor intracellular domain, resulting in *signal generation*. Signaling pathways used by growth factor receptors commonly fall into two categories: (1) those based on receptor interactions with membrane-associated G-proteins, producing intracellular second messenger molecules such as Ca^{2+} and membrane lipid breakdown products; and (2) those based on receptor kinase activity, covalently modifying intracellular protein substrates by phosphorylation, primarily on tyrosine residues (Yarden and Ullrich, 1988; Ullrich and Schlessinger, 1990; Cantley *et al.*, 1991) (see Chapter 5). It is important to note that these categories are non-exclusive; the insulin receptor and EGF receptor are examples of receptors that make use of both types of signaling pathways.

The physical mechanisms by which activation of receptor signaling occurs are not completely understood. Both intermolecular and intramolecular mechanisms have been hypothesized (see Gill *et al.* (1987) for a helpful exposition concerning the EGF receptor). In the former, signaling requires aggregation of two receptors into a dimer form, permitting one receptor to activate the other. In the latter, activation occurs through a conformational change induced in the cytoplasmic domain across the transmembrane region. It is likely that no single mechanism will be found to be general for all growth factor receptors. In any event, activation allows the receptor to couple productively with bound signal-transducing substrates such as G-proteins and/or to act as kinases on enzyme substrates, initiating an intracellular signal cascade that ultimately leads to a mitogenic response.

6.1.1.b Cell Cycle

Proliferation of eukaryotic cells in culture involves progression through four major phases, G_1, S, G_2, and M, of the *cell cycle*, as shown schematically in Figure 6–2 (see Baserga (1985) and Pardee (1989) for more detailed information). S and M are the time periods of DNA synthesis and mitosis. These possess relatively fixed lengths, of approximately 6–8 h and 1 h, respectively, for higher eukaryotic cells. G_2 is the gap between DNA synthesis and mitosis, and is also of fairly constant length, about 2–3 h. G_1 is defined functionally as the gap between the easily observed events of cell division and DNA synthesis. It lasts many hours (typically in the range 18–72 h, depending on the cell type), during which the cell grows in size. This phase is the most variable in length of the four, so the doubling time of cycling cells is largely governed by the length of G_1.

Complicating this situation is the fact that cells can remain viable for long periods in a non-proliferating, quiescent state, designated G_0. The distinction between G_0 and G_1 can be subtle, as both G_0 and G_1 cells possess an unduplicated

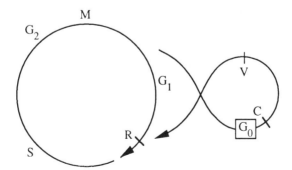

Figure 6–2 Illustration of eukaryotic cell cycle, based on 3T3 fibroblast studies. Beginning in quiescent (G_0) phase, cells pass critical points C and V to pass into G_1, then pass critical point R to enter DNA synthesis phase (S). G_2 follows before cell division (M). Redrawn from Pardee (1989) *Science*, vol. 246, pp. 603–608. Copyright 1989 by the AAAS.

DNA content. However, they appear to differ in some properties; macromolecular (both protein and RNA) synthesis rates, enzyme activity, and membrane transport levels are all relatively low in G_0 cells, whereas some RNA and protein species seem to be characteristic of the G_0 state. The specific proliferation rate of a cell population primarily depends on the fraction of the population that is actively progressing through the cell cycle, since the rate of progression through the cycle appears to be fairly constant for a given cell type except for transition into and out of G_0. In particular, growth factors are believed to influence the passage of a cell from G_0 into G_1, and through G_1. The most substantial information on this point has been obtained for fibroblastic cells in culture, especially the partially-transformed mouse cell line designated 3T3 (Pardee, 1989). Understanding of intracellular events more directly linked to control points in cell cycle passage has been advanced dramatically with studies on yeast cells and frog oocytes (Murray and Kirschner, 1989). It is presumed that these events are analogous to those stimulated by growth factor/receptor binding in mammalian blood and tissue cells.

Given our focus on receptor-mediated processes, we will consider the fibroblast studies of *cell cycle control*, based on the review by Pardee (1989). In 3T3 fibroblasts, there are three well-defined landmarks for passage of a cell from G_0 through G_1 into S. These are denoted C, V, and R (Figure 6–2). Movement from G_0 into G_1 is achieved by passing control point C, the "competence" point, and requires incubation of the cells with PDGF. Once in G_1, EGF is necessary to move past point V, the "entry" point. The length of time from V to R – the "entry" point into S phase – is approximately 6 h under adequate nutrient conditions. This is roughly the same period taken in G_1 by cycling cells, so V must be very near the border between M and G_1 in some physiological sense. The sole growth factor required by 3T3 cells after V is IGF-I, which permits the cells to begin DNA synthesis. This last stage of progression additionally requires net protein synthesis; inhibition of protein synthesis causes exponentially growing cells to stop at this point with unduplicated chromosomes.

A simplistic version of the key events involved in growth factor regulation of the cell cycle is illustrated in Figure 6–3. Formation of growth factor/receptor complexes activates receptor cytoplasmic domains, initiating the mitogenic signal. Pathways involving second messengers and covalently modified proteins are followed. Substrates of protein kinases as well as counteracting phosphatases (Fischer *et al.*, 1991) form a cascade of signal propagation leading to regulation of DNA transcription via modified DNA-binding proteins. Eventually, proliferation-associated genes are expressed, new proteins are synthesized, and the chromosomal DNA is replicated, with the ultimate result of cell division. A critical component in cell cycle regulation is the protein cyclin (Draetta, 1990), which affects the activity of a key protein kinase known as cdc2 (Pines and Hunter, 1990; Hunter and Pines, 1991). In addition, part of the phosphorylation/dephosphorylation cascade may be devoted to modification of growth-inhibiting proteins, such as the retinoblastoma gene product Rb, which appear to restrict cell proliferation until modified by phosphorylation (Sager, 1989).

In sum, though there exist multiple events and, likely, multiple regulators of

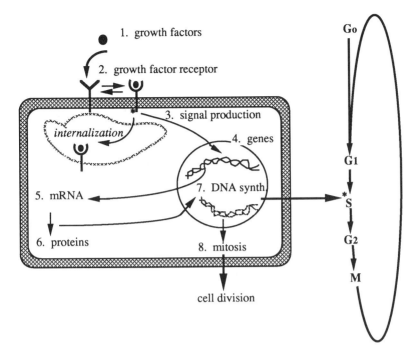

Figure 6–3 Sketch of events involved in growth factor receptor regulation of cell proliferation. Growth factor binding to receptors activates receptor signaling activity, leading to expression of growth-associated genes for production of proteins required to enter DNA synthesis phase.

cell progression through the cell cycle, growth factor receptors clearly play a central role.

6.1.1.c Oncogenes

Elucidation of the roles for some specific proteins involved in growth regulation has been closely associated with the investigation of oncogenes. *Oncogenes* are viral genes apparently derived from normal cellular genes, or proto-oncogenes, that are capable of transforming non-tumorigenic animal cells into tumor-forming cells (see Burck *et al.*, 1988; Hunter, 1991). Some oncogene products are directly related to growth factors and their receptors themselves. Examples of these include the *sis* gene product, which is essentially identical to PDGF, and the *neu* and *erbB* gene products, which are related to the EGF receptor. These oncogenes may transform normal cells to tumorigenic cells by quantitatively affecting the initial receptor/ligand event – in the first case by generating an autocrine growth factor, and in the second case by either increasing the available number of receptors for a common growth factor or by introducing a constitutively signaling receptor. The mode of action of other oncogene products such as *src* and *ras*, although known to affect signaling by growth factor receptors, is less clear. Satisfactory mathematical models for growth factor regulation of cell proliferation should provide a robust framework capable of accounting for these diverse oncogene effects.

6.1.2   Experimental Methods

Although many of the molecular components involved in the stimulation of DNA synthesis by growth factor/receptor binding have not yet been identified, substantial data exist on cell proliferation itself. Three major methods for the measurement of cell proliferation exist: cell counting, tritiated-thymidine uptake, and flow cytometry (see Baserga, 1985). *Cell counting* is clearly the simplest. It can be accomplished by direct microscopic observation of anchorage-dependent cells growing on a surface or by means of a Coulter counter for cells in suspension. From data on total cell number versus time, a specific cell population proliferation rate constant, k_g (time^{-1}) can be defined according to Eqn. (6–2). This equation requires that there is no cell death (which can be easily monitored by cell viability tests), and yields an average specific proliferation rate constant based on all cells – whether actively cycling or quiescent.

For a typical experiment using the *tritiated-thymidine uptake* method, the cell population is incubated with a pulse of [^{3}H]thymidine (about 20 min) and then observed at a sequence of later times. Tritiated-thymidine uptake occurs solely during S phase, so cell-associated radioactivity, generally quantified through autoradiography (microscopic observation of silver grains, induced by the radio-activity, in an overlying emulsion) or scintillation counting, provides a measure of the number of cells that have progressed through this stage of the cell cycle. The rate of DNA synthesis quantified in this manner can be used to estimate the specific population proliferation rate constant, since the ratio of DNA synthesis rate to total cell number should be equal to the ratio of cell division rate to total cell number.

Tritiated-thymidine uptake methods are also capable of quantifying the fraction of cells in each of the cell cycle phases. This approach is based on the fact that simple light microscopy can distinguish cells in mitosis (M phase) from those in interphase (G_1–S–G_2), whereas autoradiography distinguishes those cells currently synthesizing DNA (S phase) from those that are not (G_2–M–G_1). From measurements of percentage labeled mitoses, *i.e.*, the fraction of cells in mitosis that also have accumulated the labeled thymidine, as a function of time after the pulse (see Figure 6–4), the lengths of each cell cycle phase can be calculated. Various methods for doing this calculation have been proposed, offering different techniques for breaking down the experimental curves (*e.g.*, Hartmann *et al.*, 1975).

More recently, fluorescence-based *flow cytometry* has gained popularity for cell cycle studies. Certain fluorescent dyes, such as mithramycin, ethidium bromide, and acriflavine, bind specifically to DNA. Following laser excitation, cells treated with the dye fluoresce with an intensity proportional to their DNA content. A flow cytometer is able to count cell number as a function of fluorescence intensity. Cells in G_2 and M phases provide twice the intensity of those in G_1, and cells in S phase give intermediate intensity levels. Example DNA distributions obtained by this procedure are shown in Figure 6–5. Deconvolution of the distributions into three groups permits estimation of the fraction of cells in stages G_1, S, and G_2 + M, though this is fraught with complications (Zietz and Nicolini, 1978). Independent determination of the overall cell cycle time is necessary to calculate the corresponding lengths of these three stages. Flow cytometry is much faster

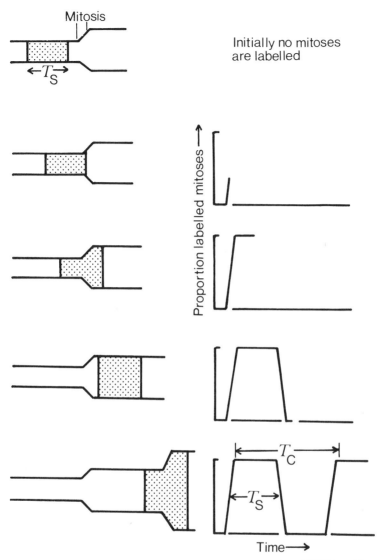

Figure 6–4 Cell cycle analysis by autoradiography. After a short pulse of [^{3}H]thymidine, which is taken up only by cells synthesizing DNA, a labeled cohort of cells traverses the series of phases in the cycle: G_2, then M, then G_1, then S. The peak width is equal to the duration of S phases (T_S), and the interval between two sequential peaks is the total cell cycle time ($T_C = T_S + T_{G_2} + T_M + T_{G_1}$). Redrawn from Baserga (1985).

than autoradiography, and avoids the use of radiolabeled compounds. It further allows simultaneous measurement of related quantities, such as levels of RNA or particular proteins. An example of the use of flow cytometry to investigate the dependence of the transition from quiescence into active cycling on growth factor concentration can be found in Bohmer and Beattie (1988).

The mathematical models for effects of growth factor and receptor properties

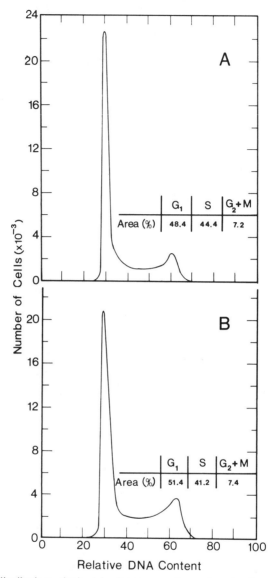

Figure 6–5 DNA distributions obtained by fluorescence flow cytometry, for proliferating Chinese hamster ovary cells. Panels A and B represent two different fluorescent dyes. The percentages of cells in each phase of the cell cycle were obtained by computer-fits of the distribution curves. G_1 phase corresponds to the farthest-left peak, with cells possessing a single copy of each chromosome. Redrawn from Baserga (1985).

on cell proliferation we discuss next treat only overall cell proliferation rates, mainly based on DNA synthesis rate data. However, we expect that in the near future such models will be enhanced toward explicit consideration of progress through the cell cycle.

6.1.3   Models

Given the preceding background discussion, it is apparent that some mechanistic understanding is available at both the front and back ends of the mitogenic process. That is, much is known about growth factors and their receptors, including binding, trafficking, and early signal transduction events. Also, a fair amount is known about the control of the cell proliferation cycle. Unfortunately, very little information currently exists that connects these two aspects of the overall behavior. Therefore, mathematical models aiming to provide insight relevant to molecular properties must for the present steer clear of the unknown intermediate events. Numerous models for the cell proliferation cycle have been proposed in the past, with primary application to cancer and hematopoiesis (Swan, 1977). None of these, however, include any reference to growth factor/receptor regulatory processes, so they are of no direct relevance to our purposes. Only in recent years have efforts begun to relate growth factor ligand and receptor properties and interactions to cell proliferation responses using mathematical models. Appropriately, these models include detail only for receptor/ligand features and for reasonable quantitative representations of the mitogenic response. Intermediate events are left for now as "black boxes", to await accumulation of experimental information.

   We will describe two models that focus on control of fibroblast proliferation by EGF/EGF-receptor interactions. The EGF/fibroblast system is ideal for modeling studies because of its comparative wealth of biochemical data (Carpenter and Wahl, 1990), its experimental convenience and relative simplicity, and its potential significance for biotechnological and medical applications (*e.g.*, cell culture media, wound healing). The first model addresses the relationship between growth factor/receptor binding and mitogenic response. Understanding gained from that is then combined with information on receptor/ligand trafficking dynamics from Chapter 3 to examine the effects of trafficking on cell proliferation.

   A third model will then be briefly introduced, treating implications of extracellular ligand diffusion effects inherent in autocrine growth factor systems. This last model combines concepts from the first two together with considerations of diffusional influences from section 4.2.

6.1.3.a   *Relationship Between Growth Factor/Receptor Complexes and Cell Proliferation*

A model attempting to relate cell proliferation rate to growth factor receptor binding was offered by Knauer *et al.* (1984) for human fibroblasts (HF) stimulated by EGF. (Similar data and analysis were also provided for mouse embryo fibroblasts.) Original data on the dependence of cell proliferation rate, as quantified by percentage maximum [^{3}H]thymidine incorporation into newly-synthesized DNA, on EGF concentration are shown in Figure 6–6. The two curves in this plot correspond to small and large volumes of growth medium, with the large volume experiments requiring a lower EGF concentration for a given mitogenic response. This observation indicates that it is not merely the EGF concentration present that governs the response, but also the amount of EGF present. Receptor-mediated endocytosis and subsequent degradation of the growth factor is responsible

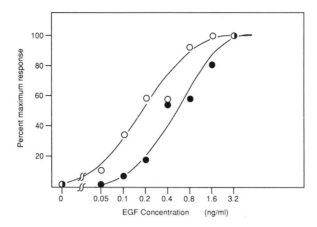

Figure 6–6 Mitogenic response of cultured human fibroblasts to EGF, comparing small and large volumes of medium. Abscissa is initial EGF concentration, ordinate is % maximum DNA synthesis rate. Open circles represent experiments using 80 ml of culture medium, filled circles represent experiments using 2 ml. For a given EGF concentration, cell proliferation rate is slower in the smaller volume of medium, because ligand is depleted (and thus its concentration is decreased) by cell binding, uptake, and degradation. Redrawn from Knauer *et al.* (1984).

for this behavior. Hence, the model developed by Knauer *et al.* incorporates endocytosis as a key aspect of the receptor-mediated response.

The central tenet of the Knauer *et al.* model is that the mitogenic response can be directly related to the number of receptor/ligand complexes present during steady-state incubation in growth factor. This may appear simplistic in light of the unfolding details of the cell cycle and the many events that likely occur between growth factor binding and DNA synthesis. However, it is not an unreasonable first attempt given the observation that, although any given round of EGF binding and trafficking occur on the time scale of minutes, cells require 6–8 h of persistent occupation of EGF-R to invoke a full mitogenic response (Aharanov *et al.*, 1978).

In order to investigate this hypothesis, a mathematical model of binding and trafficking processes was used to relate EGF concentration, L (the abscissa of Figure 6–6) to the number of EGF/EGFR complexes per cell, C. This model had been previously validated by Wiley and Cunningham (1981), and is a limiting case of the base model outlined in section 3.2.1.a. Essentially, Eqns. (3–1,a–c,e) were used, but with no recycling of either receptor or ligand (*i.e.*, $f_R = f_L = 1$).

When ligand concentration varies only slowly with time, a *quasi-steady state* for the number of free surface receptors R_s, surface complexes C_s, intracellular free receptors R_i, and intracellular complexes C_i is reached in less than one hour for this particular system (Wiley and Cunningham, 1981). The numbers of surface and intracellular complexes can be found by solving these equations at under steady-state conditions (*i.e.*, $dR_s/dt = dC_s/dt = dR_i/dt = dC_i/dt = 0$):

$$C_s = \left(\frac{K_{ss}L}{1 + K_{ss}L} \right) \frac{V_s}{k_{eC}} \tag{6–3a}$$

and

$$C_i = \left(\frac{k_{eC}}{k_{deg}}\right)C_s \qquad (6\text{–}3b)$$

where K_{ss} is the apparent cellular affinity constant:

$$K_{ss} = \frac{k_{eC}k_f}{k_{eR}(k_r + k_{eC})} \qquad (6\text{–}4)$$

k_{eC} is the complex internalization rate constant, V_s is the rate of new receptor synthesis, k_{deg} is the degradation rate constant, k_f is the association rate constants, k_r is the dissociation rate constant, and k_{eR} is the free receptor internalization rate constant. All of these rate constants are defined as in Chapter 3. Parameter values were determined from independent experiments (Knauer *et al.*, 1984), for HF: $k_{eC} = 5.3 \times 10^{-3}\,\text{s}^{-1}$, $k_{eR} = 9.3 \times 10^{-5}\,\text{s}^{-1}$, $V_s = 18\ \#/\text{s}$, $k_f = 3.1 \times 10^6\,\text{M}^{-1}\,\text{s}^{-1}$, $k_r = 2.5 \times 10^{-2}\,\text{s}^{-1}$, and $k_{deg} = 8.3 \times 10^{-4}\,\text{s}^{-1}$.

The total number of cell complexes is given by $C_T = (C_s + C_i)$, summing Eqns. (6–3a,b). C_T can be calculated easily as a function of the ligand concentration L using the parameter values specified above. By substituting these numbers into Eqns. (6–3a,b), the abscissa representing L in Figure 6–6 is translated into a corresponding abscissa in terms of C_T. The resulting plot is shown in Figure 6–7, in which the ordinate shows cell mitogenic response, quantified as percentage maximum [³H]thymidine incorporation into newly-synthesized DNA.

This plot exhibits a direct relationship, apparently linear, between mitogenic response and EGF/EGFR complex number. The slope of this line represents a

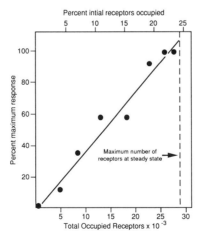

Figure 6–7 Relationship between steady-state EGF/EGF-R complex number and mitogenic response of cultured human fibroblasts to EGF. Abscissa is complex number, C_T, as calculated for a given EGF concentration using a trafficking model. Ordinate is % maximum DNA synthesis rate. These points correspond to the original data (for 80 ml culture medium) shown in Figure 6–6. Vertical dashed line represents the number of surface receptors remaining after downregulation at highest EGF concentration. Thus, maximal proliferation response is attained when all surface receptors are bound. Redrawn from Knauer *et al.* (1984).

quantity we will define as the *intrinsic mitogenic signal generation* by the complexes, γ, having units of % maximum proliferation rate per complex. The value of γ quantitatively characterizes the ability of a growth factor/receptor complex to produce a mitogenic response via the intracellular signal transduction cascade. All the processes involved in that cascade are incorporated into γ.

Mathematically, this relationship can be written:

$$\left. \begin{aligned} \% \text{ max proliferation rate} = \gamma(C - C_{\text{threshold}}), \\ C_{\text{threshold}} < C < C_{\text{maximal}} \end{aligned} \right\} \quad (6\text{--}5)$$

This formulation allows for the possibility of a threshold number of complexes, $C_{\text{threshold}}$, needed to stimulate mitogenic response above a background level. Also, it formalizes the number of complexes at which the maximum proliferation rate is attained: C_{maximal}. We have written Eqn. (6–5) in terms of a generalized complex number, C, for applicability to further work (see next section) in which the total number of cell complexes is not used as the governing quantity.

Notice from Figure 6–7 that the maximal proliferation rate occurs when only about 25% of the initial cell surface receptors are occupied at steady-state. However, this does not mean that there are "spare", or behaviorally unnecessary, receptors. Rather, endocytosis and degradation of receptor/ligand complexes leads to downregulation of EGF surface receptors, so that all remaining surface receptors are occupied at the maximal response. This result is indicated by the vertical dashed line in Figure 6–7, which represents the total number of surface receptors remaining, after downregulation, at the ligand concentration giving maximal mitogenic response.

Linear dependence of mitogenic stimulation on steady-state receptor occupancy is perhaps a surprising result, given the complicated signal transduction and response cascade following receptor binding. It suggests that continued occupancy of EGF receptors during the required 6–8 h commitment period is the overall rate-limiting step in DNA synthesis by fibroblasts, and that key intracellular messengers are generated during this period at a rate proportional to the number of receptor/growth factor complexes. This concept is consistent with the hypothesis that phosphorylation of certain intracellular proteins by the receptor tyrosine kinase is a crucial event leading to DNA synthesis. One simple attempt at accounting for the receptor occupancy dependence has appeared (Starbuck *et al.*, 1990), but it reproduces the observed DNA synthesis response only under very limited regimes of parameter values.

Most importantly, the observation of a direct relationship – whether precisely linear or not – between mitogenic response and the number of growth factor/receptor complexes means that understanding of the effects of growth factor and receptor properties on cell growth can be approached by analyzing their influence on the number of signaling complexes, C_s, as well as on the intrinsic signal generation, γ. Of immediate interest, this result forms a cornerstone for the model to be discussed in the next section for the influence of EGF-R trafficking on fibroblast proliferation responses to EGF. In the following section, it is used in a model analyzing the effect of initial cell density on the proliferation rate of cells stimulated by an autocrine growth factor.

A further implication of the direct relationship between growth factor/receptor complex levels and proliferation rate is that effects of oncogene products and other components involved in the mitogenic response may be broken down into two major classifications: (1) those explicitly related to intrinsic signal generation – the value of γ, the slope of the curve in Figure 6–7, and (2) those related to the number of complexes, C_s, the location on the abscissa of Figure 6–7. Such an approach may help investigators elucidate the mechanism for which growth regulation is altered by these components.

6.1.3.b Effect of Growth Factor/Receptor Trafficking Dynamics

Despite its seminal significance, a major shortcoming of the Knauer *et al.* work is that it does not discriminate between alternatives for the spatial compartmentation of actively-signaling EGF/EGF-R complexes. A linear relationship (albeit with different slopes, and corresponding values of the intrinsic signal generation, γ) would be obtained for these data whether the number of surface complexes C_s, the number of intracellular complexes C_i, or the total number of complexes $C_T = (C_s + C_i)$ is used as the abscissa of Figure 6–7. Some investigators (*e.g.*, Murthy *et al.*, 1986; Lauffenburger *et al.*, 1987) have suggested that intracellular EGF/EGF-R complexes could be responsible for generation of the mitogenic signal. Indeed, the majority of EGF/EGF-R complexes can be found inside the cell; Wiley and Cunningham (1981) report a steady-state ratio of $C_i/C_s = 7$ for human fibroblasts in the presence of EGF. Note also from Figure 6–7 that only about 25% of the receptors remain on the cell surface. There is some evidence that intracellular EGF/EGF-R complexes may retain active signaling capability (*e.g.*, Sorkin and Carpenter, 1991).

Another way to pose this issue is to ask whether there is an effect of receptor/ligand trafficking properties on a cell proliferation response, exclusive of any alteration of the intrinsic signal generation. Work by Wells *et al.* (1990) indicates that the effects of trafficking may, in fact, be extremely significant. These investigators transfected various EGF-R forms, including a number of mutants exhibiting altered trafficking properties, into NR6 cells (an NIH 3T3 line devoid of endogenous EGF-R). Figure 6–8 gives their data on proliferation of NR6 cells transfected with wild-type (WT) EGF-R or $\Delta973$ EGF-R, a mutant truncated in the cytoplasmic domain after amino acid residue 973 (see Figure 6–1). These two receptors forms possess apparently equivalent tyrosine kinase activity but the $\Delta973$ EGF-R is not internalized efficiently compared to the WT EGF-R. The growth curves in Figure 6–8 clearly show that the $\Delta973$ receptor shifts the dose–response curve of NR6 cell proliferation to EGF concentrations lower by approximately an order of magnitude than for the WT receptor. Morphological observations for the growth of $\Delta973$-transfectants show that these cells actually exhibit a transformed phenotype even at low EGF concentrations. NR6 cells transfected with the $\Delta973$ EGF-R mutant became refractile and overgrew the monolayer, whereas cells expressing WT receptors did not form the dense foci characteristic of transformed cells. These data suggest that this non-internalizing, kinase active EGF receptor not only transmits the growth signal from the cell surface, but it may do so unremittingly. Thus, Wells *et al.* offered the hypothesis that the mitogenic signal is generated by EGF/EGF-R complexes on the cell surface, and

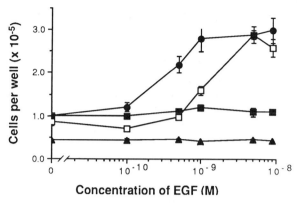

Figure 6–8 Experimental proliferation data from Wells *et al.* (1990) for NR6 cells transfected with various EGF receptor forms, as a function of EGF concentration. Ordinate is cell number after 9 days in culture. Open squares denote WT EGF-R, and closed circles represent truncated mutant Δ973 EGF-R. Closed triangles (non-functional frameshift transfectants) and filled squares (EGF-R mutant forms truncated at residue 1022 and lacking tyrosine kinase activity) show no mitogenic response to EGF. Copyright 1990 by the AAAS.

that this signal is attenuated by receptor downregulation and growth factor depletion by endocytic internalization and subsequent intracellular degradation.

In order to examine this hypothesis quantitatively, Starbuck and Lauffenburger (1992) applied an extension of the Knauer *et al.* model to analysis of the Wells *et al.* data. There are two key pieces to the extended model: (1) a more detailed mathematical model for EGF/EGF-R trafficking and (2) a linear expression for the dependence of proliferation rate on the number of signal-generating complexes similar to Eqn. (6–5) (though with different numerical values characterizing the curve). Along with these features, the crucial assumption representing the central hypothesis is that only cell surface EGF/EGF-R complexes, C_s, generate an effective mitogenic signal. That is, the expression based on Eqn. (6–5) relates C_s, instead of C_T, to proliferation rate in the Starbuck and Lauffenburger model. We will go through the development of these key model aspects in order.

First, the *trafficking model* is illustrated in Figure 6–9. As described in section 3.2.2 and shown schematically in Figure 3–16, both constitutive and induced (coated pit) EGF-R internalization are included, as is recycling of EGF/EGF-R complexes. These two features are the essential differences from the Wiley and Cunningham model used by Knauer *et al.* The model equations corresponding to Figure 6–8 are listed in Table 6–1. Values for all the parameters in these equations were determined from independent experiments on B82 cells transfected with WT and Δ973 EGF-R (see Starbuck and Lauffenburger (1992) for a summary of these procedures, with relevant data found in Lund *et al.* (1990), Starbuck *et al.* (1990), and Wiley *et al.* (1991). These values are given in Table 6–2. The only difference in these values comparing the two EGF-R types is the coated-pit affinity constant, K_{cp}. K_{cp} is essentially equal to 0 for Δ973 EGF-R, indicating that this mutant receptor exhibits no significant affinity for coupling with coated pit adaptors. It is thus incapable of using the induced endocytic internal-

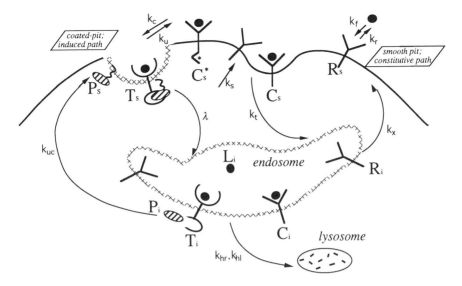

Figure 6–9 Schematic illustration of trafficking model used by Starbuck and Lauffenburger (1992), including receptor coupling with coated-pit adaptor molecules, dual pathway (coated-pit and smooth-pit) internalization, and recycling. Model equations are given in Table 6–1. Key species are surface receptor/ligand complexes, C_s, and ternary complexes (in coated pits), T_s. Details are given in section 3.2.2.a. Signaling complexes, denoted C_s^*, are considered identical to surface complexes, C_s, in this work. Reprinted with permission from Starbuck and Lauffenburger (1992). Copyright 1992 American Chemical Society.

ization pathway, so is internalized only via the constitutive pathway (Chen *et al.*, 1989; Lund *et al.*, 1990). Consequently, as discussed in section 3.2.2.a, Δ973 EGF-R has a much lower effective internalization rate constant, k_e, than does WT EGF-R.

With these specified parameters this model was validated experimentally by demonstrating successful quantitative prediction of differences in EGF-R downregulation and EGF degradation for B82 cells transfected with WT and mutant EGF-R (Starbuck and Lauffenburger, 1992). This comparison was shown in Figures 3–18a and 3–18b for receptor downregulation and ligand depletion, respectively. Cells possessing a non-internalizing receptor exhibit almost no significant EGF-R downregulation or EGF degradation compared to cells possessing the normal receptor. This result is due to the inefficient internalization of the truncated receptor, as quantitatively characterized by its coated-pit coupling affinity constant being essentially equal to zero and the substantially diminished value of k_e.

Second, the *proliferation/surface complex number relationship* (the analog to Eqn. (6–5)) must be determined for the NR6 cells of the Wells *et al.* study. In order to accomplish this task, theoretical profiles for the level of active surface complexes, C_s, as a function of EGF concentration are computed from the trafficking model. Parameter values obtained from the B82 cells are applied to the NR6 cells, because measurements have demonstrated quantitatively similar binding and trafficking properties between B82 cells and the NR6 parent NIH 3T3 line (Chen *et al.*, 1989). These computations, for cells possessing either WT or

Table 6–1 Model equations for EGF binding and trafficking. Notation used in these equations corresponds to Figure 6–9, with the exception of N_{Av}, which is Avogadro's number ($\#/$mole). Units are: R_s, C_s, T_s, P_s, R_i, C_i, P_i, L_d in $\#/$cell; L_o, L_i in M.

Surface species:

$$\frac{dR_s}{dt} = -k_f R_s L_0 + k_r C_s - k_t R_s + k_x R_i + k_s$$

$$\frac{dC_s}{dt} = k_f R_s L_0 - k_r C_s - k_c C_s P_s + k_u T_s - k_t C_s + k_x C_i$$

$$\frac{dT_s}{dt} = k_c C_s P_s - k_u T_s - \lambda T_s$$

$$\frac{dP_s}{dt} = -k_c C_s P_s + k_u T_s + k_{uc} P_i$$

$$\left(\frac{N_{Av}}{\rho}\right)\frac{dL_0}{dt} = -k_f R_s L_0 + k_r C_s$$

Intracellular species:

$$\frac{dR_i}{dt} = -k_f R_i L_i + k_r' C_i + k_t R_s - k_{hr} R_i - k_x R_i$$

$$\frac{dC_i}{dt} = k_f R_i L_i - k_r' C_i + \lambda T_s + k_t C_s - k_{hr} C_i - k_x C_i$$

$$\frac{dP_i}{dt} = \lambda T_s - k_{uc} P_i$$

$$(V_e N_e N_{Av})\frac{dL_i}{dt} = -k_f R_i L_i + k_r' C_i - k_{hl}(V_r N_e N_{Av})L_i$$

Degraded species:

$$\frac{dL_d}{dt} = k_{hr} C_i + (V_e N_e N_{Av})k_{hl} L_i$$

Δ973 EGF-R are shown in Figure 6–10. The values plotted are those obtained at a quasi-steady state, which is achieved after approximately 6 h for the cell density used, 10^6 cell/ml in culture dishes. Because incubation in EGF for about 6–8 h leads to a full cell mitogenic response characteristic of the EGF concentration used (Aharanov et al., 1978), this choice should reasonably correspond to the concentration governing typically observed proliferation data. The results in Figure 6–10 predict that cells transfected with Δ973 EGF-R reach a much higher level of surface signaling complexes for any concentration of EGF added than do cells transfected with WT EGF-R.

Next, these calculated C_s levels are translated into corresponding computations for cell proliferation rate. A *relationship between % maximum proliferation rate and C_s* was developed for NR6 cells using the WT EGF-R growth curve

Table 6–2 Model parameter values determined from binding and trafficking experiments for B82 cells transfected with WT EGF-R or Δ973 EGF-R. The Δ973 truncation mutant lacks significant affinity for coupling in coated pits, leading to a diminished internalization rate. A discussion of parameter determination is provided in Starbuck and Lauffenburger (1992).

Model parameter	Definition	B82 cell value WT EGF-R	Δ973 EGF-R
Surface binding parameter			
k_f	forward rate constant, ligand binding	$7.2 \times 10^7 \text{ M}^{-1} \text{ min}^{-1}$	
k_r	reverse rate constant, ligand binding	0.34 min^{-1}	
K_{DC}	equilibrium dissociation constant, C complex (k_r/k_f)	$4.7 \times 10^{-9} \text{ M}$	
k_c	coated-pit coupling rate constant	$2.0 \times 10^{-5} \text{ #}^{-1} \text{ min}^{-1}$	0
k_u	coated-pit uncoupling rate constant	0.10 min^{-1}	
K_{DT}	equilibrium dissociation constant, T complex (k_u/k_c)	$5.0 \times 10^3 \text{ #}$	
K_{CP}	coated pit affinity constant $[k_c/(\lambda + k_u)]$	$1.8 \times 10^{-5} \text{ #}^{-1}$	(0)
R_T	total number receptors per cell	$1.8 \times 10^5 \text{ #/cell}$	
P_T	total number of coated pit proteins per cell	$8.1 \times 10^4 \text{ #/cell}$	
P_T/R_T	ratio of # pit proteins to # receptors	0.45	
Internalization/intracellular trafficking parameter			
λ	coated-pit internalization rate constant	1.0 min^{-1}	
k_{uc}	coated-pit uncoating rate constant	(instantaneous)	
k_t	constitutive internalization rate constant	0.03 min^{-1}	
k_s	receptor synthesis rate	$130 \text{ #/cell min}^{-1}$	
k_r'	endosomal binary complex dissociation rate constant	$10 * k_r$	
k_u'	endosomal ternary complex dissociation rate constant	(instantaneous)	
k_x	recycling rate constant	$5.8 \times 10^{-2} \text{ min}^{-1}$	
k_{hr}	receptor degradation rate constant	$2.2 \times 10^{-3} \text{ min}^{-1}$	
k_{hl}	ligand degradation rate constant	$1.0 \times 10^{-2} \text{ min}^{-1}$	
$V_e N_e$	total endosomal volume	10^{-11} l/cell	
ρ	cell density	$1.0 \times 10^9 \text{ cell/l}$	

in Figure 6–8, as shown in Figure 6–11. Mitogenic response is represented by an average DNA synthesis rate. Best-fit values were obtained for the parameters in Eqn. (6–5), with C replaced by C_s. $C_{sthreshold}$ and $C_{smaximal}$ characterize complex numbers providing background and maximal proliferation rate, respectively. $C_{sthreshold}$ corresponds here to roughly 8% maximum proliferation rate instead of zero, in order to provide comparison to the data of Wells *et al.* (who used a background of 2% serum). In comparison to the values found by Knauer *et al.* for HF cells ($C_{sthreshold} = 290$, $C_{smaximal} = 3200$) higher values were obtained for the NR6 cells ($C_{sthreshold} = 600$, $C_{smaximal} = 6500$), indicative of lower intrinsic signal generation, γ, for NR6 cells relative to HF cells. Such variation among different cell types is to be expected.

Having now established the quantitative relationship between % maximum proliferation rate and C_s for NR6 cells, *proliferation predictions* can be made

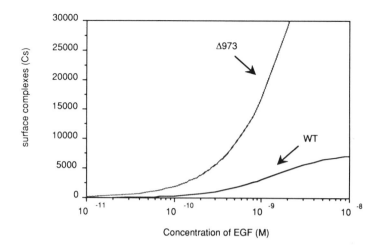

Figure 6–10 Trafficking model predictions of steady-state surface receptor/ligand complexes, C_s, for WT and Δ973 EGF-R, calculated using the equations in Table 6–1 with the parameter values in Table 6–2. The only parameter value difference between WT and Δ973 receptors is the affinity for coupling with coated-pits: $K_{cp} = 0$ for Δ973 EGF-R, $K_{cp} = 1.8 \times 10^{-5}$ for WT EGF-R. This difference causes a significant decrease in the internalization rate constant for the Δ973 receptor compared to that for the WT receptor. Reprinted with permission from Starbuck and Lauffenburger (1992). Copyright 1992 American Chemical Society.

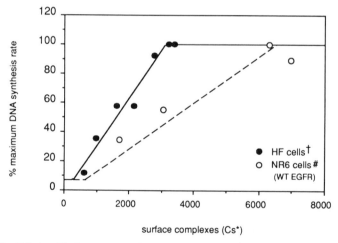

Figure 6–11 Relationship between steady-state surface complexes, C_s (assumed identical to the signaling complexes, C_s^*) and mitogenic response to EGF, for NR6 cells transfected with WT EGF-R (open circles, data from Wells *et al.*, 1990) and human fibroblasts (closed circles, data from Knauer *et al.*, 1984). The NR6 relationship was determined by plotting the cell growth rate data from Figure 6–8 versus the number of surface complexes calculated from the trafficking model (Figure 6–10) for the WT EGF-R transfectants over the range of EGF concentrations. The non-zero background proliferation corresponds to proliferation in 2% serum used as in the medium along with the specified EGF concentration. Reprinted with permission from Starbuck and Lauffenburger (1992). Copyright 1992 American Chemical Society.

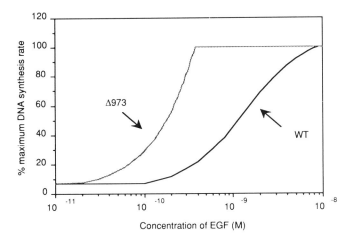

Figure 6–12   Theoretically calculated mitogenic response curves for NR6 cells transfected with WT or Δ973 EGF-R. These curves were constructed by using the values calculated for surface complex numbers at specified EGF concentrations (Figure 6–10) along with the dependence of DNA synthesis rate on surface complex numbers for NR6 cells shown in Figure 6–11. These curves can be compared to the experimental data of Wells *et al.* (1990) given in Figure 6–8. The comparison for WT EGF-R confirms the relationship determined in Figure 6–11, but the comparison for Δ973 EGF-R indicates a successful *a priori* prediction (since no additional parameter values needed to be fitted for construction of this curve). The reduced internalization rate of the Δ973 EGF receptor leads to a more sensitive proliferation response to EGF, due to diminished receptor downregulation and ligand depletion effects which would serve to attenuate the mitogenic signal. Reprinted with permission from Starbuck and Lauffenburger (1992). Copyright 1992 American Chemical Society.

from the trafficking model. Figure 6–12 shows the predictions for the Δ973 EGF-R, in comparison to the WT EGF-R, transfected into NR6 cells. This plot is obtained by using Figure 6–11 to convert Figure 6–10 into a cell growth plot (using an average % maximum DNA synthesis rate). Clearly, the Starbuck and Lauffenburger model quantitatively accounts for the WT EGF-R data of Wells *et al.* shown in Figure 6–8. More importantly, it is successful in *a priori* prediction of the Δ973 EGF-R data in that same figure; cells transfected with this non-internalizing, kinase-active receptor mutant require roughly 10-fold less EGF for half-maximal mitogenic response. This successful prediction for Δ973-transfectants arises by changing the value of only a single rate constant in the trafficking model: the affinity constant, K_{CP}, for coated-pit aggregation of Δ973 EGF-R compared to WT EGF-R. This provides quantitative support for the Wells *et al.* hypothesis, and for the concept that growth factor receptor trafficking may significantly influence a mitogenic response.

Note that it was assumed that the dependence of mitogenic response on surface complex number depends here only on the cell type but not on the EGF-R type. This is reasonable because Wells *et al.* found that the receptor tyrosine-kinase activity is essentially unchanged between the two EGF-R types considered here. However, the signaling activity (and hence the slope or thresholds of this relationship) might conceivably be altered by other means.

A mathematical model shown capable of analyzing effects of EGF/EGF-R binding and trafficking properties on fibroblast proliferation holds promise for predicting *effects of other model parameters* to identify their corresponding influence on cell growth. Along these lines, computations for the influence on proliferation rate of EGF/EGF-R binding, recycling, and degradation rate constants, EGF-R synthesis rate, and EGF-R tyrosine kinase activity have been generated (see Starbuck and Lauffenburger, 1992). Alterations in this last parameter, receptor kinase activity, influences DNA synthesis rate through changes in the slope of the linear curve of the Knauer *et al.* relation, Eqn. (6–5). The other parameters were assumed to affect proliferation rate by changing the number of EGF/EGF-R complexes for a given EGF concentration in a manner calculated from the trafficking model. These predictions are given in their paper, and should motivate further experimental tests using mutant EGF receptors with lesions in recycling or kinase activity, for instance (Chen *et al.*, 1989), and variant forms of EGF itself (Campion *et al.*, 1990).

As an example, an especially intriguing issue is the comparatively more potent effect of TGFα versus EGF on cell responses despite their similar affinities for binding to the same receptor. Ebner and Derynck (1991) have proposed that TGFα has a greater proclivity for recycling than does EGF, so that it causes less receptor downregulation and ligand depletion. This sort of question is ideally suited for the modeling approach just described.

6.1.3.c Diffusion Effects in Autocrine Systems

In the models discussed so far, the growth factor in the extracellular medium was provided exogenously at a specified concentration, with dynamics of change in concentration with time caused by binding, uptake, and degradation accounted for in the model computations. An interesting extension of modeling work on growth factor control of cell proliferation comes from considering novel implications due to *autocrine growth factors*. Recall that autocrine factors are those produced by the same cells that they bind to and act upon. Here, the number of receptor/ligand complexes is governed by combined influences of ligand secretion, binding to cell receptors, and diffusion away from the cell surface. If cell proliferation is regulated by cell surface receptor/growth factor complexes, as described above, then growth rate may be affected by growth factor diffusion properties in addition to the binding and trafficking behavior we have examined in great detail.

One feature of autocrine growth factor systems is a dependence of cell growth rate on the initial cell density, as originally observed in early days of cell culture (*e.g.*, Rein and Rubin, 1968): low cell density cultures exhibit slower specific growth rates than do higher density cultures. More recently, this "*inoculum density*" *phenomenon* appears commonly in hybridoma biotechnology for monoclonal antibody production. This dependence makes intuitive sense because the growth factor concentration will increase when a greater cell density is present. However, quantitative treatment is needed to determine what fraction of secreted ligand will, in fact, bind to cell receptors in competition with diffusive loss to the bulk medium.

In order to analyze this situation, Lauffenburger and Cozens (1989) developed a model incorporating ligand diffusion considerations (as presented in section 4.2) along with the cell proliferation framework of Knauer *et al.* (1984) discussed earlier in this chapter. These investigators recognized that cell growth rate could be predicted from a Knauer-type expression, Eqn. (6–5), for dependence of proliferation rate on the number of receptor/growth factor complexes, as shown in Figure 6–7. Hence, all that would be needed for prediction of cell density effects on growth rate are two relationships: (1) an expression for the dependence of local growth factor concentration on cell density, and (2) an expression for the number of complexes as a function of growth factor concentration. The second of these is accomplished using the Knauer *et al.* trafficking model (Eqns. (6–3) and (6–4), whereas the first requires treatment of diffusion and binding of a secreted ligand following the concepts of section 4.2.

For the case of cells growing while attached to a substratum (see Figure 6–13), Lauffenburger and Cozens derived an equation governing the *local concentration,* L_c, of *autocrine growth factor* available for binding to cell surface receptors:

$$L_c = Q \left\{ \frac{(1 + \rho_{cell})\left(1 + \dfrac{\eta}{\beta}\right)}{1 + \alpha\rho_{cell}(1 + \rho_{cell})\left(1 + \dfrac{\eta}{\beta}\right)} \right\} \tag{6-6}$$

The two central parameters in this expression are Q and η. Q is a dimensionless ligand secretion rate, equal to $[(q/V_c)/(3D_L/a^2)]$ where q is the secretion rate (moles/cell-time), V_c is a ligand sampling volume in the local cell proximity (see section 4.1.1), D_L is the ligand diffusion coefficient, and a is the cell radius. η is the fractional surface area coverage of the substratum by cells; this is

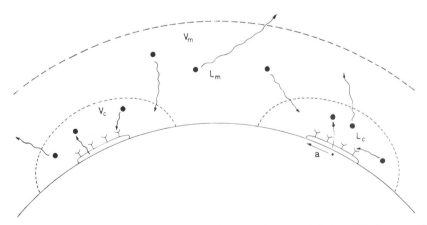

Figure 6–13 Schematic illustration of cells secreting autocrine growth factors. Kinetic competition exists between growth factor binding to cell receptors versus diffusive loss into the medium. L_c characterizes the ligand concentration with a local cell "sampling volume" V_c available to the cell for binding, while L_m represents ligand concentration in a microenvironment of volume V_m near the growth substratum from which ligand can diffuse to other cells before being lost to the bulk medium. Eqn. (6–6) of the text relates the local ligand concentration to the secretion rate along with binding and diffusion parameters. Reprinted from Lauffenburger and Cozens (1989) by permission of John Wiley & Sons, Inc. Copyright © 1989.

proportional to the cell density. ρ_{cell} is the dimensionless ligand binding rate constant, relative to diffusion, as defined in Eqn. (4–40) ($\rho_{cell} = k_{on}R/4\pi a D_L$). α is the ratio of cell volume to V_c, and β is a geometric factor characterizing the ratio of cell to substratum dimension. Details of the derivation can be found in the original reference. We point out that the expression reported there is a limiting case of Eqn. (6–6) using the simplifying assumption that $\alpha\rho_{cell} \ll 1$; i.e., cell surface binding is not diffusion-limited (see section 4.2.4).

The central finding implied by Eqn. (6–6) is that the local growth factor concentration, L_c, is linearly proportional to the cell density under conditions $\alpha\rho_{cell} \ll 1$. (It should be noted that the geometric factor β changes with substratum size so that cell growth does not scale simply with fractional cell surface coverage as system dimensions are altered (Lauffenburger and Cozens, 1989).) When the corresponding number of cell growth factor/receptors are calculated using the Knauer *et al.* trafficking model (Eqns. (6–3) and (6–4)), and the result is incorporated into the relationship between mitogenic response and number of complexes (Eqn. (6–5)), a prediction of the effect of cell density on proliferation rate can be obtained.

Qualitatively, the Lauffenburger and Cozens results show features consistent with experimental inoculum density observations reported in the literature. Most importantly, the cell proliferation rate increases for greater cell densities. However, quantitative comparisons are not presently available because values for one of the most crucial model parameters – the ligand secretion rate q – are difficult to obtain. Attempts to determine q by measuring the rate at which a secreted ligand appears in extracellular culture medium are compromised by the complicating interference of binding to cell surface receptors. Approaches overcoming this problem, for instance the use of receptor-blocking antibodies may be helpful for future investigations.

6.2   CELL ADHESION

6.2.1   Background

In the previous chapters, we have primarily discussed the binding of cell surface receptors to ligands free in solution. Cell surface receptors can also bind to ligands located on another cell or on a surface. In these cases, specific (*i.e.*, receptor-mediated) cell–cell adhesion or cell–substrate adhesion may result, as shown schematically in Figure 6–14. The ability of cells to recognize and adhere specifically to other cells or to insoluble extracellular tissue matrix proteins is critical to many physiological processes. Cells flowing through the body's circulatory systems (blood and lymph) adhere to vessel endothelia in particular organs. Prominent examples here are homing of lymphocytes to specific lymphoid organs, margination of leukocytes at inflammatory regions, platelet deposition at sites of injury, and metastatic spread of tumor cells. T helper lymphocytes must reversibly adhere to antigen presenting cells, and cytotoxic T lymphocytes must reversibly adhere to their target cells. Cell adhesion is also critical to embryonic development, maintenance of proper tissue growth and differentiation states, and

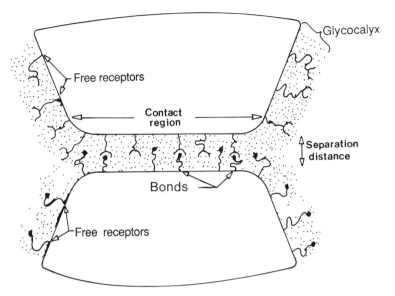

Figure 6–14 Cell–cell and cell–substrate adhesion. An adherent cell can be quantitatively characterized by the number of bonds or bridges linking it with another cell or surface, the separation distance s, the contact area A_c, and the number of free receptors within and outside of the contact area. Adapted from Bell *et al.* (1984). Reproduced from the *Biophysical Journal*, 1984, Vol. 45, pp. 1051–1064 by copyright permission of the Biophysical Society.

generation of traction for migration and/or tissue organization. Excellent reviews of these physiological processes are available (*e.g.*, Trinkaus, 1984; McClay and Ettensohn, 1987; Springer, 1990; Singer, 1992). Finally, cell/surface adhesion can be exploited for biotechnological purposes, such as affinity-based separation of cell populations (Sharma and Mahendroo, 1980; Berenson *et al.*, 1986) and anchorage-dependent cell culture (Farmer and Dike, 1989).

We will focus in this section on the relationship between receptor/ligand binding and cell–surface or cell–cell adhesion. In particular, we want to develop models to quantitate the various physical interactions underlying adhesion so that we can ask under what conditions cell adhesion will and will not occur. Models will be particularly useful for analyzing phenomena in which a cell modulates its interactions with other cells and surfaces by changing the number and/or affinity of its receptors, for models can indicate the quantitative alterations in adhesiveness that are a consequence of these changes. Such understanding may guide attempts to alter cell adhesion for biotechnological or medical purposes. In addition, receptors acting to promote cell adhesion, *i.e.*, adhesion receptors, may play a signaling role as well. In these cases, we are interested in models to predict the number of bound receptors and hence the initial "signal" that the adhering cell receives.

6.2.1.a Adhesion Receptors

Over the last decade, tremendous efforts have been made to identify the receptors involved in cell–substratum and cell–cell adhesion. At least four major superfamilies

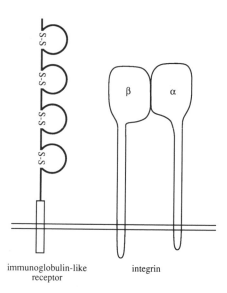

immunoglobulin-like   integrin
receptor

Figure 6–15 Basic structure of adhesion receptors of the integrin and immunoglobulin families. Integrins are heterodimers of α and β subunits. Immunoglobulins are monomers (shown) or dimers containing one or more loops closed with a disulfide bond.

of structurally related cell receptors have been found: the integrin, immunoglobulin, cadherin, and selectin superfamilies. Receptors belonging to the *integrin superfamily* are heterodimers of non-covalently associated α and β subunits (Figure 6–15) (Albelda and Buck, 1990; Hemler, 1990; Hynes, 1992). Some integrins are involved in binding extracellular matrix (ECM) proteins, including fibronectin, laminin, and collagen. Other integrins function as binding proteins for molecules on cell surfaces. For example, the T lymphocyte integrin commonly known as LFA-1 binds the ligands ICAM-1 and ICAM-2, which are members of the immuno-globulin family. The ligand specificity of an integrin is determined by the particular combination of α and β subunits, as several varieties of each exist. In several cases, integrins have been shown to bind to proteins containing the tripeptide sequence arginine–glycine–aspartate (RGD in single letter code).

Although the cytoplasmic domain of integrins is short (typically no more than 50 amino acids), it is apparently an important function of integrins to convey information into the cell interior. For instance, cell binding to ECM proteins strongly influences not only cell morphological and migration properties but also cell differentiation state. One mechanism for this information transmission is the reversible binding of the cytoplasmic domain of the β subunit to talin, a cytoskeletal component associated with the actin microfilaments involved in intracellular mechanical force generation (see Figure 6–16). The aggregation of many receptor/ligand bonds in association with talin and actin is known as a *focal adhesion* (Burridge *et al.*, 1988). Hence, integrins are transmembrane proteins which are known to interact simultaneously with both extracellular ligands and *cytoskeletal components*. Integrins may also be involved in other facts of signal transduction, including phosphorylation and cytoplasmic alkalinization,

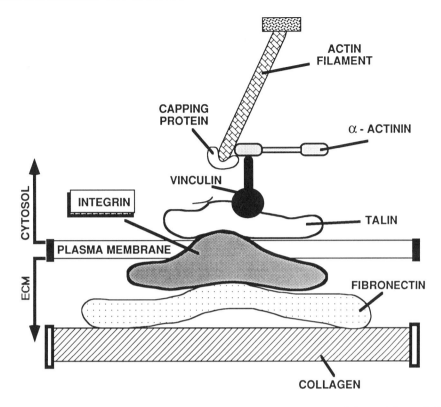

Figure 6–16 Interaction of a receptor of the integrin family with extracellular and intracellular components. Many integrins are known to bind components of the extracellular matrix (ECM) such as fibronectin. Integrins can also interact with talin, and, indirectly, actin. The precise details of linkage between actin and integrins are not completely understood but may occur as shown here.

although information on this is only now beginning to appear (Dustin, 1990; Hynes, 1992).

Individual bonds between integrins and their ECM ligands appear to have an unusually low ligand binding affinity, with K_D of the order of $10^{-6}-10^{-7}$ M. Whether this easy reversibility of binding has significant functional consequences is an open question, although recent models for cell motility predict that such a low affinity for binding to an adhesion mediator on movement substrata may be helpful in facilitating locomotion (Lauffenburger, 1989; DiMilla et al., 1991). Furthermore, the aggregation of many low affinity bonds in one area of the cell, as is found in a focal adhesion, may provide an extremely high avidity interaction.

Recent evidence suggests that the affinity of integrins for their ligands can be modulated, so that appropriately stimulated cells can increase the affinity of their integrin molecules for ligands. This *affinity modulation* may provide a way for a cell actively to control its adhesiveness to a substrate or to another cell on a relatively short time scale (Hynes, 1992). As one example, T-cells recognizing antigen via their T cell receptors are found to have increased avidity of their LFA-1 molecules for ICAM molecules on another cell (Dustin and Springer, 1989). This increased avidity may be responsible for a successful cell–cell coupling and resulting T cell activation.

The *immunoglobulin superfamily* consists of receptors with the characteristic structure shown schematically in Figure 6–15 (Hunkapiller and Hood, 1986; Springer, 1990). Immunoglobulins, the T cell receptor, and the class I and class II major histocompatibility complex (MHC) molecules, all of which are involved in immune cell interactions, belong to this superfamily. These members of the family possess variable regions as a result of somatic diversification. The family also includes the neural cell adhesion molecules (N-CAMs).

Several other cell adhesion molecules are members of the *cadherin superfamily* or the *selectin superfamily*. Cadherin binding is calcium-dependent and homophilic, *i.e.*, a receptor on one cell binds a like receptor on another cell just as if it were a complementary ligand (Takeichi, 1990). Cadherins are thought to play a role in development and in intercellular junctional structures. The newly identified selectin superfamily of adhesion receptors, also known as LEC-CAMS, have lectin-like N-terminal domains and bind to carbohydrates (Brandley *et al.*, 1990; Foxall *et al.*, 1992). Two identified members of this family are CD62 and ELAM-1, which are expressed on endothelial cells and play a role in the binding of these cells to neutrophils at sites of tissue injury and inflammation.

Although integrins have been implicated in active signaling to the cell interior, as mentioned above, in general it is not known to what extent adhesion receptors participate in such signaling. The many and varied effects of cell adhesion on growth, differentiation, and migration could be due to merely mechanical interactions, but it is more likely that active signaling on ligand binding, for example through covalent modification, is an important mechanism here as it is for other receptors.

6.2.1.b   Modeling Issues

In all of the previous text, our models for receptor-mediated processes have involved equations that described explicit relationships between receptors, ligands, and molecules that interact with receptors (G-proteins, coated pit proteins, etc.) or are generated as a result of receptors binding (IP_3, Ca^{2+}, etc.). In developing models for cell adhesion, we will find that additional physical interactions must be taken into account. Table 6–3 lists the different aspects of the cell adhesion problem.

We use the term *specific interactions* to refer to the involvement of receptor/ligand bonds in cell adhesion. In many cases, it is believed that adhesion would not occur without these interactions, and thus the expression of receptors on a cell and/or the modulation of receptor affinity or number with time serves to control the types of surface and cells with which it will interact.

Non-specific interactions are defined as interactions between a cell and surface or a cell and another cell that do not involve receptors, *i.e.* are not biochemically specific, but do act to increase or decrease the overall strength of the interaction. The three relevant types of non-specific forces for cell–cell and cell–surface adhesion are electrostatic forces, steric stabilization, and van der Waals forces. These will be discussed in more detail later.

Table 6–3 Aspects of cell adhesion

Specific interactions (receptor-mediated)
Non-specific interactions
   Electrostatic forces
   Steric stabilization
   van der Waals (electrodynamic) forces
Mechanics of cell deformation
Forces acting on a cell
Metabolic effects
   Focal adhesions
   Active movement (actin polymerizations)

Additional considerations in modeling cell adhesion include the *mechanics of cell deformation*. Cells adhering to other cells or to a surface are deformed by the interaction, and the equations of mechanics describe this deformation and the resulting forces that may act to allow or break receptor/ligand bonds. Any consideration of the mechanics of cell deformation will require an accurate description of the *forces acting on a cell*; these forces may result from receptor/ligand bonds pulling on the membrane, from fluid forces or centrifugal forces acting on the cell, or even from the force applied to a cell with a micropipette.

Furthermore, *metabolic effects*, such as the development of focal adhesions, may play a role in cell adhesion. The time course of focal adhesion formation is of the order of 10 min (Mueller *et al.*, 1989). Generally it is thought that such effects serve to strengthen the adhesion of an already adherent cell. In addition, motile cells may exhibit pseudopod protrusion and retraction driven by active mechanisms within the cell itself.

Two major classes of mathematical models for receptor-mediated cell adhesion have been developed. The first class, *equilibrium models*, is directed toward finding the equilibrium behavior of a system in which a cell possessing membrane adhesion receptors is located near another cell or surface bearing complementary ligands. This class of model has as its primary goal the prediction of whether adhesion will occur, the strength of adhesion, the interfacial contact area facilitating adhesion, and/or the amount of force necessary to disrupt the adhesion, as a function of receptor and ligand properties (number densities, binding equilibrium constant, mobilities) and cell mechanical properties (*e.g.*, viscosity, elasticity). Within this class two approaches have been used, representing thermodynamic and mechanical equilibria.

The second class of models, *kinetic models*, attempts to describe cell attachment to or detachment from another cell or surface as a function of time. The goals here are to predict the effects of receptor and ligand properties (now necessarily including association and dissociation rate constants) and cell mechanical properties on the probability of attachment during a given encounter of a cell with the surface or the time course of attachment or detachment in the presence of particular external force conditions (such as fluid shear flow). After a few preliminaries, we will deal with both equilibrium and kinetic classes of models in turn.

6.2.2 Experimental Methods

Cell adhesion has been studied experimentally with numerous assays (see Hubbe, 1981; Bongrand and Bell, 1984; Curtis and Lackie, 1991 for summaries). The simplest is the *sedimentation assay*. In this assay, a culture dish surface is treated to present a particular substrate for cell adhesion, a cell suspension is incubated in the culture dish for a given period of time, and then a gentle washing is used to remove cells not adhering to the dish surface. Data are typically presented as number of cells remaining attached after washing, as measured by either microscopic count or radioactivity for radiolabeled cells.

A slightly more complex cell adhesion assay is the *centrifugation* assay developed by McClay *et al.* (1981). In sealed microtitre wells, cells are brought into contact with a test surface using a low centrifugal force. The wells are then inverted in the centrifuge, and the force required to detach the cells is determined; alternatively, cells can be radiolabeled and the number of cells remaining attached as a function of the detachment force can be measured by radioactivity levels.

Another group of experimental assays exploits fluid mechanical forces to study cell attachment and/or detachment. These assays allow a much more rigorous estimation of the relevant physical forces than the sedimentation and centrifugation assays. Most commonly used is the *parallel-plate flow chamber*, in which a uniform level of fluid shear stress is generated down the length of a rectangular chamber. Set-ups which generate position-dependent shear stresses are the *rotating disk assay* and the *radial flow chamber*; the former has an increasing shear stress, and the latter a decreasing shear stress, with radial distance from the axis. The parallel-plate and radial flow chambers possess a special advantage in that individual cells can be observed microscopically during the course of attachment and detachment. Because of the uniform shear stress, the parallel-plate assay is particularly well-suited for attachment studies (Wattenbarger *et al.*, 1990). In contrast, the radial flow chamber is best suited for detachment studies, for a range of shear stresses can be examined in a single experiment (Cozens-Roberts *et al.*, 1990b).

A final type of adhesion assay is the *micromechanical assay*, in which a single cell held through suction pressure at the tip of a micropipette is brought into contact with an adhesive surface, possibly another cell. Detachment is accomplished by increasing the suction pressure on the cell. Adhesion energy is estimated from the curvature of the cell membrane at the edge of the contact zone, using viscoelasticity models for cell mechanical properties (*e.g.*, Dong *et al.*, 1988).

In the sedimentation, centrifugation, and radial flow chamber assays for cell adhesion, the time that adherent cells are allowed to incubate prior to attempts to remove them, or the *incubation time*, is an important variable. As the incubation time increases, *cell spreading* and flattening against the adhesive surface may occur. Spreading typically occurs on a time scale of minutes to hours, is temperature-dependent, and increases the adhesive force dramatically. Cell spreading is likely the result of the active process (requiring metabolism) of focal contact formation and the passive process of receptor diffusion and accumulation in the contact area, *i.e.*, the area of cell membrane in sufficiently close apposition to the surface containing the ligand to permit receptor/ligand bond formation (McClay *et al.*, 1981).

Finally, it should be noted that for these diverse experimental approaches, not only the magnitudes of the contact and distraction *forces* but also the qualitative nature of those forces are different. For instance, the distraction force in the fluid mechanical assays (parallel-plate flow chamber, rotating disk, and radial flow chamber assays) is primarily a shear stress containing a significant torque component, whereas the micropipette assay induces a distraction force which is essentially normal. Thus the results obtained in different assay systems are not necessarily comparable. Quantitative estimates of the forces likely in different assay systems can be found in Hubbe (1981) and Bongrand and Bell (1984).

6.2.3 Physical Aspects

6.2.3.a *Non-specific Interactions*

There are three major types of non-specific interaction important for cell adhesion processes: electrostatic forces, steric stabilization, and van der Waals or electro-dynamic forces (Bongrand and Bell, 1984; Israelachvili and McGuiggan, 1988). Although all are present at once, each is dominant at a different cell–cell or cell–substrate separation distance.

Electrostatic forces, which may be attractive or repulsive, result when two charged surfaces are brought together. The surface of a cell consists not only of an approximately 70 Å thick lipid bilayer containing receptors and other embedded molecules but also what has been termed the *glycocalyx*. The glycocalyx is made up of short oligosaccharide chains bound to glycoproteins and glycolipids and high-molecular-weight proteoglycans. Surface receptor binding sites appear within this coat. The glycocalyx is of the order of 100 Å thick and is negatively charged due to the numerous sialic acid residues present.

For cell–cell adhesion, then, the bringing together of two negatively charged surfaces results in an overall repulsive electrostatic force between the cells. For cell–surface adhesion, the electrostatic forces may be positive or negative, depending on the composition of the surface. For example, a polylysine-treated surface presents a positively charged surface that will attract most cells, so non-specific cell–surface adhesion may result regardless of the availability of receptors and complementary ligands.

The glycocalyx consists of polymers in a hydrated environment. As two polymer coats approach each other, the layers overlap and some of the water of hydration is pushed out. A repulsive force termed *steric stabilization* results because of the osmotic tendency of water to return and because of the steric compression of the polymer chains. This force dominates at small separation distances, and likely acts to prevent significant interpenetration of the glycocalyxes on two adhering cells.

The remaining non-specific forces to consider are the *van der Waals*, or *electrodynamic*, forces. These attractive forces result from the charge interactions of polarizable molecules, including molecules with no net charge, on the cell surface and in the solvent.

Bongrand and Bell (1984) calculate the magnitudes of the three non-specific forces for the case of cell–cell adhesion by using a simplified physical model for

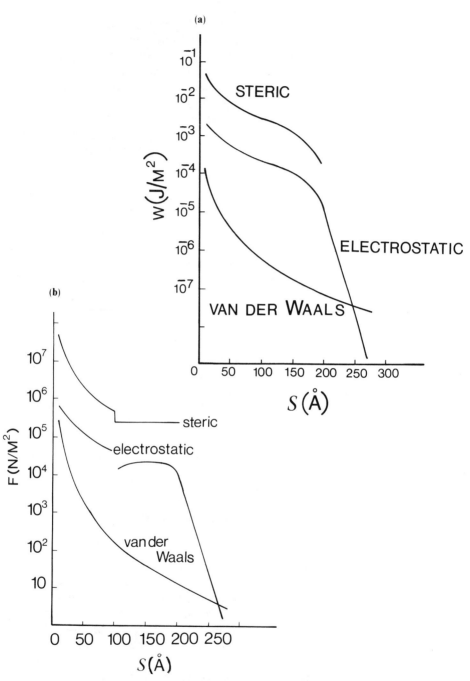

Figure 6–17 Non-specific forces in cell adhesion. Electrostatic, steric stabilization, and van der Waals forces are shown. (a) The interaction potential between two like cells is plotted as a function of the cell–cell separation distance. Recall that van der Waals interactions are attractive; electrostatic and steric stabilization interactions are repulsive. (b) The force per unit area required to separate two like cells is plotted as a function of the cell–cell separation distance. Reprinted from Bongrand and Bell (1984) by courtesy of Marcel Dekker, Inc.

both cell membranes and mathematical descriptions of the different forces. They obtained the results shown in Figure 6–17 for the interaction potential per unit area, W, between two like cells and the force per unit area, F/A, required to separate the two cells. In both figures, s is the distance between the lipid bilayers of the two cells, or the separation distance. The two plots are related by the relationship $F/A = -\partial W/\partial s$; in other words, the energy required to separate two adherent cells is obtained by integrating the force from the distance s to infinity.

Note from Figure 6–17a that at small separation distances, repulsive potentials dominate the cell–cell interaction; these diminish after cell–cell separation distances of the order of 200 Å (*i.e.*, much beyond glycocalyx interpenetration). Because cell–cell and cell–surface adhesion are observed to occur with separation distances of roughly 100–300 Å, these repulsive forces are important to consider in a treatment of cell adhesion.

As the separation distance increases, repulsive potentials fall off and the attractive van der Waals forces act to increase the likelihood of cell–cell adhesion. This is shown schematically in Figure 6–18. With respect to the typical situation under consideration here, that of cell–cell adhesion, the chief message of Figure 6–18 is that steric stabilization dominates once cell surfaces and their associated glycocalyxes begin to come into contact, at a distance of less than about 200 Å. Further, van der Waals attraction only very weakly overcomes electrostatic and steric repulsion at separation distances of the order of 200–250 Å. This corresponds to a small minimum in the diagram of potential energy versus separation distance. In classical colloidal systems, this is termed the secondary minimum

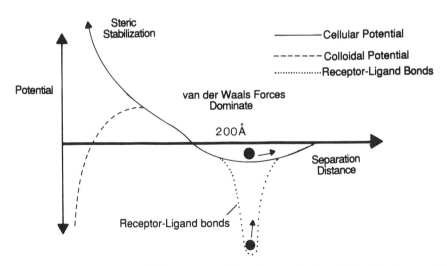

Figure 6–18 Interaction potential between two like cells as a function of the cell–cell separation distance. The net potential due to the combination of non-specific interactions (electrostatic, steric stabilization, and van der Waals) and specific (receptor-mediated) interactions is shown. Negative potential corresponds to a net attraction between the two cells. Note that at a separation distance of about 200 Å, receptor/ligand bonds strengthen the weak adhesion due to van der Waals forces.

and there is a strong primary minimum at smaller separation distance. In cell–cell systems, steric repulsion prevents the primary minimum.

We can obtain from Figure 6–17b the force per unit area necessary to uncouple cells that are separated by a distance s. At a typical separation distance for non-specifically adherent cells of 250 Å, approximately 10^2 N/m^2 (10^3 dyn/cm^2) applied force per unit area will easily detach the cells. Because physiological fluid shear forces present in venule circulation are estimated at 1–10 dyn/cm^2 (Lawrence *et al.*, 1990) and cell-generated contractive forces present in tissue are of the order of 10^3–10^5 dyn/cm^2 (Felder and Elson, 1990; Harris *et al.*, 1980; Kolodney and Wysolmerski, 1992), the motivation for a stronger – but still reversible – adhesive interaction between cells is evident. This is the role played by receptors, to be discussed shortly.

6.2.3.b Mechanics of Cell Deformation

A cell can act to resist adhesion by virtue of its own mechanical structure. A great deal of work has focused on the influence of cell mechanical properties on a spectrum of cellular phenomena over the past two decades (Elson, 1988). The approach is to develop quantitative *models for cell rheological behavior* (*e.g.*, erythrocytes (Evans, 1989), lipid vesicles (Evans and Needham, 1987), and leukocytes (Dong *et al.*, 1988; Yeung and Evans, 1989)).

When developing a model for cell mechanics, a choice must be made regarding the description of the complex and inhomogeneous structure of the cell. To do so, one generally conceptualizes the cell as a combination of *viscous* and *elastic elements*. Viscous elements are characterized by a resistance to flow or the rate of deformation. They deform irreversibly, because the displaced molecules do not "remember" their original location. Elastic elements, on the other hand, have memory and are characterized by a resistance to the extent of deformation. When the shape of an elastic element is changed, the energy of deformation is stored and can act to reform the original shape of the element. Thus elastic elements deform reversibly. Purely viscous behavior is typical of liquids; purely elastic behavior is typical of solids. Other materials, such as eukaryotic cells, may exhibit a combination of both behaviors.

Even the cell membrane can exhibit both viscous and elastic characteristics. *Three independent modes of deformation* can take place in a thin membrane: area dilation, shear at constant area, and bending (Figure 6–19). For a membrane with some elasticity, the three material properties describing these modes are the elastic area compressibility modulus K, the elastic shear modulus μ, and the bending elastic modulus B. These parameters are defined by the relationships:

$$\tau = K\alpha \tag{6–7a}$$

$$\tau_s = \tfrac{1}{2}\mu\left(\left[\frac{\lambda_1}{\lambda_2}\right]^2 - \left[\frac{\lambda_2}{\lambda_1}\right]^2\right) \tag{6–7b}$$

$$M = B\Delta\left(\frac{1}{R_1} + \frac{1}{R_2}\right) \tag{6–7c}$$

where τ is the isotropic tension (dyn/cm), τ_s is the surface shear resultant

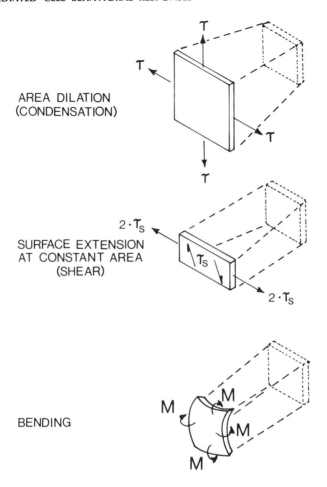

Figure 6-19 Three independent modes of membrane deformation. In each case, the initial shape of the membrane element is a flat square. Area dilation is due to an isotropic tension τ acting to increase the membrane area. Shear at constant area is due to a surface shear resultant τ_s acting to change the length or width of the membrane. Bending is due to a bending moment M acting to bend the membrane without changing its length or width. Reprinted with permission from Evans (1988). Copyright CRC Press, Inc., Boca Raton, FL.

(dyn/cm), and M is the bending moment (dyn). The ratio of the instantaneous dimension of a deforming membrane element to its reference state is given by λ_1 and λ_2 for the two sides of an element. The dimensionless parameter α is the fractional change in area given by $\lambda_1\lambda_2 - 1$. R_1 and R_2 are the radii of curvature of the deforming element, and Δ indicates that the change in the quantity $(1/R_1 + 1/R_2)$ should be evaluated. Qualitatively, small values of K, μ, and B mean that small forces will induce in a membrane element large area changes, significant extension at constant area, and significant bending, respectively, as compared with large values of these parameters. For a membrane with some viscous properties as well, the corresponding material properties that characterize resistance to flow are the surface dilational, shear, and bending viscosities.

It may be helpful to consider the quantitative implications of these equations together with estimates for the elasticity parameters in lipid bilayers and red blood cells (Evans, 1988). Because of the relatively large energy needed to expose hydrophobic portions of a membrane lipid bilayer to aqueous media, the area compressibility modulus K is large, of the order of 10^2 dyn/cm. A stress of 10^4 dyn/cm^2 acting only to increase membrane area would yield for a membrane/ cortex layer 10^{-5} cm in thickness the result $\alpha = 10^{-3}$, or only a 0.1% change in membrane area (10^4 dyn/cm$^2 \times 10^{-5}$ cm/10^2 dyn/cm). In contrast, the elastic shear modulus μ is very small. In red blood cells, which are known to have a substantial subsurface protein network, μ is of the order of 10^{-2} dyn/cm; for lipid bilayers, μ is equal to zero and thus the bilayer behaves as a true two-dimensional liquid. The same stress of 10^4 dyn/cm^2 acting only to shear the membrane of a red blood cell would cause the ratio λ_1/λ_2 to change from 1 to about 4.5 according to Eqn. (6–7b). Smaller values of μ would result in more dramatic changes in the shape of a membrane element. Finally, the bending elastic modulus B is also very small, of the order of 10^{-12} dyn-cm, permitting sharp bending very easily. Thus one may reasonably model the cell as having a constant area, no resistance to shear, and significant bending due to applied forces; in fact, in a later section (6.2.4.b) we examine an equilibrium model for adhesion which, for the reasons just stated, includes only the deformation of bending.

The viscous and elastic properties of the cell play an important role in adhesion. The extent of cell spreading, for example, is governed by a balance of mechanical forces: membrane tension induced by formation of energetically-favorable receptor/ligand bonds tends to promote spreading whereas membrane elasticity tends to oppose it. The non-specific cell/substrate repulsive forces discussed earlier will also act to oppose spreading. In addition, constraints on spreading are imposed by the requirements of constant total membrane surface area (the cell membrane is not believed to stretch appreciably, as shown by our calculation above) and constant cell volume.

There are two popular models for the mechanical structure of the cell, both of which have been shown to describe adequately some aspects of cell behavior. Evans (1988) and Yeung and Evans (1989) suggest that most nucleated cells can be modeled by assuming that the cytoplasmic core acts as a Newtonian viscous fluid and that the cell membrane and associated cortical cytoskeleton act as a viscoelastic solid. This model is shown schematically in Figure 6–20a; pistons represent viscous behavior and springs represent elastic behavior. A viscous fluid is described by the relationship:

$$\tau_s = \eta \nabla v \tag{6–8}$$

where τ_s is the shear stress, v is the vector of fluid flow rates, ∇v is the rate of fluid strain, and η is the shear viscosity (Newtonian fluids have a constant viscosity at constant temperature and pressure). The value measured for the cytoplasmic shear viscosity varies with the method of measurement (due to differences in the length scale of measurement compared to length scales of cytoskeletal components; see Elson (1988)), with values at 37 °C having been found in the range of about 1–10^3 poise (dyn-sec/cm^2) – that is, 10^2–10^5 times that of water (Evans and Yeung, 1989). In this model of cell rheology (Figure 6–20a), the equilibrium

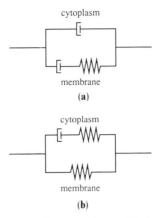

Figure 6–20 Schematic representation of two popular models for the mechanical structure of a cell. Pistons represent viscous behavior; springs represent elastic behavior. (a) The model of Evans (1988) and Yeung and Evans (1989) describes the cell as a viscous liquid, the cytoplasm, inside a viscoelastic solid, the cell membrane. (b) The model of Schmid-Schonbein *et al.* (1981) describes the cell as a viscoelastic solid, the cytoplasm, inside an elastic membrane.

shape and stress distribution for a cell are determined by the membrane/cortex elastic properties subject to a uniform internal pressure due to the liquid core. Transient shapes and stresses for cells far from equilibrium are significantly affected by the viscous properties of both the cytoplasm and the membrane/cortex.

A second and alternative rheological model for the description of the cell has been offered by other researchers (Schmid-Schonbein *et al.*, 1981). In their model, shown schematically in Figure 6–20b, the membrane/cortex is considered to be an ideal elastic element and the cytoplasm to be a viscoelastic Maxwell solid.

Each of these two models for cell rheology capture some, but not all, of the relevant physics of cell deformation. The second model (Figure 6–20b) does well in describing short time scale (∼seconds) behavior because the non-Newtonian behavior of the cytoplasm is included. For example, this model accurately describes the micropipette studies of Schmid-Schonbein *et al.* (1981). In addition, Elson (1988) has noted that the viscosity of the cytoplasm decreases as the shear rate increases, clearly non-Newtonian behavior. On the other hand, the second model predicts that under conditions of constant suction pressure within a micropipette, a cell should stop flowing into the pipette within seconds to minutes. The experimental data of Evans and Yeung (1989), however, show a continuous flow under these conditions. Furthermore, experimental recovery upon diminution of the suction pressure is much less than predicted by the second model. The model of Evans and Yeung more accurately describes this long time-scale behavior than the second model.

Other cell rheological models have been suggested (*e.g.*, Skalak *et al.*, 1990). Furthermore, more detailed models of the cell, including aspects of actin polymerization and cytoskeletal networks, have been developed by Dembo *et al.* (1984) and Nossal (1988), although these are not yet ready for use in this context.

Finally, most extensive cell spreading is generated only when active, metabolism-dependent intracellular contractile forces exist, as in motile cells; otherwise, the contact area remains a small fraction of the total cell surface area. This aspect must be included in order to achieve a satisfactory model for adhesion and spreading for such cell types.

6.2.3.c Receptor-Mediated Interactions

The non-specific forces described earlier provide only a weak attractive force, of the order of 10^3 dyn/cm^2 (10^{-5} dyn/μm^2) for typical cell–cell separation distances. In order to strengthen adhesive interactions as well as provide specificity, receptors must play a role. One can examine the *strength of a receptor/ligand bond* from both equilibrium and kinetic standpoints. From the *equilibrium* perspective, Bell (1978) estimates the strength of a single receptor/ligand bond from the relation $f_c = \Delta G/r_0$, where f_c is the force required to break the bond, ΔG is the free energy of bond formation, and r_0 is the range of the bond potential energy minimum. For $r_0 = 10$ Å and $\Delta G = 13$ kcal/mole (corresponding to an equilibrium dissociation constant $K_D = 10^{-9}$ M) we obtain $f_c = 9 \times 10^{-6}$ dyn/bond. A covalent bond with $\Delta G \sim 70$ kcal/mole and $r_0 \sim 1$ Å, would require $f_c \sim 4 \times 10^{-4}$ dyn/bond. Non-covalent receptor/ligand bonds with $K_D < 10^{-9}$ M (*i.e.*, higher affinity bonds) would fall somewhere in between, with a logarithmic dependence of f_c on K_D^{-1} (*cf.* Eqn. (2–3)).

A related approach to analyzing the strength of a receptor/ligand bond is offered by a *kinetic* point of view. This approach was introduced by Bell (1978) and is based on the kinetic theory of isotropic materials (Zhurkov, 1965). Considering the forward and reverse rate constants, k_f and k_r, for receptor/ligand association and dissociation, Bell proposed that the *dissociation rate constant is increased by a physical stress*:

$$k_r = k_{r0} \exp\left\{\frac{\gamma f}{K_B T}\right\} \tag{6–9}$$

where k_{r0} is the unstressed dissociation rate constant, f is the applied force stressing a bond, and γ is a parameter loosely defined as the bond interaction range and likely of the order of r_0. T is absolute temperature and K_B is Boltzmann's constant. Bell used Eqn. (6–9) to determine the force needed to detach a cell initially attached via multiple receptor/ligand bonds, by computing the force able to accelerate bond disruption to a sufficient extent that the system dynamically tends toward a state of complete dissociation. He obtained an approximate expression for the total detachment force divided by the initial number of bonds, or the *adhesion strength per bond*, F_{bond}:

$$F_{bond} \approx 0.7\left(\frac{k_B T}{\gamma}\right) \ln\left\{\frac{n_s}{K_D}\right\} \tag{6–10}$$

where n_s is the surface ligand density and K_D is the surface equilibrium dissociation constant. For $\gamma = 10$ Å, $n_s = 10^{11}$ #/cm^2, and $K_D = 10^5$ #/cm^2 (corresponding to a solution value of 10^{-9} M, using an effective volume with height of 200 Å) Eqn. (6–10) yields $F_{bond} = 4 \times 10^{-6}$ dyn. Note that the kinetic approach thus gives a similar but lower estimate for the bond strength than does the equilibrium approach, because it does not require all bonds to break simultaneously. An advantage to the kinetic approach is that it permits dynamical modeling, and we will use this expression for k_r later in this chapter. It is typically assumed that k_f is unaffected by stress, but some analyses have suggested how it could vary with strain (see Dembo *et al.* (1988) or Eqn. (6–33a)).

Let us compare these estimates of specific bond interactions to our previous estimates of non-specific interactions. We stated earlier the estimate that a force per unit area of about 10^{-5} dyn/μm^2 is sufficient to detach a cell held by only non-specific forces to form another cell. This is equivalent to a single high-affinity receptor/ligand bond per μm^2 of cell/cell contact area. Given that the cell surface receptor number density will usually be about 10–$100/\mu$m^2, receptor/ligand bonds can be expected to provide at least an order-of magnitude stronger adhesive strength than non-specific interactions. Furthermore, because receptors can accumulate within a contact zone by diffusion in the cell membrane, the net enhancement can be even greater than this (*e.g.*, Chan *et al.*, 1991).

Recently, Kuo and Lauffenburger (1993) have provided the first *experimental test of the relationship between chemical bond affinity and mechanical bond strength.* Using the radial flow chamber, they measured the critical shear stress for detachment of receptor-coated beads from a ligand-coated substratum as a function of receptor density, n_T (see Figure 6–21). The slope of this curve should represent the detachment forced carried by the receptor/ligand bonds, while the intercept should give the non-specific bead/surface interaction force. Using immunoglobulin G as the receptor and protein A as the ligand, values for K_D,

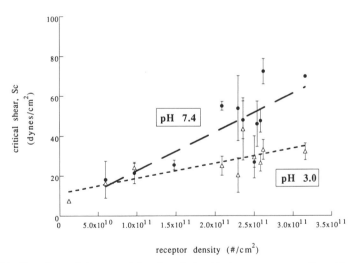

Figure 6–21 Experimental data examining the relationship between receptor/ligand bond chemical affinity and mechanical adhesion strength (from Kuo and Lauffenburger, 1993). The critical shear stress, S_c, necessary for detachment of 10 μm latex polystyrene beads coated with rabbit immunoglobulin G (IgG) as "receptors" from a glass surface coated with protein A "ligand", both via covalent linkages, is measured as a function of the bead "receptor" density for two pH values. The slope characterizes specific, receptor/ligand bond adhesion strength. At pH = 7.4, the specific adhesion strength is about 2.5 times greater at pH = 7.4 than at pH = 3.0. Equilibrium binding data show that the protein A/IgG binding affinity is approximately 40 times greater at pH = 7.4 than at pH = 3.0.

n_T, and n_s were measured using separate binding experiments and could be varied individually. K_D was altered by changing the fluid pH from 7.4 to 3.0, causing a decrease in affinity by a factor of 40, from 4.6×10^{-9} M to 1.8×10^{-7} M. This pH change correspondingly caused a decrease in the bond force by a factor of only about 2.5 comparing the two slopes in Figure 6–21. Eqn. (6–10) predicts that the bond force should be decreased by a factor of about 1.4 when the K_D values measured at the two pH values (5×10^5 #/cm² at pH = 7.4 and 2×10^7 #/cm² at pH = 3.0) are substituted along with the measured value of n_s, 3.5×10^{11} #/cm². This comparison was taken by the investigators to provide evidence in support of Bell's theoretical analyses, because of the apparent logarithmic dependence of bond force on affinity. Absolute values of F_{bond} were calculated to be about 6.7×10^{-6} dyn/bond and 2.6×10^{-6} dyn/bond for pH = 7.4 and 3.0, respectively, using estimates of the bond density and contact area, consistent with the theoretical predictions. Additional data for various IgG species which exhibit differing binding affinities for protein A support the validity of this logarithmic relationship (Kuo and Lauffenburger, 1993).

An alternative way of viewing the influence of specific receptor/ligand bonds is on the potential energy diagram shown in Figure 6–18. Although steric repulsion should still keep the minimum separation distance of about 200 Å between cell membranes (Mege et al., 1987; Bongrand and Bell, 1984), the depth of the secondary minimum at this separation distance should be increased from about 10^{-4} ergs/cm² to about 10^{-3} ergs/cm² for a density of 100 bonds/μm². Helm et al. (1991) used the surface-force apparatus to measure the strength of adhesion between two lipid bilayers containing avidin and biotin, respectively (recall that these molecules bind with extremely high affinity, with $K_D \sim 10^{-15}$ M). These investigators found an energy vs. separation distance diagram similar to that sketched in Figure 6–18, with a well width of about 0.5 nm and a well depth of about 7.4×10^{-4} erg/cm² at a separation distance of approximately 100 Å. The bond density was not known. Detachment of the two surfaces occurred by extraction of the avidin/biotin complex from the lipid bilayers instead of bond dissociation, indicating that the bond strength is at least as great as the force needed to remove proteins from the lipid membrane. This technique may be useful for measuring the binding strength of lower affinity bonds, however, once difficulties in determining the actual number of bonds present are overcome.

We have been assuming that receptor/ligand bonds simply dissociate to their free receptor and ligand states after sufficient physical stress. However, as suggested by the Helm et al. (1991) experiments, uprooting the receptor/ligand complex from the cell membrane might also occur. Bell (1978) estimates that the force needed to pull a glycophorin molecule through a lipid bilayer is approximately 10^{-5} dyn/molecule, of the same order as the force needed to break a moderate affinity bond. More force may be needed for receptors anchored to the cytoskeleton. The recent experimental work of Evans et al. (1991a) suggests that uprooting of receptors is indeed a feasible possibility. In their micropipette experiments, cells that are pulled apart appear to extract receptors from the cell membrane. For at least two reasons, this result must still be shown to be general if it is to be used in models for detachment. First, the receptor/ligand combination used by Evans et al. is likely to produce considerable aggregation or crosslinking, generating a linkage of sufficiently great avidity to make uprooting of receptors more likely. Second, in micropipette experiments (as in the surface force apparatus) a normal force acts over the entire contact area, whereas in physiological situations the force modes experienced by bonds may be quite different.

6.2.3.d Contact Area

A final preliminary consideration has been mentioned frequently but only superficially in the preceding discussion. The *contact zone* is defined as the region of cell surface in sufficiently close proximity to the adhesive surface to permit receptor/ligand bond formation. Mege *et al.* (1987), using electron microscopic observations of macrophage/erythrocyte adhesion, concluded that an apparent separation distance of 500 Å was a valid functional definition of the contact zone. The amount of membrane surface area within the contact zone is called the *contact area*, although this quantity is ambiguous because of difficulties in distinguishing between areas in true contact and those not in the presence of numerous membrane invaginations and protrusions. Practically all theoretical treatments of cell adhesion use an apparent contact area, defined simply by πa^2, where a is the radius circumscribing the region containing membrane in "true" (*i.e.*, within receptor/ligand binding distance) contact; it is a projected area.

The size of the contact area is a crucial parameter in receptor-mediated cell adhesion, since it will help determine the number of cell receptors available for bond formation. What factors govern the size of the contact area? At equilibrium, the size is determined by non-specific interactions, specific interactions, cell mechanical properties, cytoskeletal arrangement, and any external forces acting on the cell. Under non-equilibrium conditions, forces present during incubation and distraction forces will also play a role and the contact area may change with time. For instance, incubation of cells with a surface under centrifugation should generate a greater initial contact area than incubation under sedimentation.

6.2.4 Equilibrium Models

Both for biotechnological purposes and for understanding the basic physiological event of cell adhesion, there are several key questions that mathematical models aim to answer. First, under what conditions will cells adhere to other cells and to surfaces? Second, how difficult is it to disrupt an adhesive interaction? We have seen in our earlier discussion that it is likely that receptor/ligand interactions are necessary to produce adhesive interactions of significant strength, and we would like to quantitate the strength of adhesion as a function of, for example, bond number and affinity. Finally, what is the contact area when cells adhere to other cells or to surfaces, and how many bonds are in that contact area? Two approaches have been used to begin to quantify these physical aspects of cell adhesion at equilibrium, a thermodynamic approach and a mechanical approach. In the models we will discuss below, the *thermodynamic approach* offers the advantage of explicit calculation of the contact area, number of bonds, and cell separation distance, but neglects consideration of cell mechanics in its current formulation. The *mechanical approach* offers the advantage of explicit consideration of the mechanics of cell deformation and the ability to calculate the shape of the edge of the contact area, but does not calculate the contact area or the effect of receptor/ligand affinity on cell adhesion. Both are offered here, however, for their foundational nature; most later models of cell adhesion, including some still under development, rely on the basic physical aspects that are elegantly elucidated in these two models.

6.2.4.a Thermodynamic Prediction of Cell–Cell Contact Area and Bond Number

Bell et al. (1984) pioneered the equilibrium thermodynamic approach to cell–cell adhesion. The key postulate of their model is that the process of receptor diffusion, formation of bonds, and development of the contact area are assumed to be more rapid than metabolic events such as receptor synthesis, development of focal adhesions, and redistribution of the glycocalyx. Under these conditions, then, a cell may reach a quasi-equilibrium which may be calculated from physical properties of the cell. The role of their model is to calculate this equilibrium state and thus find the contact area, cell–cell separation distance, and number of bonds as a function of receptor properties and non-specific interactions. The mechanics of cell deformation are not included in this analysis.

According to thermodynamics, the equilibrium state is defined by the *minimum Gibbs free energy* at constant temperature and pressure. Thus Bell et al. formulate an expression for the change in Gibbs free energy ΔG of a closed system containing two free deformable cells which change from a state of no interaction to a state of adhesion. For such a system, ΔG is equal to the sum of the free energy change due to bond formation and the free energy change due to the appearance of non-specific forces as the two cells are brought together. If it is found that ΔG is positive for a change from a state of no interaction to a state of adhesion, then adhesion does not represent a minimum in Gibbs free energy and adhesion will not be a stable equilibrium state. The opposite is true for the case when adhesion is thermodynamically favored. In other words, the thermodynamically favored interaction between the two cells will be found at the minimum value of ΔG, that is, the largest negative value of ΔG.

The expression for ΔG is developed as follows. The two cells are considered to have R_{1T} and R_{2T} total surface receptors, respectively, capable of binding each other. In other words, R_1 is the ligand for R_2 and R_2 is the ligand for R_1. B is the number of bridges (receptor/ligand complexes acting to bridge one cell to another), and R_{ic} and R_{io} ($i = 1, 2$) are the numbers of free receptors within and outside the contact area, respectively, for cells 1 and 2. Assuming a constant number of total surface receptors on each cell, conservation gives $R_{iT} = B + R_{ic} + R_{io}$. Corresponding receptor densities are $n_{iT} = R_{iT}/A_i$, $n_b = B/A_c$, $n_{ic} = R_{ic}/A_c$, and $n_{io} = R_{io}/(A_i - A_c)$, where A_i is the total surface area of cell i and A_c is the contact area. The cell–cell separation distance, the distance between the cell membranes (see Figure 6–14), is assumed to be uniform across the contact area and equal to s. s is important in quantitating the non-specific forces, which vary with separation distance (see Figure 6–17), and in assessing the stress on individual receptor/ligand bonds.

The change in Gibbs free energy for the system can be expressed as:

$$\Delta G(A_c, s, B) = \sum_j N_j \mu_j(n_j) + A_c \Gamma(s) \qquad (6\text{–}11)$$

Note that ΔG is a function of the contact area, the separation distance, and the number of bonds. N_j is the number of the various receptor species (*i.e.*, N_j should be replaced by B, R_{ic}, R_{io} for $j = 1, 2$, and 3, respectively), n_j is the density of the jth receptor state (*i.e.*, n_j should be replaced by n_b, n_{ic}, and n_{io} for $j = 1, 2$, and 3, respectively), and μ_j is the chemical potential of the jth receptor state, written in ideal solution form as a function of the density of that receptor species:

$$\mu_j = \mu_j^0 + K_B T \ln n_j \qquad (6\text{--}12a)$$

It is conceivable that non-ideal behavior should be accounted for at high number densities (Ryan *et al.*, 1988), although this is not done here. $\Gamma(s)$ is the free energy of non-specific repulsion between the cells per unit area of contact, or, in other words, the mechanical work that must be done against non-specific repulsive forces to bring a unit area of membrane from an infinite separation to a cell–cell separation distance of s.

Two terms requiring special comment are the *bond* or *bridge chemical potential* μ_b ($\mu_b = \mu_b^0 + K_B T \ln n_b$) and the non-specific interaction energy. Bell and coworkers treat bonds as mechanical springs which possess minimum energy at some optimal cell/cell separation distance, L, consistent with the natural bond length (see Figure 6–22). When $s < L$ the bond "spring" is compressed, whereas when $s > L$ the bond "spring" is stretched – in both cases increasing the bond energy. Accordingly, the standard chemical potential for a bond is expressed as a series around the natural bond length:

$$\mu_b^0(s) = \mu_b^0(L) + \tfrac{1}{2}\kappa_b(s - L)^2 \qquad (6\text{--}12b)$$

where κ_b is the mechanical spring constant for the bond. The way in which κ_b might vary with receptor and ligand biochemical properties is unknown. Next, the *non-specific interaction energy* $\Gamma(s)$ is modeled empirically as a separation distance-dependent net repulsion:

$$\Gamma(s) = \frac{\sigma}{s} \exp\left\{-\frac{s}{\delta_g}\right\} \qquad (6\text{--}13)$$

where σ is a measure of the stiffness of the glycocalyx and δ_g characterizes its thickness. This simple relationship is essentially a description of the repulsive steric stabilization forces and electrostatic forces. Van der Waals attractive forces are assumed to be less significant, as is seen to be plausible from Figure 6–17b, and are therefore not included in the model.

Substitution of Eqns. (6–12) and (6–13) into Eqn. (6–11) yields a final expression for the change in Gibbs free energy ΔG as a function of the state variables A_c, s, and B. Minimization of ΔG with respect to each of these state variables permits their equilibrium values to be determined. This minimization requires the statements of total receptor conservation on each cell given earlier as well as several *physical constraints*. These constraints are simply that separation distance must be non-negative, bond number cannot exceed total receptor numbers, and contact area cannot exceed some value A_{max} set *a priori* based on geometric and morphological constraints. For example, A_{max} can be no larger than the surface area of the smaller cell, which would correspond to total engulfment of the smaller cell by the larger cell; it is likely that A_{max} is less than

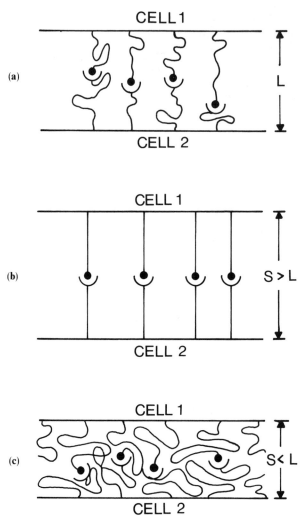

Figure 6–22 Receptors/ligand bonds between adherent cells. Bonds are modeled as springs which possess minimum energy at the optimal length L. Both stretching and compressing the bond increase the chemical potential of the bond and thus the Gibbs free energy. Reprinted from Bell *et al.* (1984). Reproduced from the *Biophysical Journal*, 1984, Vol. 45, pp. 1051–1064 by copyright permission of the Biophysical Society.

this value. The minimizations are performed by setting to zero the partial derivatives of ΔG with respect to each of the state variables in turn and solving the resulting three algebraic equations.

The key prediction of this equilibrium thermodynamic model for cell–cell adhesion is shown in the *phase diagram* of Figure 6–23a. Three regions for cell–cell adhesion are found and are placed in the figure according to the values of R_{1T}, the total number of receptors on cell 1, and a parameter ξ. $\xi(s)$ is a dimensionless quantity characterizing a modified ratio of non-specific repulsion to binding

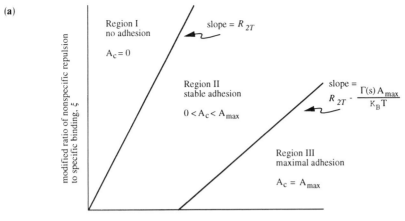

(a)

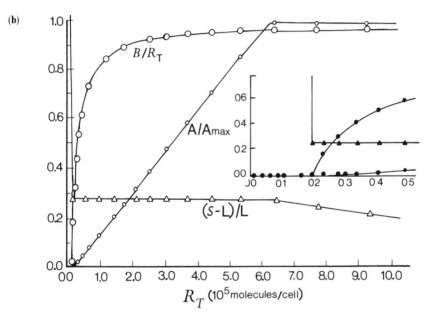

(b)

Figure 6–23 (a) Predicted phase diagram for cell–cell adhesion. Cells do not adhere in region I, adhere with less than maximum contact area in region II, and with contact area A_{max} in region III. Cells are more likely to adhere as the number of receptors R_{1T} increases, as the association equilibrium constant K_A increases, or as the non-specific repulsion energy Γ decreases. Redrawn from Bell (1984). (b) Predicted dependence of the number of cell–cell bridges B, the contact area A_c, and the separation distance s on the total receptor number R_{1T}. Reprinted from Bell *et al.* (1984). Reproduced from the *Biophysical Journal*, 1984, Vol. 45, pp. 1051–1064 by copyright permission of the Biophysical Society.

according to:

$$\xi(s) = \frac{A_1 A_2 \Gamma(s)}{K_B T K_A(s)} \qquad (6\text{–}14)$$

where $K_A(s)$ is the separation distance-dependent binding equilibrium constant:

$$K_A(s) = K_D^{-1} \exp\left\{ -\frac{\kappa_b(s - L)^2}{2\mathrm{K}_B T} \right\} \tag{6-15}$$

In region I, no stable adhesive interaction occurs. The attractive forces offered by receptor/ligand bonds are simply not great enough to overcome the repulsive forces resulting from close cell–cell contact. In region II, stable cell–cell adhesion exists with a contact area less then A_{max}; in other words, a balance exists between attractive and repulsive forces such that the contact area is less than its allowed maximum. In region III, stable cell–cell adhesion results and the contact area is A_{max}. In this region, then, attractive forces far outweigh repulsive forces and the area of contact is limited only by geometric and morphological constraints.

As shown in Figure 6–23a, a *threshold* in receptor number exists for stable cell–cell adhesion. This can be seen by observing that an increase in R_{1T} can allow the transition from no adhesion (region I) to stable adhesion (region II) with no other changes in model parameters. This theoretical prediction is consistent with a variety of experimental data (*e.g.*, Weigel *et al.*, 1979). Similarly, a threshold in receptor affinity (through the parameter $\kappa_b(s)$) also exists. The implication of this theoretical result is that if a cell can alter the number of its receptors or the affinity of its receptors it can thereby modulate its ability to interact with other cells; recall our earlier mention in section 6.2.1.a of the modulation of the avidity of LFA-1 for ICAM (Dustin, 1990).

In Figure 6–23b, sample computational results of the model are shown for the dependence of B, A_c, and s on total receptor number, $R_{1T} = R_{2T} = R_T$, for equilibrium adhesion between two identical cells. There are three key insights to be drawn from this figure. First, once adhesion is thermodynamically possible, the contact area increases almost linearly with total receptor number until the maximum possible area A_{max} is reached. This is consistent with the data of Capo *et al.* (1982), who measured A_c versus R_T for the agglutination of thymocytes at various cell surface concentrations of the adhesive lectin (*i.e.*, a protein capable of specific binding to sugar residues) concanavalin A. Second, the cell–cell separation distance is basically constant at a value not too different from the glycocalyx thickness and about 30% above the unstressed bond length L, consistent with the data of Mege *et al.* (1987), until the maximum contact area regime is reached. Third, most of the receptors are predicted to be accumulated as bonds in the contact area, depleting the remainder of the cell surface by diffusion into this reacting domain. Substantial accumulation of receptors in contact zones has, in fact, been observed experimentally (*e.g.*, Kupfer and Singer, 1989; Singer, 1992).

The Bell *et al.* model, then, is successful in explaining a number of the physical features of cell–cell adhesion. The model framework has more recently been extended to include a number of other physical possibilities, including glycocalyx molecules that may move out of the contact area (Torney *et al.*, 1986) and cell–substrate binding and the existence of essentially immobile receptors (Bell, 1988).

Finally, a significant omission of the Bell *et al.* (1984) equilibrium thermodynamic analysis is that the contribution of cell membrane deformation energy is neglected in the minimization of free energy, except for the assumed constraint that the contact area cannot be greater than A_{max}. This concern is addressed below, using a mechanical approach to model cell adhesion.

6.2.4.b   Mechanical Relationship Between Membrane Tension and Receptor/Ligand Bonds

An alternative approach to an equilibrium model of cell adhesion is taken by Evans (1985a, 1988), emphasizing the mechanics of cell formation. Useful background on cell mechanics and modeling of the cell is given in section 6.2.3.b and in Evans and Skalak (1980) and Evans (1988).

The complete mechanical description of the cell requires the choice of a model for cell rheology, determination of the forces applied to the cell by receptor/ligand bonds, and determination of the forces applied to the cell by outside sources (such as fluid flow or a micropipette). Evans' model is an abbreviated mechanical approach to the description of cell–substrate adhesion. In this model, only the mechanics of a fraction of the cell membrane are described explicitly. Whereas analysis of the shape of entire cells requires coupling of the mechanical models for both the cell interior and the cell membrane/cortex, as described earlier, for purposes of investigating local behavior of the cell–substrate or cell–cell contact area only the cell membrane/cortex mechanics are necessary and the problem is mathematically simpler.

The *relationship between an applied membrane tension* T_m^0 *and bond density* n_b at equilibrium is calculated by this mechanical model. The membrane tension T_m^0, which is assumed constant, is that expected to be applied by the rest of the cell or by outside forces. At equilibrium, then, one expects that the forces due to "pulling" on the membrane will be balanced by the resistance offered by receptor/ligand bonds. Clearly the greater the applied tension, the greater the bond density necessary to resist that tension. The model, shown schematically in Figure 6–24, describes this relationship mathematically.

This abbreviated mechanical approach is limited in that only the membrane contour in the contact area and the membrane tension are calculated. The role of receptor/ligand affinity is not included explicitly, the size of the contact area cannot be calculated, and non-specific forces are neglected. However, this approach does provide the foundation for more sophisticated analyses of adhesion, including dynamic models.

As shown in Figure 6–24, the cell membrane is divided into two regions: that within the contact zone and that immediately outside. Only the membrane curvature shown is modeled; curvature in the plane perpendicular to the figure is considered too small to be significant. Within each of the two membrane regions a local mechanical equilibrium balance is written for the forces tangent to the membrane and for the forces normal to the membrane. These force balances simply require that both the sum of the forces acting normal to the membrane and the sum of the forces acting tangential to the membrane must each be zero at equilibrium.

To write these force balances one must consider the contributions of the local tension in the plane of the membrane element, T_m, which varies with position or arc length s. Note that this tension T_m is the local result of the macroscopic applied membrane tension T_m^0. We must also consider the action of bending the membrane.

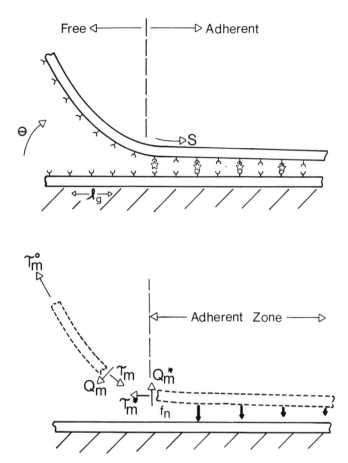

Figure 6–24 Mechanical model schematic. An applied constant membrane tension of T_m^0 acts to lift the membrane away from the substrate; receptor/ligand bonds act to hold the membrane close to the surface. For each small section of membrane or membrane element, a tension T_m and transverse shear resultant Q_m act to displace the element. In addition, normal forces due to receptor/ligand bonds must be considered for the region of the membrane in the contact area. The arc length is given by s. θ is the angle that a membrane element forms with the substrate surface, and θ_0 is the macroscopic cell/substrate contact angle. Adapted from Evans (1985a).

As the cell membrane bends it forms an angle θ with the substrate surface, so $d\theta/ds$ is the local membrane curvature. If there is some stiffness or integrity to the membrane, then a binding deformation can result in a force perpendicular to the membrane. In other words, pulling down on one edge of a somewhat stiff membrane element will cause points far from the edge to be lifted. This lifting is due to a transverse shear resultant Q_m which is shown schematically in Figure 6–24 and related to the bending moment M by

$$Q_m = -\frac{dM}{ds} \tag{6–16}$$

The *constitutive equation* used to relate the change in membrane curvature to the elastic bending modulus B and the bending moment M is given by:

$$M = B\left[\frac{d\theta}{ds} - \left(\frac{d\theta}{ds}\right)_0\right] \tag{6-17}$$

where $(d\theta/ds)$ gives the current membrane curvature and $(d\theta/ds)_0$ its stress-free value (see also Eqn. (6–7c)). Thus $Q_m = -dM/ds = -B(d^2\theta/ds^2)$. Hence, the greater the bending modulus B (*i.e.*, the elastic resistance to bending deformation), the greater the transverse shear stress required to produce a given membrane curvature. Area dilation and in-plane shear stresses are assumed to be negligible for the applied stress, a reasonable assumption based on our discussion in section 6.2.3.b, so other constitutive equations are not needed.

After careful consideration of geometry (see Evans and Skalak, 1980), *mechanical force balances* can now be written. Outside the contact zone the force balances are:

$$\frac{dT_m}{ds} - Q_m\frac{d\theta}{ds} = 0 \tag{6-18a}$$

$$T_m\frac{d\theta}{ds} + \frac{dQ_m}{ds} = 0 \tag{6-18b}$$

while within the contact zone they are:

$$\frac{dT_m}{ds} - Q_m\frac{d\theta}{ds} = 0 \tag{6-19a}$$

$$T_m\frac{d\theta}{ds} + \frac{dQ_m}{ds} = -\sigma_n \tag{6-19b}$$

Eqns. (6–18a) and (6–19a) represent tangential force balances; Eqns. (6–18b) and (6–19b) represent normal force balances. The relationship of the membrane mechanics to bonds is given by σ_n, the *local normal stress due to receptor/ligand bonds*.

Of particular interest to us, σ_n must be modeled explicitly. This aspect is analogous to the modeling of a bond as a spring in the thermodynamic analysis by Bell and coworkers examined earlier. The force a receptor/ligand bond generates is characterized by two parameters: f_c, the force that will break the bond; and l_c, the bond length at which this force will be produced by bond stretching. The normal stress generated by a single bond is then modeled by:

$$f = \frac{f_c}{l_c}l_b \qquad 0 < l_b < l_c \tag{6-20a}$$

$$f = 0 \qquad l_c < l_b \tag{6-20b}$$

where l_b is the actual bond length. If the bond number density within the contact zone can be considered as a continuum, then the local normal stress is simply the product of the local bond density, n_b, and the normal force per bond, f, so that $\sigma_n = n_n f$. Note that we are assuming a *continuum of normal stresses* contributed by the bonds, rather than stresses offered at discrete locations. This is reasonable for large enough bond densities.

To solve the model, one specifies the externally applied tension T_m^0, the membrane and bond properties B, f_c, and l_c, and the bond density n_b. The position of the membrane is then changed until the calculated local values of T_m and Q_m satisfy the force balances given in Eqns. (6–18) and (6–19) for all points on the membrane; this position is the equilibrium configuration of the membrane.

Evans finds that the relationship between membrane tension T_m^0, adhesion energy per unit area of contact (resulting from receptor/ligand binding) ψ, and the macroscopic cell/substrate contact angle θ_0, is:

$$\psi = T_m^0 (1 - \cos \theta_0) \qquad (6-21)$$

where ψ depends on the bond density by $\psi = n_b f l_c / 2$. This equation, then, gives the relationship between applied tension and required bond density. In fact, the equation is simply the classical Young equation (Adamson, 1976), which was not known *a priori* to apply to this biological system. Although this result in itself does not answer the questions likely posed about cell adhesion, such as the probability of adhesion as a function of various cell properties (receptor number and affinity, rheological properties, etc.), it does provide a framework for future models that do so. Evans is also able to predict the shape of the membrane at the edge of the contact area. This is an improvement over the thermodynamic model of Bell *et al.* (1984), which assumed that the separation distance between two cells was constant over the contact area.

This mechanical treatment of cell–surface adhesion involves receptor/ligand bond formation only implicitly, through the parameter ψ. Recall that ψ is proportional to n_b, the density of bonds present, but n_b is not directly calculated from a binding affinity constant and receptor and ligand number densities. Since the membrane tension may affect the binding constant, as in the thermodynamic model of Bell and coworkers discussed earlier, the problem can be "closed" only when calculation of n_b is included explicitly. It is primarily this omission that prevents this model from being used to predict conditions necessary for adhesion, as was done with the thermodynamic model.

An intriguing observation is often made in cell adhesion experiments which cannot be explained by either of the thermodynamic or mechanical models just discussed. A limiting membrane tension is reached at which there is no further spreading of the contact zone during adhesion, but during separation of an adhesive contact a much greater tension may be required for complete detachment of the cell from the substrate. In fact, sometimes the required tension is so great as to rupture the cell membrane. This *difference in membrane tension for spreading and detachment* conflicts with the equilibrium models in which the work of adhesion is completely reversible, so that the tension limiting spreading is equal to that necessary for detachment. In an attempt to account for this, Evans (1985b) successfully used the mechanical equilibrium approach described above but permitted the bonds to be present at discrete spatial positions instead of as a continuum. However, the basic underlying assumption for such an analysis, that bonds are widely-spaced on the length scale of the contact zone, has not been substantiated. At least three other explanations may be offered to explain the experimental observation of a difference in membrane tension for spreading and detachment. First, the effect may be caused by receptor accumulation and uneven distribution in the contact area, or even "dragging" of bonds toward a still-adherent region as the cell is pulled away from the surface (Tozeren et al., 1989). Second, there may be active metabolic cell processes such as focal contact formation that do not contribute to cell spreading but act to strengthen the adhesive interaction. The presence of active cell processes is in direct conflict with the approaches outlined above, which treat the cell as a passive bundle of fluid, but are certainly not unexpected. Third, as an adherent cell is detached, the fraction of surface area in intimate contact with the ligand-bearing surface may increase and allow additional bond formation. This idea is based on a mechanical argument that requires membrane being pulled away from the surface to act as a lever to push adjacent membrane toward the surface (Evans et al., 1991b).

6.2.5 Dynamic Models

By developing dynamic models of cell–cell or cell–substrate adhesion we can investigate not only the equilibrium state of the cell but also the kinetics of formation of that state. This is particularly appropriate for the situation of cell adhesion under conditions of flow. Here, a predicted equilibrium solution of adhesion may be irrelevant for comparison with the results of an attachment experiment. In this case, the equilibrium state may never be reached because the cell may simply not be able to form sufficient bonds fast enough to prevent detachment. Kinetic analyses of cell adhesion may therefore be especially important for the biotechnological application of cell affinity chromatography and for the adhesion of cells to blood vessel walls.

6.2.5.a *Point Attachment Model*

To examine the situation of a cell transiently encountering the surface in the presence of fluid shear flow, Hammer and Lauffenburger (1987) developed a model for the dynamics of receptor-mediated cell adhesion to a ligand-coated surface (see Figure 6–25). The model is termed the *point attachment model* because it focuses on the initial attachment of the cell via a small contact area determined by mechanical and non-specific forces before significant cell spreading induced by receptor/ligand binding (whether active or passive) can occur. That is, it is a non-equilibrium model with regard to the eventual contact area that may develop after a time period long compared to the duration of initial encounter. The time during which cells come into contact with the surface and have a chance to adhere is typically of the order of seconds or fractions of seconds for cells contacting surfaces under common flow conditions, such as in the microcirculation or in experimental assays or cell separation devices *in vitro*. The objective of this model is to elucidate criteria governing the outcome of a transient cell/surface encounter; that is, under what conditions will such an encounter result in attachment?

Consider a cell possessing a single class of receptors, with total number R_{1T} and diffusivity D_R, for the ligand present on the surface at number density n_s. Bond formation and dissociation take place with two-dimensional rate constants k_f and k_r, respectively; note that these are rate constants for a two-dimensional reaction,

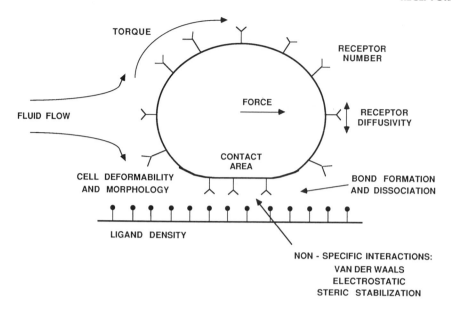

Figure 6–25 Point attachment model for analysis of the dynamics of cell–substrate adhesion. A cell is assumed to attach initially via a small and constant contact area. If the cell forms enough bonds within the time it is in contact with the substrate, it will adhere; otherwise, it will be moved away from the substrate by fluid flow. Reprinted from Hammer and Lauffenburger (1987).

because both receptor and ligand are confined to surfaces. It is likely that this two-dimensional association reaction will demonstrate significant diffusion-limitation, as discussed in section 4.2.

Hammer and Lauffenburger assume that the contact area A_c is the region at which the cell and surface approach within a distance s appropriate for receptor/ligand binding, about 100–300 Å, during the initial contact of the cell with the surface. They postulate that this distance and the size of the contact area are determined by non-specific forces, cell mechanical properties, and depositional forces (causing sedimentation from the bulk fluid). For reasonable parameter estimates, such as those proposed by Bongrand and Bell (1984), initial contact areas of the order of 1 μm^2 are calculated; these represent a small fraction ($<1\%$) of typical total cell surface area, hence the name "point attachment". Furthermore, the contact area is assumed to remain constant during this initial attachment of the cell in this model.

A critical aspect of the model is then to view processes occurring at this contact zone chronologically during the encounter. In Figure 6–25, it is shown that the fluid force on the cell can be transduced into a physical stress on bonds that form. Initially this stress will be small, as the first bonds form and are translated to the rear edge of the cell due to dragging or cell rolling. Once cell motion is arrested, however, the stress on the cell, and therefore on the bonds, increases to a maximum value. Hammer and Lauffenburger model this change in stress with a step-function, so that the bonds are completely unstressed for time $t < t_c$ and are subject to maximum stress for $t > t_c$. t_c is termed the *contact time* and is calculated as the time needed for a bond to form and be translated to the rearward edge of the contact zone:

$$t_c = \frac{1}{k_f n_s} \ln\left[1 - \frac{1}{A_c n_{1T}}\right]^{-1} + \frac{\sqrt{A_c/\pi}}{v} \tag{6-22}$$

where n_{1T} is the initial cell receptor density on the cell and v is the fluid velocity at the cell center-of-mass. (A similar treatment can also be applied to incubation for a specified contact time in the absence of fluid shear stress; in such a case t_c is simply equal to the experimental incubation time.) All bonds are assumed to be stressed equally.

The maximum force on the bonds is calculated by considering the forces and torques acting on the cell as a result of fluid flow and is based on a low-Reynolds number hydrodynamic analysis for a non-deformable particle near a surface in shear flow due to Goldman *et al.* (1967a,b):

$$F = 6\pi\eta_s R^2 Sg(S, R, A_c, \eta_s, s) \tag{6-23}$$

where η_s is the fluid viscosity, R is the cell radius, S is the fluid shear rate at the surface, and g is a complicated function additionally dependent on the cell/surface separation distance s. A major assumption here is that all the fluid stress is transmitted to the bonds. In reality, it is likely that some portion of this stress will be dissipated by viscoelastic deformation of the cell, so Eqn. (6-23) should be modified by a factor less than unity characterizing this dissipation.

Within the contact zone, receptor/ligand bond formation and dissociation occur according to the following kinetic species balances on the number densities of bonds n_b and free receptors in the contact area n_{1c}:

$$\frac{dn_b}{dt} = k_f n_s n_{1c} - k_r n_b \tag{6-24a}$$

$$\frac{dn_{1c}}{dt} = -k_f n_s n_{1c} + k_r n_b + k_t(n_{1T} - n_{1c}) \tag{6-24b}$$

where $k_t = (2D_R/[A_c/\pi]) + (2\Omega R/\pi[A_c/\pi]^{1/2})$ is the transport rate constant for receptor movement into and out of the contact zone by diffusion and effective "convection"; Ω is the angular velocity of the cell, assumed equal to zero for $t > t_c$ when the bonds are fully stressed. The bond dissociation rate constant k_r depends on the bond stress f ($f = F/A_c n_b$) according to Eqn. (6-9) for $t > t_c$, and is simply equal to k_{r0} for $t < t_c$. The association rate constant k_f is constant.

Eqns. (6-24a,b) can be numerically integrated from $t = 0$, using initial conditions that specify that a cell first contacting a surface has no bonds:

$$n_b(0) = 0 \tag{6-25a}$$

$$n_{1c}(0) = \frac{R_{1T}}{4\pi R^2} \tag{6-25b}$$

with the goal of discovering the long-time dynamic outcome. That is, if $n_b = 0$ for large t, the cell must have failed to adhere during the encounter; on the other hand, if n_b approaches a non-zero steady-state value for large t, the cell must have become stably adherent. Key computational results can be expressed in terms of five *dimensionless parameters*, the bond breakage energy α_b, the fractional contact area β, the equilibrium dissociation constant for receptor/ligand bond formation κ, the bond formation rate θ_b, and the receptor accumulation rate δ:

$$\alpha_b = \frac{\gamma F}{K_B T} \qquad (6\text{-}26a)$$

$$\beta = \frac{A_c}{4\pi R^2} \qquad (6\text{-}26b)$$

$$\kappa = \frac{k_{r0}}{k_f n_s} \qquad (6\text{-}26c)$$

$$\theta = k_f n_s t_c \qquad (6\text{-}26d)$$

$$\delta = \frac{k_t}{k_f n_s} \qquad (6\text{-}26e)$$

The most important result of the point attachment model is shown in Figure 6–26. The total cell receptor number required for adhesion is plotted versus the dimensionless equilibrium dissociation constant κ for one set of values of the dimensionless parameter α_b, β, and δ. Each curve represents a different value of θ, the dimensionless bond formation rate. Notice that as κ increases (corresponding to lower binding affinity and bond strength), the required number of receptors increases. Most importantly, note that all the curves merge into a single curve

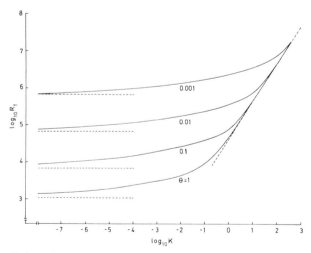

Figure 6–26 Predictions of the point attachment model. The minimum number of receptors required for adhesion is plotted as a function of the scaled dissociation constant κ for several values of the bond formation rate θ. For all curves, $\alpha_b = 25$, $\beta = 1.5 \times 10^{-3}$, and $\delta \ll 1$. Redrawn from Hammer and Lauffenburger (1987). Reproduced from the *Biophysical Journal*, 1987, Vol. 52, pp. 475–487 by copyright permission of the Biophysical Society.

at sufficiently low binding affinity; this curve represents an *equilibrium-controlled adhesion regime*. Here, the *critical number of receptors required for adhesion* R_{1T}^c is governed by the balance between distractive fluid shear force and the bond strength. This critical number of receptors can be shown to be equal to $\alpha_b \kappa e / \beta$, where e is the exponential constant ≈ 2.7. Notice that this result depends on the equilibrium constant but not any individual kinetic rates.

In contrast, as κ decreases (corresponding to higher binding affinity and bond strength) the required number of receptors decreases, but its value now becomes strongly dependent on the bond formation rate θ and independent of the binding affinity κ. This behavior represents the *rate-controlled adhesion regime*, in which the number of receptors required for adhesion is governed by the balance between bond formation rate and contact time. Here, the R_{1T}^c can be calculated from the mean time necessary to form a single bond and is equal to $1/\beta(1 - e^{-\theta})$. Once a bond is formed, the affinity is sufficiently high that an adhesive equilibrium state will essentially be assured before that bond is disrupted. To determine which of the regimes is applicable in a given situation, it is sufficient to evaluate the quantity $\alpha_b \kappa e (1 - e^{-\theta})$. For $\alpha_b \kappa e (1 - e^{-\theta}) < 1$, the rate-controlled regime is operative. That is, rate processes rather than equilibrium processes will be the primary influence for small enough dimensionless distractive forces and binding dissociation constant.

The existence of these two behavioral regimes may help to explain a variety of experimental observations. For instance, consider the effect of receptor diffusivity on cell adhesion. Increasing the receptor diffusivity D_R will primarily serve to increase the value of k_f, the receptor/ligand bond formation rate constant, which is likely to be near its diffusion limit. Hence, only cell adhesion experiments operating under the rate-controlled regime should exhibit a dependence on D_R, as shown in Figure 6–27. As an example, consider a typical cell affinity chromatography column (Hammer *et al.*, 1987). Assuming 1 cm min^{-1} linear fluid velocity with 250 μm beads and $\eta_s = 1$ centipoise yields a force per unit area of the cell of $\sim 5 \times 10^4$ dyn/cm^2, so that $\alpha_b \sim 600$. If the surface ligand is at about 10% of an effective monolayer, then $n_s \sim 10^{11}$ #/cm^2. With $k_f \sim 10^{-8}$ cm^2/#-s and $t_c \sim 0.1$ s, we find $\theta \sim 100$. Under these circumstances, cell/bead contact will be occurring under rate-controlled conditions for binding affinities such that $\kappa < 6 \times 10^{-4}$; with our value for n_s, this corresponds to $K_D < 6 \times 10^7$ #/cm^2 (for two-dimensional binding reactions). A typical receptor/ligand bond K_D of 10^{-8} M in three dimensions is equivalent to $(10^{-8}$ moles/l$) \times (6 \times 10^{23}$ #/mole$) \times (10^{-3}$/cm$^3) \times (10^{-6}$ cm membrane thickness$) = 6 \times 10^6$ #/cm^2, so that most typical systems should, indeed, be rate-controlled. Thus, it is not surprising that receptor diffusivity plays a significant role in cell/surface adhesion in such columns (*e.g.*, Rutishauser and Sachs, 1975).

Rigorous quantitative experiments investigating the dynamics of cell attachment are much fewer than experiments directed toward equilibrium adhesion behavior or the dynamics of detachment of previously adhered cells. One exception is the study by Wattenbarger *et al.* (1990) on the attachment of liposomes to lectin-coated glass surfaces in a parallel-plate flow chamber. The liposomes contained glycophorin molecules acting as receptors for the lectin. By tracking individual liposomes microscopically, these investigators were able to measure the

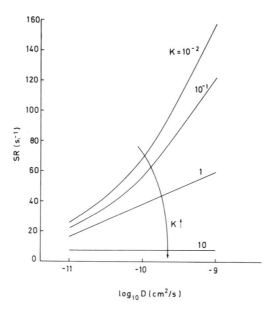

Figure 6-27 Predictions of the point attachment model. The maximum allowable shear rate, SR, for cell adhesion to occur is plotted as a function of the receptor diffusion coefficient D_R for several values of the scaled dissociation constant κ. An increase in receptor mobility affects cell adhesiveness only for low values of κ. Redrawn from Hammer and Lauffenburger (1987). Reproduced from the *Biophysical Journal*, 1987, Vol. 52, pp. 475–487 by copyright permission of the Biophysical Society.

probability of attachment during a particular surface encounter, or *sticking probability* p_{exp}, as a function of fluid shear rate and receptor number. At low shear rates nearly all encounters led to stable adhesion, independent of receptor level. As shear rate increased, the sticking probability decreased significantly for lower receptor levels but not for higher receptors levels. Wattenbarger *et al.* calculated the ratio R^c_{1T}/R_{1T} for their experiments, where the critical receptor number $R^c_{1T} = 1/\beta(1 - e^{-\theta})$ is that predicted by the model for the rate-controlled regime of adhesion and R_{1T} is the actual receptor number estimated to be available on the liposomes. They found $p_{exp} = 1.0$ for cases in which the ratio is greater than 1, whereas $p_{exp} < 1$ for cases in which the ratio is less than 1. This is in semiquantitative agreement with the Hammer and Lauffenburger model, for they predict adhesion ($p_{exp} = 1.0$) for $R^c_{1T}/R_{1T} < 1$ and no adhesion ($p_{exp} = 0$) for $R^c_{1T}/R_{1T} > 1$. The model is not able to predict intermediate values of p_{exp} because it is a deterministic model: all cells are assumed to behave identically. In order to predict non-uniform cell behavior, two approaches could be considered. Cells could be assumed to have varying numbers of receptors or other parameters, or the event of adhesion could be assumed to be governed by probabilistic or stochastic equations. The latter approach is examined next.

6.2.5.b Probabilistic Attachment and Detachment

Cozens-Roberts *et al.* (1990a) extended the dynamic model discussed above in order to interpret the kinetics of cell attachment and detachment. For attachment studies, they also assumed that a small contact area forms relatively quickly upon an initial cell/surface encounter. For their analysis of detachment, it is still assumed that a fairly small contact zone exists initially and does not decrease in size following application of stress. Further, all bonds present in the contact zone are presumed to be stressed equally, as in the Hammer and Lauffenburger model.

The key difference from the earlier model is that the kinetic bond formation equations are cast in probabilistic instead of deterministic form (see section 4.1.2). Eqn. (6–24a) for rate of bond formation is replaced by a system of equations governing the probabilities, $P_B(t)$, that B bonds exist between the cell and surface at time t:

$$P_B(t + \Delta t) = P_B(t) + k_f n_s [R_c - (B - 1)]P_B(t)\Delta t - k_f n_s [R_c - B]P_B(t)\Delta t$$

$$- k_r B P_B(t)\Delta t + k_r(B + 1)P_{B+1}(t)\Delta t \qquad B = 1, 2, 3, \ldots, R_c \quad (6\text{–}27)$$

where Δt is a time step sufficiently small that only one kinetic bond formation or dissociation event can take place and higher order terms have been neglected. Eqn. (6–24b) is eliminated by the simplifying assumption of surplus ligand: $n_s \gg R_c/A_c$, where R_c is the total number of receptors in the contact area. It is assumed that there is no recruitment of receptors into the contact area, so R_c is constant. In the limit as Δt goes to 0 relative to the experimental observation time and $R_c \gg 1$, this system of difference equations can be approximated by a single partial differential equation of Fokker-Planck form (Gardiner, 1983):

$$\frac{\partial p}{\partial \tau} = -\frac{\partial}{\partial u}[A(u)p] + \frac{1}{2}\frac{\partial^2}{\partial u^2}[B(u)p] \qquad (6\text{–}28)$$

where

$$A(u) = (1 - u) - \kappa u \exp\left[\frac{\alpha_b}{R_c u}\right]$$

$$B(u) = \left\{(1 - u) + \kappa u \exp\left[\frac{\alpha_b}{R_c u}\right]\right\}\frac{1}{R_c}$$

with p a continuously-distributed probability (probability density function) over the range of dimensionless bond number $u = B/R_c$ and dimensionless time $\tau = k_f n_s t$. The dimensionless parameters α_b and κ are as defined by Eqns. (6–26a,c). Common techniques for solving partial differential equations can be found in Pinsky (1991).

With appropriate boundary and initial conditions, Eqn. (6–28) is solved numerically for the distribution of bond number probabilities $p(u, \tau)$. Some sample results for cell *detachment* are plotted in Figure 6–28. The initial condition is that predicted by the analogous probabilistic attachment model and is shown on this plot as the curve marked $\tau = 0$. (We note that in the original reference, the initial condition for the detachment model incorrectly neglected a factor of R_T in the denominator of its Eqn. (9).) Notice that even though the cells are homogeneous with respect to all parameter values (*e.g.*, receptor number and binding rate constants), detachment occurs over a continuous time range. A significant feature of the predicted behavior is the seemingly stable level of adherent cell number

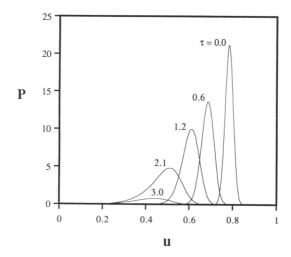

Figure 6–28 Predictions of the probabilistic model for cell detachment. The probability density function p is plotted as a function of the scaled bond number u for several values of the scaled time τ. The initial condition for the detachment simulation is the curve marked $\tau = 0$ and is the result of a probabilistic analysis of cell attachment. Parameters values used were $k_f = 10^{-14} \, \text{cm}^2/\#\text{-s}$, $k_{r0} = 10^{-4} \, \text{s}^{-1}$, $n_s = 10^{11} \, \#/\text{cm}^2$, $R_c = 500$, $\alpha_b = 1.3$.

at long times, despite the fairly rapid detachment of a substantial number of identical cells. This results from formation of a strong peak in the distribution $p(u, t)$ for small values of α_b that decays only slowly.

For comparison to experimental data, p_a, the probability that at least one bond exists (*i.e.*, that a cell is adherent), is defined by:

$$p_a(\tau) = \int_\delta^1 p(u, \tau) \, du \qquad (6\text{–}29)$$

$p_a(t)$ is proportional to the number of cells adhered to the surface at a given time. In an accompanying experimental paper, Cozens-Roberts *et al.* (1990c) measured the fraction of antibody-coated latex beads remaining attached, after 60 min in shear flow, to a glass surface coated with complementary antibody, as a function of shear rate in the radial flow assay. Figure 6–29 shows some of these data together with calculations from the model (Saterbak *et al.*, 1993). In this figure, the dashed curve represents the model results in the absence of probabilistic effects, using an experimental measurement of heterogeneity in bead receptor number, the dotted curve represents model predictions for purely probabilistic effects neglecting heterogeneity, and the solid curve represents both effects combined. If both probabilistic and heterogeneity influences were neglected, detachment behavior would be ideal, 'all or none' in nature. That is, all the beads would detach simultaneously under identical conditions, so that an ideal curve would be a step change from 1 to 0 in p_a at the critical force α_{bc}. In this particular case, then, probabilistic effects account for about 10% of the deviation of observed behavior from ideal, whereas heterogeneity effects account for about 90%.

Finally, several shortcomings of the model, both in its deterministic and

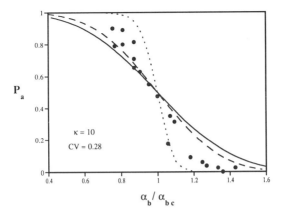

Figure 6-29 Experimental data (Cozens-Roberts *et al.*, 1990c) and model predictions (Saterbak *et al.*, 1993) for bead detachment. Plotted is the fraction of initially adherent beads remaining on a ligand-coated surface after 60 min exposure to a fluid shear stress in the radial flow chamber. The shear stress is scaled by the shear stress at which 50% of the beads are removed. Filled circles are experimental data, dashed curve is model prediction for pure heterogeneity effects, dotted curve is model prediction for pure probabilistic effects, and solid curve is model prediction combining heterogeneity and probabilistic effects. Measured coefficient of variation for bead receptor number is 0.28, estimated dimensionless binding equilibrium dissociation constant is $\kappa = k_{r0}/k_f n_s = 10$, and $R_c = 500$.

probabilistic forms, exist with regard to application to living cells: the assumption that bonds in the contact zone are stressed uniformly, the neglect of non-linear association among bonds (as in focal contact formation), the assumption of a constant contact area, and the assumption of no recruitment of receptors into the contact area. The first of these issues is addressed in the next model to be discussed.

6.2.5.c *Mechanical Approach to Cell Attachment and Detachment*
In the dynamic models discussed above, the mechanics of cell deformation have not been explicitly included. In contrast, Dembo *et al.* (1988) combine a cell mechanical picture with receptor/ligand bond formation dynamics. A central goal of this work is to elucidate how the *critical membrane tension for local detachment* of an edge of the contact zone depends on cell, receptor, and ligand properties. The model is further capable of describing the transient behavior of contact zone spreading or shrinking during cell attachment or detachment, including the local geometry of the cell/surface interface and the velocity of edge of the contact zone.

Consider the situation illustrated in Figure 6-30. The ligand-covered surface which the cell encounters is the x-axis of a Cartesian coordinate system, and the distance above this surface is marked on the y-axis. The location of the membrane is described by an arc length s so that the shape of the membrane at any time t is given by the curve $\{x(s, t), y(s, t)\}$. For large positive s, the membrane is considered to be deep within the contact zone, whereas for large negative s, the membrane approaches its free orientation outside the contact zone. A tension T_f is assumed to act in this latter region at the orientation angle θ_f. Bonds between

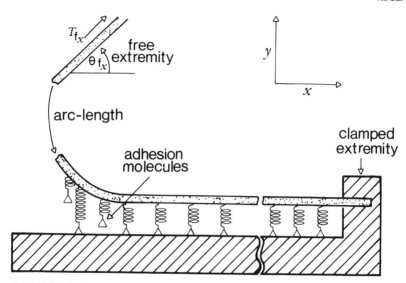

Figure 6–30 Model schematic. A two-dimensional membrane is subjected to an externally applied tension T_f. The position along the membrane is described by the arc length s; tension is assumed to be applied at $s = -\infty$ and the membrane is assumed to be clamped at $s = +\infty$. Given material properties of the membrane and an idealization of receptor/ligand bonds as springs, the goal of the model is to calculate the velocity v_c of the edge of the contact area. If the velocity is greater than zero, the cell is detached from the surface; if the velocity is less than zero, the cell is spreading onto the surface. Redrawn from Dembo et al. (1988).

cell receptors and surface ligands can form at any location, with forward rate constant $k_f(y)$ and reverse rate constant $k_r(y)$. That is, both the association and dissociation rate constants are postulated to be functions of separation distance between the cell and surface. Under the condition that receptors and ligands do not freely diffuse within their respective planes, a species balance equation for the local bond density n_b can be written:

$$\frac{\partial n_b(s, t)}{\partial t} = k_f(y)[n_{1T} - n_b(s, t)]n_s - k_r(y)n_b(s, t) \qquad (6\text{–}30)$$

where n_{1T} is the total cell receptor density and n_s is the density of ligand on the surface. No recruitment of receptors into the contact area is allowed (this assumption has been relaxed in the model of Ward and Hammer, 1993). Surface ligand molecules are assumed to be in stoichiometric excess and thus there is no need to account for the depletion of free ligand.

The contact zone is defined to be the locus of points at which the bond density is greater than or equal to a threshold value n_{bc}. The smallest value of s at which the contact threshold is attained or exceeded is the edge of the contact zone and defined by $s = s_c(t)$. Hence, the local *velocity of the edge of the contact zone* is

$$v_c = \frac{\partial s_c}{\partial t} \qquad (6\text{–}31)$$

Velocities greater than zero correspond to cell detachment, and velocities less than zero correspond to cell attachment or spreading. Adopting a reference frame with respect to the edge of the contact zone, so that $s' = s - s_c$, the balance equation for bond density can be rewritten with a convective term as

$$\frac{\partial n_b}{\partial t} = v_c \frac{\partial n_b}{\partial s'} + k_f [n_{1T} - n_b] n_s - k_r n_b \qquad (6\text{–}32a)$$

The local contact zone edge velocity (the velocity at $s' = 0$) can then be written explicitly:

$$v_c = - \left[\frac{k_f n_{1T} n_s - (k_f n_s + k_r) n_b}{\dfrac{\partial n_b}{\partial s'}} \right]_{s'=0} \qquad (6\text{–}32b)$$

since $n_b(s' = 0) = n_{bc}$ and thus has a zero time derivative.

Once $k_f(y)$ and $k_r(y)$ are specified, Eqns. (6–32a and b) will relate receptor/ligand binding kinetics and local membrane geometry. Dembo et al. utilize a view of *bonds as springs*, with equilibrium length L and spring constant κ_b (see section 6.2.4.a). In addition, they confer upon the molecular transition state between the individual receptor/ligand pair and the resulting complex similar spring-like properties except with its own characteristic spring constant κ_{ts}. Their assumed *rate constant functionality* is:

$$k_f(y) = k_f(L) \exp \left[-\frac{\kappa_{ts}(y - L)^2}{2 K_B T} \right] \qquad (6\text{–}33a)$$

$$k_r(y) = k_r(L) \exp \left[\frac{(\kappa_b - \kappa_{ts})(y - L)^2}{2 K_B T} \right] \qquad (6\text{–}33b)$$

with the resulting association equilibrium constant:

$$K_A(y) = \frac{k_f(y)}{k_r(y)} = \frac{1}{K_D} \exp \left[-\frac{\kappa_b(y - L)^2}{2 K_B T} \right] \qquad (6\text{–}33c)$$

$k_f(L)$ and $k_r(L)$ are the normal, unstressed values for the chemical reaction rate constants (in two dimensions), previously denoted k_f and k_{r0}. K_D is then the unstressed equilibrium dissociation constant, $k_r(L)/k_f(L)$. The forward rate constant can depend on the separation distance because, in order to react, the receptor and ligand may need to stretch or compress, and thus possess energetically distorted states. The reverse rate constant may vary with separation distance because of the input of energy into the bond by stretching or compression. Notice that Eqn. (6–33c) is identical to Eqn. (6–15) used earlier in the equilibrium model of Bell and coworkers.

Two radically different types of dynamic behavior can be found depending on the comparative stiffness of the transition state and the bond. For $\kappa_b > \kappa_{ts}$, the bond is stiffer than the transition state, so bond dissociation is accelerated by the application of stress, i.e., increasing or decreasing y from L. This is the behavior arising from the dynamic model originally due to Bell (1978) and used by Hammer and Lauffenburger (1987); these types of bonds are termed *slip bonds*. In contrast, if $\kappa_b < \kappa_{ts}$, so that the bond is more flexible than the transition state, then bond dissociation is actually slower when the bond is stressed than when it is unstressed! Dembo *et al.* term these *catch bonds*. At present their existence is speculative, but their properties could explain confounding experimental observations such as the apparent irreversibility of adhesive work mentioned in section 6.2.4.b.

In addition to the equations given above to relate binding kinetics and membrane geometry, mechanical equations governing the membrane geometry must be specified. These equations incorporate the role of membrane material properties (*e.g.*, bending elastic modulus B) into force balances. Stresses due to receptor/ligand bonds are included, but forces due to external fluid flow are not considered in this formulation of the model.

The full system of equations is solved by Dembo *et al.* using a numerical finite-difference algorithm to determine the bond density, membrane displacement (x and y coordinates), and membrane tension as functions of time and arc length s. Parameter value ranges used were $k_r(L)/k_f(L)n_s = 10^{-9}$ to 10^4, $n_{1T} = 10^9$ to 10^{12} #/cm^2, $\kappa_b = 10^{-2}$ to 10 dyn/cm, $L = 1 \times 10^{-6}$–4×10^{-6} cm, and $B = 10^{-12}$–10^{-13} dyn-cm. An important implication of these values is that the ratio of membrane and bond flexibilities, $n_{1T}(K_B T)^2/B\kappa_b$, is much smaller than unity, so the membrane is extremely stiff relative to the bonds. Thus most of the applied membrane tension is transmitted to the bonds rather than stored or dissipated in membrane deformation.

In Figure 6–31a,b, the steady-state (*i.e.*, constant velocity v_c) values of cell/surface separation distance y, membrane tension T_f, and bond number density n_b, are plotted as functions of spatial position (parameterized by the arc length, s). The function $y(s)$ represents the cell contour at the edge of the contact zone. Notice the bond density distributions in Figure 6–31a for slip bonds and in Figure 6–31b for catch bonds. For the case of slip bonds, in which dissociation rates increase under stress, the bond density falls quickly to zero at the edge of the contact zone. In contrast, for the case of catch bonds there is a very small number of bonds, highly strained, at a substantial distance from the apparent edge of the contact zone. These bonds serve to maintain the cell position against the applied membrane tension, virtually by themselves.

Probably the most important result of the analysis is the analytical expression Dembo *et al.* derive for the *critical tension* T_f^c. This is the applied membrane tension required to overcome the tendency of the membrane to spread over the surface, and hence to peel the membrane away from the surface. It is formally calculated to be the tension at which the contact zone edge velocity v_c is equal to zero:

$$T_f^c = \frac{n_{1T} K_B T \ln(1 + n_s/K_D)}{1 + \cos \theta_f} \tag{6-34}$$

By comparison to Eqn. (6–21), it can be seen that this has a form similar to that of the classical Young equation. A major improvement is the explicit appearance of the total bond number density n_{1T} and binding equilibrium dissociation constant K_D. As the bond density and affinity (K_D^{-1}) increase, the critical membrane tension also increases. Although Eqn. (6–34) is derived with neglect of non-specific forces (except, perhaps, for steric stabilization interactions inherent in the value of L, the equilibrium unstressed bond length), Dembo *et al.*

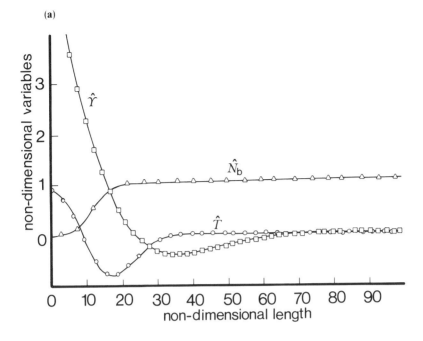

(a)

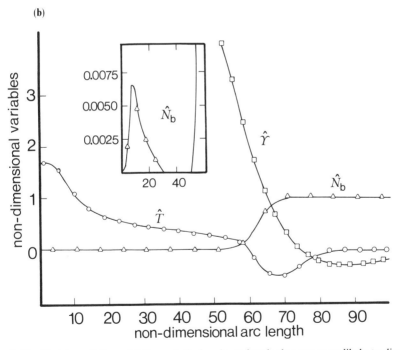

(b)

Figure 6–31 Model predictions. (a) Case of slip-bonds, or bonds that are more likely to dissociate under stress. For a constant value of the velocity v_c of the edge of the contact area, the scaled y coordinate of the membrane $(y - L)(\kappa_b/K_B T)^{1/2}$, scaled tension T_f/T_f^c, and scaled bond density (n_b/equilibrium value of n_b) are plotted as a function of the scaled arc length $(s - s_c)(\kappa_b/K_B T)^{1/2}$. (b) Case of catch-bonds, or bonds that are less likely to dissociate under stress. Scaling as in (a).

(*continued*)

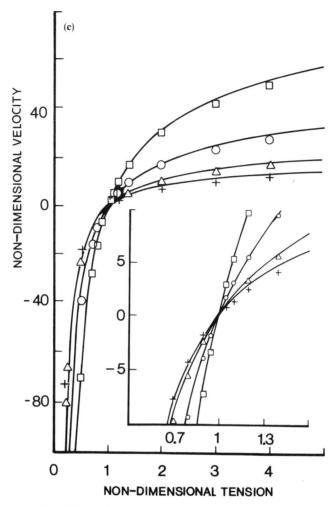

Figure 6–31 (*continued*)   (c) The scaled velocity $(v_c/k_r(L))(\kappa_b/K \cdot T)^{1/2}$ of the edge of the contact area is plotted as a function of the scaled tension T_f/T_f^c. Negative values of the velocity correspond to cell spreading; positive values of the velocity correspond to cell detachment. See original reference for parameter values. Redrawn from Dembo *et al.* (1988).

also provide a more complicated expression that includes a net repulsive potential.

A second major result of the analysis is an approximate formula for v_c, the velocity of the contact zone edge. An example set of calculations is shown in Figure 6–31c, with v_c plotted as a function of the applied membrane tension T_f for several values of K_D. For applied tensions below critical, there is typically a rapid spreading of the contact zone (indicated by negative values of v_c), whereas for larger applied tensions there is a more gradual increase in peeling of the cell from the surface.

The analysis presented by Dembo *et al.* (1988), as well as the models presented earlier in this section, do not incorporate the possibility of *focal contact formation*. Very recently, the approach of Dembo *et al.* has been adopted in an equilibrium analysis of the effect of focal contacts on the force needed to detach an adherent cell (Ward and Hammer, 1993). These investigators find that detachment of cells following focal contact formation may require forces 1000 times greater than with no focal contacts, forces of the order of those necessary to rupture the cell membrane itself. Future extensions of the Ward and Hammer model are likely to include the dynamics of focal contact formation and the mechanics of cell deformation.

Before leaving our discussion of cell adhesion, we note the models described in this section fail to account for a number of other aspects of adhesion. First, many cell–cell interactions are likely to involve *multiple types of receptor/ligand interactions*. It is likely that the unstressed lengths of the bonds vary with the identity of the receptor and ligand, which suggests that the separation distance s and distribution of types of bonds may change with position in the contact area. For example, Springer (1990) suggests that T cell receptor/MHC and CD2/LFA-3 interactions occur within intermembrane distances of 13 nm, whereas LFA-1/ICAM interactions occur with intermembrane distances of about 30 nm; these interactions may occur within the same macroscopic contact area. In addition, some receptors may be multivalent. An approach to incorporating multiple receptor species and valencies, as well as both thermodynamic and mechanical aspects of adhesion, has recently been proposed by Zhu (1991).

As a second example, in our discussions of this section we have admitted two possibilities for cell behavior: adhesion and complete detachment. The many observations of *cell rolling* clearly present an intermediate situation. In an attempt to describe cell rolling in fluid flow, Hammer and Apte (1992) formulate a probabilistic model in which the cell body is pictured as a sphere with numerous rigid microvilli on the surface. Receptors at the tips of the microvilli interact with a ligand-covered surface. The rate constants for association and dissociation are given by Eqns. (6–33a,b), but in probabilistic form as in Eqn. (6–27). Computer simulations based on this model are able to account for the entire spectrum of cell adhesion: no adhesion, rolling, and stable attachment. In particular, the model is able to account for the observations by Lawrence and Springer (1991), who observed the rolling and attachment of neutrophils during shear flow on artificial lipid bilayers containing CD62 adhesion molecules of the selectin family but not on bilayers containing ICAM-1 of the immunoglobulin family. Hammer and Apte suggest that the key property of selectins allowing them to mediate neutrophil attachment in shear flow is a very low value of the *bond slippage coefficient*, defined as the relative difference in spring constants for the bond and the receptor/ligand complex transition state in Eqns. (6–33a,b): $(1 - \kappa_{ts}/\kappa_b)$. When this quantity is near 0, force applied to a bond does not accelerate dissociation to any appreciable degree. Model computations indicate that the slippage coefficient for selectin/ligand family bonds must be less than 0.01 in order to account for the neutrophil rolling velocities observed by Lawrence and Springer. This would presumably be an unusually small value, compared to that for integrin/immunoglobulin family bonds, for instance. At the present time, however, no method is available for measuring the slippage coefficient.

6.3 CELL MIGRATION

6.3.1 Background

Active migration of eukaryotic cells through three-dimensional matrices and over two-dimensional substrata occurs in a wide spectrum of physiological and biotechnological situations. White blood cells of the inflammatory and immune receptors, including neutrophils, macrophages, and lymphocytes, must migrate through tissue spaces in order to carry out host defense functions such as phagocytosis of invading bacteria, killing of tumor and virus-infected cells, and adhesion-mediated cell–cell activation. Endothelial cells of microcirculatory blood vessel walls can be induced to migrate out into tissue to form new vascular networks in a phenomenon termed angiogenesis, and migration of epidermal cells and dermal fibroblasts into cutaneous wound spaces is needed for closure and healing of these wounds. Among other numerous examples, migration of tumor cells into organs is a key feature of tumor metastasis, nerve network formation appears to require migration of neural axons, and migration of embryonic cells to specific locations is a critical aspect of proper embryonic development. Overviews of these applications (Curtis and Pitts, 1980; Wilkinson, 1982; Trinkaus, 1984; Kater and Letourneau, 1985; Hudlicka and Tyler, 1986) and excellent expositions of information concerning biochemical and biophysical mechanisms underlying cell migration (Lackie, 1986; Singer and Kupfer, 1986; Devroetes and Zigmond, 1988; Harris, 1989; Bray, 1992) are available.

As in the previous two sections, our purpose here is to discuss mathematical models that relate cell function to receptor and ligand properties. In particular, we wish to develop quantitative understanding of the role of receptor/ligand binding, trafficking, and signaling processes in governing the speed and direction of cell movement. An ability to modulate cell migration behavior by manipulation of receptor and ligand properties, either pharmacologically or genetically, may be useful in developing improved therapies directed toward the applications mentioned above. Also, the new field of tissue engineering, where *in vivo* enhancement or even *in vitro* reconstruction of tissue function is desired, requires that cell populations distribute themselves appropriately within extracellular matrix environments.

6.3.1.a *Experimental Phenomena*

The basic form of eukaroytic cell migration in isotropic environments as observed *in vitro*, termed *random motility*, can be described as a *persistent random walk* (Gail and Boone, 1970; Dunn, 1983). That is, over short time periods a cell is typically found to move along a relatively straight path, whereas over long time periods movement appears similar to Brownian motion of inert particles, characterized by many direction changes (see Figure 6–32a,b). Over short time intervals the apparent cell velocity is constant. Over long time intervals, on the other hand, the rate of area coverage is constant.

"Short" and "long" time periods are not absolute, but merely relative to a characteristic of movement known as the *persistence time*, which we will denote with the symbol P (time). Although this quantity can be defined in rigorous, mathematical fashion (in a variety of ways, *e.g.* Dunn (1983); Othmer *et al.* (1988);

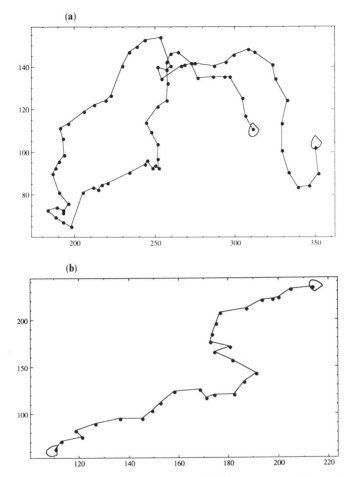

Figure 6–32 Experimental paths of individual neutrophil leukocytes undergoing random motility in uniform environments. Points represent cell centroid positions at 1-minute intervals, axis numbers correspond to position in μm. Note the persistence in movement direction for short time periods (a few minutes) and randomness in direction for longer time periods.

Tranquillo and Lauffenburger (1987)), an intuitive interpretation is simply the average time between significant direction changes. The magnitude of P can differ greatly among cell types, having been measured at about 4 min for rabbit neutrophils (Zigmond *et al.*, 1985), about 30 min for rat alveolar macrophages (Farrell *et al.*, 1990), 1 h for mouse fibroblasts (Gail and Boone, 1970), and 4–5 h for human microvessel endothelial cells and smooth muscle cells (Stokes *et al.*, 1991; DiMilla *et al.*, 1993).

    Cell speed, denoted by S (distance/time), can similarly be defined either by rigorous mathematical formula or intuitively as a centroid displacement per time (in the limit that time be short enough that movement is in a constant direction). S also varies among cell types, being observed on the order of 20 μm min^{-1} for neutrophils, 2 μm min^{-1} for alveolar macrophages, 30 μm h^{-1} for fibroblasts, and

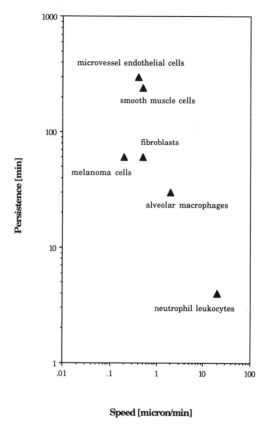

Figure 6–33 Plot of experimental measurements for cell speed and persistence time pairs for various cell types moving on two-dimensional substrata *in vitro*. All cell types are as mentioned in the text with corresponding references, except for the melanoma cells for which data are courtesy of R. Dickinson and R. Tranquillo, University of Minnesota (personal communication).

and 25–30 μm h^{-1} for microvessel endothelial cells and smooth muscle cells (with the same corresponding references as above).

Notice that a rough inverse correlation appears to exist between speed and persistence time (Figure 6–33). A rationale for such a relationship may be found by considering that the product of these two quantities is the cell's analog to a "mean free path-length", which should be neither too short nor too long compared to a characteristic size of the cell or its surroundings. If a cell had a very short mean free path-length it would effectively get nowhere, whereas if it had a very long mean free path-length it would risk moving great distances in improper directions.

An essential point to make about the speed and persistence time values plotted in Figure 6–33 is that they are merely representative measurements under particular assay conditions. For all cell types considered on this plot both speed and persistence time will exhibit a distribution of values depending on the cellular environment, as will be discussed in great detail as we proceed. Just as an example, speed and persistence time are likely to vary with the movement substratum for a given cell type, such as between a two-dimensional surface and a three-dimensional matrix. For instance, melanoma cells exhibit a much smaller persistence time when migrating within a three-dimensional collagen gel than when migrating over a two-dimensional collagen-coated surface (Dickinson and Tranquillo, 1993a,b).

Persistence is related to *cell polarity*, which allows locomoting cells to retain memory of their current direction (Devroetes and Zigmond, 1988; Singer and Kupfer, 1986). Most locomoting cells exhibit front (lamellipodal) and rear (uropodal) regions, recognizable through morphology, that are stable over times of the order of the directional persistence time. Both translocation and direction changes are accomplished primarily via active membrane protrusions termed lamellipodia extended from the cell front, with little similar activity at the rear.

Migration generally can be considered as a continual *movement cycle* of lamellipod extension and subsequent cell body translocation into an extended lamellipod (Figure 6–34). Mechanical forces needed to achieve locomotion are generally believed to be generated through polymer filaments of the protein actin (Egelhoff and Spudich, 1991; Heath and Holifield, 1991; Bray, 1992), but the fundamental mechanisms seem to be different for the two phases of the movement cycle. *Lamellipodal extension* appears to rely on polymerization of actin filaments from actin monomers attached to the local membrane, with the likely involvement of myosin-I. *Cell body translocation* appears to arise from contractile force

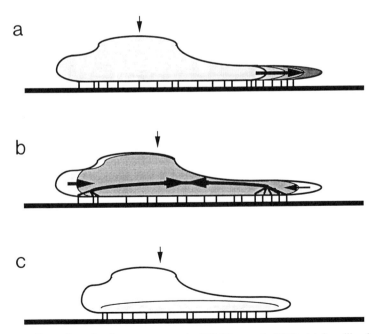

Figure 6–34 Illustration of cell movement cycle (DiMilla *et al.*, 1991). (a) lamellipod extension; (b) contractile force generation; (c) relaxation. Net cell translocation arises from displacement of the cell body toward the newly attached lamellipod with detachment of the uropod.

produced within the submembrane cortex consisting of actin filaments and myosin-I and/or myosin-II, with transmission of force into traction on the extracellular matrix or substratum via cell membrane interactions with the surroundings (Stopak and Harris, 1982). Both facets of actin-based force generation – polymerization and contraction – are active, energy-dependent processes. A *relaxation* phase may follow before the next round of lamellipod extension. These phases overlap in a continuous fashion, and we do not mean to imply a strongly discrete cycle.

Key characteristics of cell migration can be affected by environmental factors, primary among which are chemical stimuli acting through receptor-mediated signaling pathways. Figure 6–35 illustrates the major classes of behavior, following the categorization by Tranquillo and Alt (1990). The term orthokinesis refers to variations in speed, and klinokinesis to changes in persistence time (whether due to alterations in direction-change frequency or magnitude). When either of these

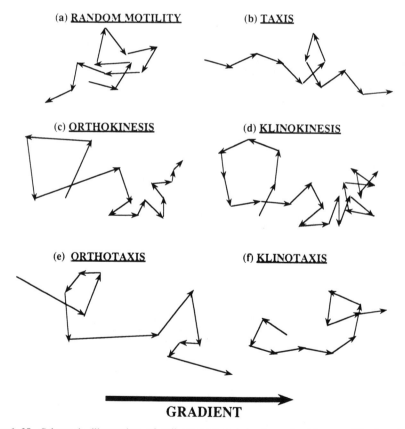

Figure 6–35 Schematic illustration of cell movement behavior categories possibly found in a concentration gradient of a chemical stimulus ligand (from Dickinson and Tranquillo, 1993a). Orthokinesis is a change in movement speed as concentration changes, klinokinesis is a change in turning frequency as concentration changes. Orthotaxis is speed affected by movement direction, and klinotaxis is random turning affected by movement direction. Taxis, or topotaxis, is directed turning in the gradient direction.

effects is due to specific chemical stimuli, the cell behavior is termed *chemokinesis*, typically without regard to the nature of the modulated property (*i.e.*, speed or persistence time). The direction of locomotion can be influenced by spatial gradients of environmental factors in *tactic* responses. Orthotaxis and klinotaxis denote speed and (random) turning frequency varying with direction relative to the gradient, respectively, whereas topotaxis (simply "taxis" in the figure) is true biased turning toward the gradient. A directional response to a concentration gradient of a soluble chemical stimulus is termed *chemotaxis*, and a directional response to a gradient of substratum adhesiveness is termed *haptotaxis*. Tactic responses to other types of stimuli may also be observed (Lackie, 1986).

A chemotactic response is typically a function of both the absolute attractant concentration and the steepness of the attractant concentration gradient. Example data for neutrophil leukocytes (Zigmond, 1977) and the slime mold amoeba *Dictyostelium discoideum* (Fisher *et al.*, 1989) are shown in Figures 6–36 and 6–37.

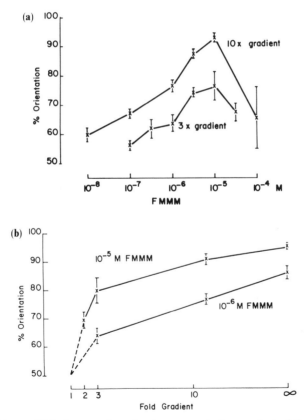

Figure 6–36 Neutrophil leukocyte orientation in a concentration gradient of the chemoattractant peptide fMMM (from Zigmond, 1977). (a) Fraction of cells oriented toward higher concentrations as a function of peptide concentration for a constant relative gradient steepness, either 3-fold or 10-fold across a 1 mm distance. (b) Fraction of cells oriented toward higher concentrations as a function of gradient steepness for a constant midpoint peptide concentration, either 10^{-5} or 10^{-6} M. Reproduced from the *Journal of Cell Biology*, 1977, Vol. 75, pp. 606–616 by copyright permission of the Rockefeller University Press.

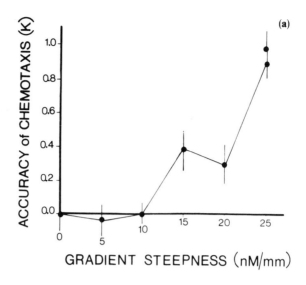

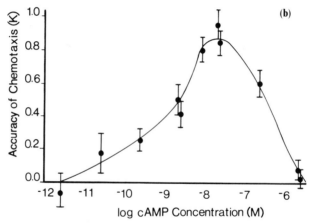

Figure 6–37 *Dictyostelium discoideum* orientation in a concentration gradient of the chemoattractant cAMP (from Fisher *et al.*, 1989). (a) Orientation accuracy as a function of gradient steepness, for a constant midpoint cAMP concentration of 25 nM. (b) Orientation accuracy as a function of cAMP concentration for a constant relative gradient steepness. Accuracy is expressed in terms of the von Mises distribution parameter K that measures how strongly individual cell directions cluster around the mean; in the range observed here K is roughly equal to the chemotactic index, CI. Reproduced from the *Journal of Cell Biology*, 1989, Vol. 108, pp. 973–984 by copyright permission of the Rockefeller University Press.

Chemotaxis is quantified in these plots as the fraction of a cell population moving toward higher attractant concentrations (Figure 6–36), and by the directional accuracy of this movement (Figure 6–37). *Directional orientation bias* increases with concentration gradient steepness, asymptotically approaching a maximal level as steepness increases. The dependence on attractant concentration is biphasic, increasing at low concentrations to a maximum, then decreasing as concentration increases further.

A related measure of the magnitude of a tactic response is the *tactic index* (McCutcheon, 1946): the ratio of net distance moved along the stimulus gradient to the total distance moved by a cell. When averaged over many cells, or over long time periods for a single cell, this quantity should vary between 0 and 1, the behavioral extremes of purely random and perfectly directed movement. The tactic index also depends on the attractant concentration and the attractant concentration gradient since it reflects both the magnitude of the stimulus and the response sensitivity. Unlike cell speed, persistence time, and directional orientation bias, then, the tactic index is not an intrinsic parameter characterizing cell behavior solely.

6.3.1.b Receptor-mediated Aspects

Cell surface receptors are involved in at least two ways in the process of cell migration. First, *adhesion receptors* (see section 6.2) appear to provide for conversion of intracellular contractile forces into traction on the substratum for cell locomotion. Second, *chemosensory receptors* transduce ligand binding events into chemokinetic and/or chemotactic movement responses.

It has been demonstrated that members of the *integrin* family (Buck and Horwitz, 1987; Albelda and Buck, 1990; Hemler, 1990) are required for migration of most cell types, at least *in vitro* over two-dimensional surfaces coated with extracellular matrix (ECM) proteins such as fibronectin, laminin, and collagen. Integrins are heterodimeric receptors (see section 6.2.1.a) consisting of an α and β subunit. Each subunit is a transmembrane protein possessing a large extracellular region capable of binding to specific sites on ECM proteins or related peptides, and a small intracellular domain that can form linkages with cytoskeletal elements, such as talin and α-actinin. The particular combination of α and β chains determines the ECM binding repertoire of a given integrin. For example, $\alpha_5\beta_1$ binds exclusively to fibronectin, whereas $\alpha_1\beta_1$ can bind to either laminin or collagen-IV. ECM binding requires association of both subunits, whereas the precise requirements for association with the cytoskeleton linkage are variable.

The ability of these adhesion receptors to reversibly interact with both ECM proteins and cytoskeletal structures via their extracellular and intracellular domains (see Figure 6–16), respectively, provides a mechanism for generating a net mechanical force on the substratum. At the same time, active signaling by these receptors may be involved in regulation of cell motility responses in both directions: ECM protein binding can influence integrin coupling with cytoskeletal elements, and cytoskeletal linkages can modulate extracellular ligand binding properties (Dustin, 1990; Hynes, 1992). The role of focal contacts (Burridge *et al.*, 1988), relatively large (roughly $1–3\,\mu m^2$) aggregates of linked integrins and cytoskeletal components located in membrane regions in very close proximity to the substratum, in locomotion is not clear at present.

Integrins may be able to mediate both haptotactic and chemotactic responses when the ECM ligands are either substratum-bound or in free solution (Aznavoorian *et al.*, 1990). The dependence of cell movement speed on the level of substratum-bound ligand is often biphasic, increasing initially with ligand density before reaching a maximum and then decreasing (*e.g.*, see Figure 6–38).

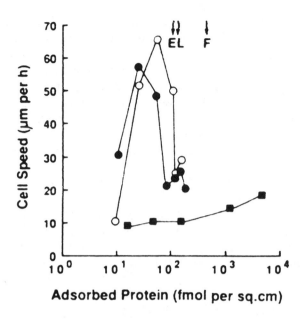

Figure 6–38 Cell movement speed versus substratum ligand density for murine myoblasts on laminin (open circles), laminin fragment E8 (filled circles), and fibronectin (filled squares). Redrawn from Goodman *et al.* (1989). Reproduced from the *Journal of Cell Biology*, 1989, Vol. 109, pp. 799–809 by copyright permission of the Rockefeller University Press.

However, monotonically increasing or decreasing plots of speed versus ligand density are sometimes found.

Chemosensory receptors for various soluble ligands that stimulate chemokinetic and/or chemotactic cell movement behavior have been found on a range of cell types. Phagocytic white blood cells (neutrophils and macrophages) possess receptors for formylated peptides (commonly found as bacterial metabolic products), activated complement component C5a and membrane lipid metabolite LTB$_4$ (both inflammatory response mediators, the former generated from complement interaction with antigens and the latter released by the phagocytic cells themselves), and exhibit chemokinetic and chemotactic responses to these chemical stimuli (Wilkinson, 1982). T-lymphocytes possess receptors for interleukin-8 (IL-8), which appears to help induce these cells to move from the bloodstream into tissue during immune responses (Oppenheim *et al.*, 1991). Endothelial cells possess receptors for acidic and basic fibroblast growth factor (FGF), which induce chemokinesis and chemotaxis in these cells during angiogenesis (Gospodarowicz *et al.*, 1989). As a further example, certain tumor cells possess receptors for a peptide obtained from degradation of the tissue protein elastin (Blood *et al.*, 1988); this peptide stimulates chemotaxis and chemokinesis by these cells, a behavior possibly relevant for *in vivo* metastasis into tissue spaces. Receptors responsible for transducing chemosensory cell migration responses generally work via interactions with G-proteins leading to membrane phospholipid turnover, elevation of intracellular Ca^{2+}, and activation of protein kinase C, all providing candidates for intracellular second messengers (Zigmond, 1989) (see Chapter 5). Some chemo-attractant receptors have been cloned and sequenced, and have been found to possess seven transmembrane segments like other receptor families that signal via G-proteins (Thomas *et al.*, 1990; Gerard and Gerard, 1991; Holmes *et al.*, 1991).

Understanding of the mechanism for *attractant gradient perception* is incomplete. Lackie (1986) provides a lucid analysis of major possibilities, including spatial comparison of attractant concentrations across a cell dimension and temporal comparison of attractant concentrations averaged over a cell as the cell moves. A spatial perception mechanism would compare receptor/attractant signals generated at two separate locations on a cell. In contrast, a temporal perception mechanism would compare receptor/attractant signals generated at two separate points in time at a particular location, perhaps the whole cell or only a portion of it. Phenomenologically, dependence of cell orientation bias in a gradient has been found experimentally to be proportional to a spatial difference in attractant-bound receptors across cell dimensions (Zigmond, 1977, 1981) (Figure 6–39). This relationship corresponds to the biphasic dependence of directional orientation bias on attractant concentration for a constant concentration gradient steepness seen in Figures 6–36 and 6–37. At low attractant concentrations very few receptors are bound so only a small difference results, and at high attractant concentrations almost all receptors are bound so again only a small difference is yielded. Maximum bias is found for attractant concentrations near the equilibrium dissociation constant, K_D, where the number of bound receptors is significant and most sensitive to differences in attractant concentration (Lackie, 1986).

Although this behavior, illustrated in Figure 6–39, does not rule out a

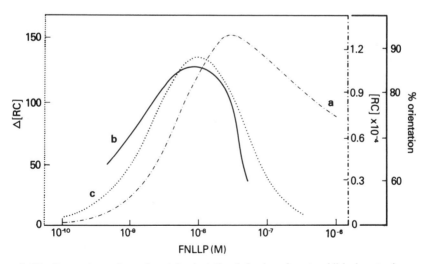

Figure 6–39 Comparison of experimental orientation behavior of neutrophil leukocytes in concentration gradients of the chemoattractant peptide fNLLP (solid line, b) with various quantities related to receptor/ligand complexes (Zigmond, 1981). Dashed line, a, is the number of total cell surface complexes after receptor downregulation. Dotted line, c, is the calculated absolute difference in complex numbers between points separated by 10 μm, *i.e.*, across the cell length along the gradient direction. The implication from this comparison is that these cells appear to exhibit orientation bias proportional to the absolute difference in complex numbers across a cell dimension in the gradient direction.

temporal sensing mechanism, experimental observations, mainly of directional orientation in steady gradients, support the ability of eukaryotic cells to perceive attractant concentration differences spatially (Zigmond, 1974; Lauffenburger et al., 1987; Fisher et al., 1989; Zicha et al., 1991). It is conceivable that spatial gradients are perceived by a "pseudo-spatial" mechanism (Dunn, 1981; Zigmond, 1982), comparing local temporal attractant changes generated from active cell processes such as extending lamellipodia in various regions of the cell. Even with this mechanism, however, a cell must compare intracellular signals generated by receptor/ligand complexes at different locations. Temporal variations in attractant concentration may contribute significantly in some circumstances along with spatial gradients to the net cell movement response. The slime mold amoeba, *Dictyostelium discoideum*, in particular, appears to respond directionally to temporal changes as well as spatial attractant concentration gradients (Fisher et al., 1989); this behavior may be peculiarly useful to these cells in their reponse to waves of attractant propagating from centers of aggregation (Lackie, 1986).

6.3.2 Experimental Methods

There exist two major classes of experimental approaches for measuring cell migration behavior: individual-cell assays and cell-population assays (Zigmond and Lauffenburger, 1986). The former are generally capable of generating more detailed data but are more difficult and time-consuming to carry out. Analysis of both types of assays requires mathematical modeling in order to yield objective parameters characterizing intrinsic cell motility properties.

6.3.2.a *Individual-cell Measurements*

Most individual-cell assays involve *visual tracking* of movement paths of a small number of cells, typically by time-lapse videomicroscopy methods. Cell migration on a protein-coated surface, beneath a fluid or gel medium, is followed by time-lapse observation. The medium may contain attractant at a specified concentration. When an attractant concentration gradient is desired, a gel (through which the chemical stimulus can diffuse freely) is necessary to suppress fluid convection. Agarose is the gel most commonly used. If the stimulus source is restricted to a particular location (*e.g.*, a well in the gel (Lauffenburger et al., 1983) or a hollow-fiber membrane (Fisher et al., 1989)), a spatial concentration gradient can be obtained. Transient and steady-state characteristics of this gradient are governed by assay geometry, molecular diffusivity, and cell uptake of the attractant (Lauffenburger et al., 1988). In all these cases, cell movement takes place over a two-dimensional substratum, which may be less physiological than assays involving three-dimensional matrices. Underlying mechanisms for migration may, in fact, differ between these two fundamentally different environments (Lackie and Wilkinson, 1984). Recently, procedures allowing tracking of cells within three-dimensional collagen matrices have been developed (Shields and Noble, 1987; Parkhurst and Saltzman, 1992; Dickinson and Tranquillo, 1993b), and these should grow in importance in the near future.

Techniques for quantifying cell behavioral responses in individual cell assays have improved markedly in recent years. Mathematical analysis permits rigorous determination of *root-mean-square speed*, S, and *persistence time*, P, from measurements of cell centroid position versus time along a cell movement path. Dunn (1983) and Othmer *et al.* (1988), although making different assumptions concerning the details of movement paths, obtain similar expressions relating S and P to the *mean-square cell displacement*, $\langle d^2 \rangle$, in n dimensions during a time interval, t:

$$\langle d^2 \rangle = nS^2[Pt - P^2(1 - e^{-t/P})] \tag{6-35a}$$

This expression successfully describes data such as those portrayed in Figure 6–32, as shown in Figure 6–40. Numerical values of S and P are best obtained by non-linear parameter estimation algorithms using Eqn. (6–34a) (*e.g.*, Farrell *et al.*, 1990), because graphical techniques (Dunn, 1983) lead to results dependent on the size of the time interval, t (Lackie *et al.*, 1987). Notice that for time intervals long compared to the cell persistence time, *i.e.*, $t \gg P$, this formula simplifies to the simpler expression

$$\langle d \rangle = nS^2 Pt \tag{6-35b}$$

This limit yields pure diffusion behavior, in which the mean-square displacement

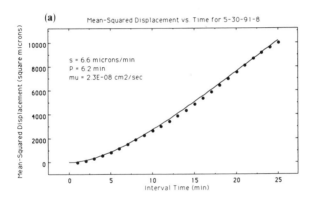

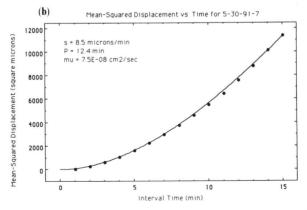

Figure 6–40 Plot of individual cell path data from Figure 6–32a,b according to Eqn. (6–34a), allowing determination of speed, S, and persistence, P. For cell a $S = 6.6\ \mu m/min$ and $P = 6.2$ min; for cell b $S = 8.5\ \mu m/min$ and $P = 12.4$ min.

is linearly proportional to the time interval with the diffusion coefficient being the constant of proportionality; this corresponds to the "long time" portion of Figure 6–40. We note that the Othmer *et al.* analysis gives P as a persistence in direction whereas that of Dunn uses P as a persistence in velocity (combining direction and speed); readers interested in this distinction are referred to the original papers.

The *chemotactic index*, CI, can be calculated approximately from the ratio of *mean cell displacement*, $\langle d \rangle$, up a chemical attractant concentration gradient, to the total cell path length, L_{path}, during a specified time interval, t: $CI = \langle d \rangle / L_{path}$. However, this definition neglects the influence of cell directional persistence at relatively short times. Othmer *et al.* (1988) derived an expression accounting for this complication:

$$CI = \frac{\langle d \rangle}{L_{path}} \left\{ 1 - \left(\frac{t}{P} \right)^{-1} [1 - e^{-t/P}] \right\}^{-1} \qquad (6\text{–}36a)$$

Again, for time intervals long compared to the cell persistence time $(t \gg P)$ this formula simplifies to the simpler expression

$$CI = \frac{\langle d \rangle}{L_{path}} \qquad (6\text{–}36b)$$

It should be emphasized that the values of S, P, and CI are not generally constant for a given cell type but rather can be influenced by a variety of environmental factors. Chemokinetic factors affect the value of S and/or P, by orthokinesis and/or klinokinesis, respectively. CI depends on the chemotactic factor concentration as well as on the concentration gradient steepness (see Figures 6–36, 6–37, and 6–39).

6.3.2.b Cell-population Measurements

Cell-population assays do not track the paths of individual cells, but instead examine number density profiles of many cells (typically 10^5 or greater) resulting after a given time period of migration. Similar in operation to the individual-cell assays is the *under-agarose assay*. A layer of agarose gel is created on a glass or plastic surface (microscope slide or culture dish), and cells are placed in a well from which they migrate underneath the gel. For random motility and chemokinesis experiments, the chemical stimulus is incorporated into the gel (and in the cell well medium) at a uniform concentration, whereas for chemotaxis experiments the stimulus is placed in a separate well from which it forms a concentration gradient by diffusing through the gel.

The most commonly-used assay, however, is the *filter assay*. Cells are placed on the surface of a porous filter separating two volumes of medium and are allowed to migrate into the filter for a given period of time. For chemokinesis experiments the stimulus is present in both medium volumes at equal concentrations, and for chemotaxis experiments it is present at greater concentration in the volume far from the cells. Gels prepared from physiological polymers, such as collagen, may be used instead of a filter.

Typically measured quantities in population assays are the so-called "leading front" (the distance moved by some arbitrary number of advance cells), the total number of cells migrating away from the initial source location, and the number of cells reaching some specified location (such as the bottom of the filter in that assay). These quantities are influenced by many factors besides intrinsic cell behavioral properties, such as assay geometry and distance scales, experimental time period, attractant diffusivity, and initial cell number. Thus, they are difficult to extrapolate to behavior under conditions any different from the particular circumstances under which they are measured. Another complication is that population migration in a gradient can be simultaneously affected by chemokinetic and chemotactic behavior. Without rigorous mathematical analysis, quantitative elucidation of the relative contributions of these two phenomena to population dispersal remains unsatisfactory.

A useful approach is the application of a mathematical model describing the cell population number density profile as a function of spatial position and time. Such a model permits determination of *model parameters* quantifying chemokinesis and chemotaxis, as demonstrated for both the under-agarose assay (Tranquillo *et al.*, 1988c) and the filter assay (Buettner *et al.*, 1989). Extensions of a phenomenological model originally proposed by Keller and Segel (1971) have served as a useful basis for this sort of approach. As reformulated by Lauffenburger (1983) from the derivation by Alt (1980), the cell flux $\mathbf{J}_c$ (cells/distance-time or cells/area-time in two- or three-dimensional assays, respectively) in the presence of a small chemoattractant concentration gradient ($\partial L/\partial x$) (concentration/distance) can be described reasonably well by the *cell flux expression* for movement along the gradient axis x:

$$\mathbf{J}_c = -\mu \frac{\partial c}{\partial x} + c \left\{ -\frac{1}{2} \frac{d\mu}{dL} + \chi \right\} \frac{\partial L}{\partial x} \qquad (6\text{--}37a)$$

where c is the cell number density (#/area or #/volume, in two- or three-dimensions, respectively), L is the attractant concentration, μ is the *random motility coefficient* (distance^2/time), and χ is the *chemotaxis coefficient* ((distance)^2/time-M). Both can vary with L, as will be described shortly. Basically, μ characterizes the amount of population dispersion, equivalent to a diffusion coefficient, and χ characterizes the amount of biased drift, similar to a velocity term for convection (Figure 6–41). Notice that two distinct processes can lead to biased migration in a gradient, as seen in the bracketed terms of Eqn. (6–37a). Only the bracketed term containing χ represents a true chemotactic contribution, due to directional bias in movement. The first term in the brackets is the chemokinetic contribution, resulting from changes in cell speed with concentration; for instance, if μ decreases as L increases, cells will accumulate at the higher concentration end of a gradient even if they possess no directional bias mechanism.

Experimentally, the *cell number density profile* $c(x, t)$, is obtained; that is, c is measured as a function of position x in the assay at a given time t. Model predictions for this profile are obtained by first substituting the cell expression (Eqn. (6–37a)) into the cell number density conservation equation:

$$\frac{\partial c}{\partial t} = -\frac{\partial \mathbf{J}_c}{\partial x} \qquad (6\text{--}37b)$$

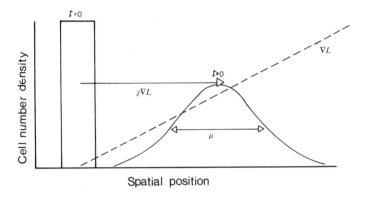

Figure 6–41 Schematic illustration of cell population migration in an attractant gradient. The cell number density profile spreads during a period of time t related to the random motility coefficient, μ. The profile peak moves during this period a distance related to the product of the chemotaxis coefficient χ and the attractant gradient ∇L.

resulting in a *partial differential equation* for $c(x, t)$ (a good background text is Pinsky (1991)). A solution can then be found either analytically or numerically (see Rothman and Lauffenburger, 1983). Eqn. (6–37b) merely says that the change in the number of cells at a given position, x, with time, t, is governed by the difference in cell flux toward and away from that position. Comparison of the model predictions for $c(x, t)$ to the experimental profiles allows determination of μ in uniform attractant concentrations and χ in attractant gradient conditions. However, the contributions of chemokinesis (variation of μ with L) to cell movement in a gradient must be separated out quantitatively first. The function $\mu(L)$ can be obtained from a set of random motility experiments at a series of uniform attractant concentrations, for which $(\partial L/\partial x) = 0$ and the bracketed term in Eqn. (6–37a) disappears. Figure 6–42 gives an example result for $\mu(L)$, showing a biphasic chemokinetic response of alveolar macrophages to the chemoattractants C5a (Farrell *et al.*, 1990).

Quantitative determination of the chemokinetic response, $\mu(L)$, then leaves χ as the only unknown in Eqn. (6–37a). This permits unambiguous determination of a value for χ in an attractant concentration gradient by applying the combination of Eqns. (6–37a and b).

It should be noted that Eqn. (6–37a) was derived under the assumption that chemokinetic behavior is due to orthokinetic effects (*i.e.*, variation of cell speed with chemical stimulus concentration) while neglecting klinokinetic effects (variation of persistence with concentration). When both effects are included it is not sufficient to know only $\mu(L)$ from cell-population experiments. Rather, one must know both $S(L)$ and $P(L)$ from single-cell experiments because the flux expression becomes more complicated, with explicit dependence of S and P on concentration involved in the bracketed chemokinetic term (Farrell *et al.*, 1990).

The chemotaxis coefficient χ is likewise not a simple constant, but rather a function of attractant concentration. Using the empirical result from Zigmond (1977, 1981) that the fraction of cells oriented up an attractant concentration gradient is proportional to the gradient in attractor/receptor complex number, C,

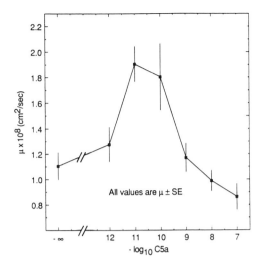

Figure 6–42 Plot of variation of random motility coefficient, μ, with concentration of chemoattractant C5a for rat alveolar macrophages (from Farrell *et al.*, 1990). This plot quantitatively captures the chemokinetic response of these cells to this stimulus.

across a cell dimension (see Figure 6–39), Tranquillo *et al.* (1988c) obtained the expression:

$$\chi = \chi_0 \frac{dC}{dL} S = \chi_0 \frac{R_T(L)K_D}{(K_D + L)^2} S \tag{6–38}$$

χ_0 is the *chemotactic sensitivity*, a simple constant possessing units of $(\#/\text{complexes}/\text{distance})^{-1}$; this reflects orientation bias per difference in complexes across the cell length along the gradient. The variation of total cell surface receptor number, R_T, with attractant concentration accounts for receptor downregulation due to endocytic trafficking. While the sensitivity χ_0 stays constant, χ decreases with increasing attractant concentration due to downregulation and saturation of receptor binding (Figure 6–43). The product of χ and the attractant concentration gradient (which is proportional to the attractant concentration) thus gives a chemotactic flux component with biphasic dependence on attractant concentration, as shown in Figure 6–39.

Allowing this functional dependence of χ on L provides excellent agreement between predictions from the cell population flux model and experimental cell number density profiles (Tranquillo *et al.*, 1988c). Rigorous experimental verification of Eqn. (6–38) was obtained by Farrell *et al.* (1990) (see Figure 6–43), who also made further improvement by modifying the cell flux expression to include the non-linear dependence of the chemotactic response on the magnitude of the attractant gradient (see Figures 6–36 and 6–37). The dependence of χ on the receptor and ligand properties L, R_T, and K_D is fundamentally related to signal transduction phenomena, likely involving considerations discussed in Chapter 5.

Comparing movement over a two-dimensional surface and within a three-dimensional matrix, the value of the random motility coefficient μ is approximately

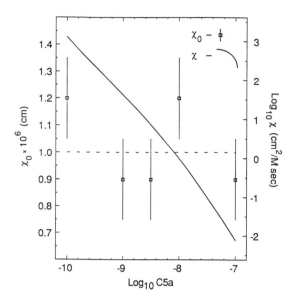

Figure 6–43 Plot of results from experiments by Farrell *et al.* (1990) for alveolar macrophage population migration in response to concentration gradients of C5a. Over three decades of C5a concentration the chemotactic sensitivity χ_0 remains constant at a value of 1.0×10^{-6} #/cm)$^{-1}$ while the chemotaxis coefficient χ changes by five orders of magnitude, consistent with Eqn. (6–38). This result implies that the sensitivity to a difference in receptor/attractant complex numbers remains constant, while the overall response to the attractant concentration gradient is modulated by binding, downregulation, and speed effects.

an order of magnitude smaller for neutrophil leukocytes in the filter assay than in the under-agarose assay (Rivero *et al.*, 1989), probably reflecting a greater resistance to movement in the porous matrix. At the same time, values of the chemotactic sensitivity χ_0 are of similar magnitude in the two assays, apparently indicating that the ability to orient the direction of movement is not noticeably altered by differences in the physical environment.

6.3.2.c *Relationship Between Individual-cell and Cell-population Parameters*

Because cell migration properties can be characterized at these two levels of description, it is important to investigate relationships between the individual-cell parameters (S, P, and CI) and the cell-population parameters (μ and χ). The model of Alt (1980) underlying Eqn. (6–37a) offers theoretical expressions for these relationships by deriving cell-population flux expressions derived from individual-cell properties. These expressions can be written as follows:

$$\mu = \frac{1}{n} S^2 P \tag{6–39a}$$

$$\chi = \frac{S \cdot \mathrm{CI}}{\nabla L} - \frac{1}{n}\left(\frac{d \ln P}{dL} - \frac{d \ln S}{dL} \right) \tag{6–39b}$$

where n is the number of dimensions in which cell movement occurs. Rivero *et al.* (1989) derive a similar result, based on a one-dimensional approximation for cell movement. The random motility coefficient, μ, is seen to be essentially the product of root-mean-square cell speed, S, and "mean free path-length", SP. The chemotaxis coefficient, χ, can be thought of as a fraction of the cell speed (effectively, the net speed in the direction of the stimulus gradient) per magnitude of the gradient, with a correction for the apparent bias in cell movement in the gradient direction due to chemokinetic effects rather than true chemotactic directional orientation.

Clearly, Eqn. (6–39a) predicts that the cell-population random motility coefficient, μ, is related to the individual-cell speed, S, and the directional persistence time, P. Similarly, Eqn. (6–39b) indicates that the cell-population chemotaxis coefficient, χ, should be related to S and the individual-cell chemotactic index, CI. Farrell *et al.* (1990) have provided *experimental verification* of these theoretical relationships for the case of aveolar macrophages responding to the activated complement factor C5a by measuring the individual-cell and cell-population parameter simultaneously as functions of C5a concentration and gradient magnitude. Their flux expression was actually slightly more complicated than Eqn. (6–37a), accounting for both orthokinesis (variation in S) and klino-kinesis (variation in P). Doing this required measuring the variations of both S and P with L in individual-cell chemokinesis experiments, instead of just measuring the overall variation of μ with L in cell-population chemokinesis experiments. By measuring all three individual-cell parameters and both cell-population parameters as functions of attractant concentration, these investigators demonstrated that Eqns. (6–39a,b) hold over many orders of magnitude of concentration, at least for this particular system.

6.3.3 Mathematical Models

We now proceed to a look at mathematical models for receptor-mediated cell migration behavior, limiting our interests to a mechanistic point of view. We will focus on the connection between receptor/ligand interactions and the behavioral response of individual cells, as for all the other cell functions we consider in this chapter.

A small but growing body of literature exists on mathematical models addressing two major questions concerning effects of receptor and ligand properties on cell migration behavior. One concerns effects of interactions between adhesion receptors and substratum-bound ligands on cell migration speed (Lauffenburger, 1989; DiMilla *et al.*, 1991). Another examines dependence of cell persistence time and directional orientation bias on processes involving chemosensory receptors and soluble chemoattractant ligands (Tranquillo and Lauffenburger, 1986, 1987; Tranquillo *et al.*, 1988a,b; Tranquillo, 1990). Finally, an effort has recently appeared that relates cell speed and direction simultaneously, investigating the role of adhesion receptor properties in haptotaxis (Dickinson and Tranquillo, 1993a). We will discuss each of these topics in turn.

6.3.3.a Adhesion Receptors and Cell Movement Speed

Experimental data demonstrate a dependence of cell speed on the adhesive interaction between cell adhesion receptors and substratum-bound ligands. An

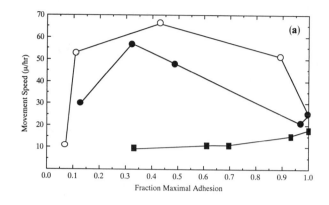

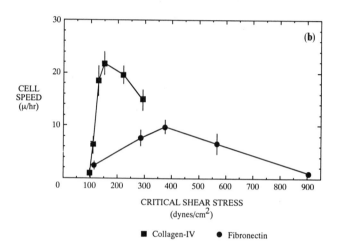

■ Collagen-IV   ● Fibronectin

Figure 6–44 Cell movement speed versus cell/substratum adhesiveness. (a) Data for murine myoblasts on laminin (open circles), laminin fragment E8 (filled circles), and fibronectin (filled squares). Plot was formulated by DiMilla *et al.* (1991) based on experimental data from Goodman *et al.* (1989) (see Figure 6–38) for migration and Goodman *et al.* (1987) for adhesion. Adhesion was determined by a sedimentation assay. (b) Data for movement speed of human smooth muscle cells as a function of cell/substratum adhesiveness on collagen-IV (squares) and fibronectin (circles), from DiMilla *et al.* (1993). Adhesiveness was measured using the radial flow chamber. (a) Reproduced from the *Biophysical Journal*, 1991, Vol. 60, pp. 15–37 by copyright permission of the Biophysical Society.

example is shown in Figure 6–44a, for skeletal muscle myoblasts migrating over polystyrene surfaces coated with the ECM proteins laminin and fibronectin, and an elastase-digestion fragment of laminin termed E8 (Goodman *et al.*, 1989). The cell speed data are the same as in Figure 6–38, but here the abscissa axis has been changed to cell/substratum adhesiveness from substratum ligand density, based on corresponding adhesion experiments by the same group of investigators using a sedimentation assay (section 6.2.2) (Goodman *et al.*, 1987). For laminin and the laminin-fragment E8, there is a *biphasic dependence of movement speed on the density of surface ligand*, with maxima occurring at densities approximately one-tenth to one-third of those at which cell attachment to the surface reaches a maximum. Low cell speed occurs on ligand densities giving relatively weak and strong extremes of adhesive strength, and high cell speed on ligand densities yielding intermediate adhesive strength. Fibronectin, on the other hand, gives a low, monotonically increasing cell speed even up through the regime representing greatest adhesive strength in the sedimentation assay (which involves only relatively weak distractive forces). Reasons for this contrasting behavior are not clear; it could arise from differences in a variety of receptor/ligand properties such as bond strength, production of signals for motile force generation, or organization of receptors by the cytoskeleton. Indeed, it is one aim of a mathematical modeling approach to try to elucidate the key underlying difference(s).

DiMilla *et al.* (1993) measured the speed of human smooth muscle cells migrating on fibronectin and collagen-IV. In Figure 6–44b cell migration speed is plotted directly as a function of the shear stress needed to detach cells from these surfaces as measured using a radial-flow chamber assay (section 6.2.2) capable of strong distractive forces. Each point represents a different protein density on the coated surface, corresponding between the migration and adhesion assays. This figure shows a clear biphasic dependence of speed on adhesive strength mediated by integrin/ECM protein interactions for both ligands in this case.

Mechanistic explanation of this kind of data is the goal of the mathematical models developed by Lauffenburger (1989) and DiMilla *et al.* (1991). A major issue is the elucidation of the effects of receptor/adhesion-ligand number densities and binding/dissociation rate constants on cell speed. These models are based on the underlying concept of a cell migration cycle, consisting of sequential stages of lamellipodal extension, cell body translocation, and relaxation (see Figure 6–35). A central presumption is that migrating cells generate the motile force necessary for cell body translocation within the cortical network of actin/myocin cytoskeleton and transmit this force as traction onto their substratum through dynamic adhesion interactions.

Observations like those in Figure 6–44 can be understood intuitively, at least in qualitative fashion. The key quantity, according to our viewpoint, is the *ratio of cell motile force to cell/substratum adhesive strength*, as illustrated in Figure 6–45. When this ratio is large, cell speed should be low because motile force generated within the cell is not transmitted effectively as traction. On the other hand, when this ratio is small the cell cannot produce sufficient force to break its attachments. It is only in an intermediate regime, where the motile force roughly balances the adhesive strength, that the cell can exhibit significant locomotion.

The magnitude of the velocity will depend on a number of additional factors, including cell mechanical properties. It also seems necessary that either the motile force or the adhesive interactions be asymmetric with respect to cell length, so that it does not merely exert isometric traction on the substratum and stay in place. At least three alternative *mechanisms for generating asymmetry* in adhesive interactions have been proposed: (1) spatial variation of adhesion receptor number due to polarized receptor trafficking (Bretscher, 1984); (2) localized proteolysis of cell/substratum attachments (Grinnell, 1986); and (3) localized covalent modification of adhesion receptors that may alter cell/substratum attachment strength (Tapley *et al.*, 1989). The common feature of these mechanisms is that they all can

INTRACELLULAR MOTILE FORCE / CELL-SUBSTRATUM ADHESIVENESS

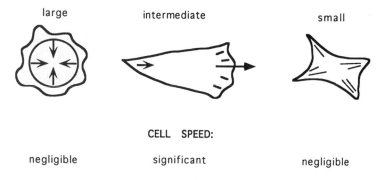

large   intermediate   small

CELL   SPEED:

negligible   significant   negligible

Figure 6–45 Conceptual foundation for dependence of cell movement speed on the ratio of intracellular motile force to cell-substratum adhesiveness. When adhesiveness is relatively small, force results in detachment of cell-substratum attachments; when it is comparatively large, force is insufficient to break attachments. Only when force and adhesiveness are in balance can attachments be disrupted dynamically; cell translocation also requires an asymmetry in adhesiveness.

generate spatial asymmetry in adhesion bond number comparing the front and back of the cell.

With this foundation, we can now discuss the mathematical model of DiMilla *et al.* (1991) for cell locomotion along a surface in one dimension. This model has its roots in a simpler model by Lauffenburger (1989) but provides a more rigorous description and analysis; actually, we will present a slightly modified version of the DiMilla *et al.* model for increased clarity. It contains three essential aspects: (1) a movement cycle framework; (2) kinetic balance equations governing the spatial distribution of adhesion–receptor/ligand bonds along the cell; and (3) a mechanical force balance relating intracellular motile force to cell displacement by means of spatially asymmetric adhesion-bond interactions with the substratum. We will briefly summarize each of these aspects in order.

During the *first stage* of the movement cycle, a period of duration t_1, the (polarized) cell extends a lamellipod in the forward direction. As the lamellipod extends, adhesion receptors on the ventral side of the extension form bonds with ligand immobilized on the underlying substratum. We assume that by the end of this process receptor/ligand interactions reach a steady state locally.

During the *second stage* of the movement cycle (with period t_c) the cortical cytoskeleton generates force that can be transmitted to the substratum as traction. Estimates of this force fall in the range of approximately 10^{-5} to 10^{-3} dynes/μm^2 for fibroblasts and endothelial cells (Harris *et al.*, 1980; Felder and Elson, 1990; Kolodney and Wysolmerski, 1992). Some portion of this force may serve to disrupt cell/substratum attachments at either the receptor/ligand or the cytoskeleton connections; in the latter case receptors and associated membrane may be left with the substratum (Regen and Horwitz, 1992). Disruption of attachments in some manner is, of course, necessary for cell body translocation. When there is asymmetry in the number of cell/substratum bonds in the lamellipodal and uropodal regions a net traction force on the substratum will be generated.

Translocation of the cell mass opposing this traction force, or cell locomotion, will occur if the substratum is fixed. When it is not fixed, the substratum itself can also be deformed (Harris *et al.*, 1980).

Finally, during the *third stage* of the cycle (period = t_r), the cytoskeleton relaxes, the motile apparatus ceases to transmit force to the bonds, and the receptors are permitted to reach an unstressed distribution once again. Altogether we have an entire *cycle duration* $t_m = t_1 + t_c + t_r$.

Within this framework, *adhesion receptor dynamics* can be analyzed using kinetic balances over the cell length, including association/dissociation, free receptor diffusion in the membrane, and free receptor endocytosis and reinsertion. Also, a viscoelastic-solid model can be used to describe both how the force generated by cytoskeletal elements is transmitted to the adhesion bonds at either end of the cell and how this force affects bond dynamics and the net deformation of the cell. Because the amount of force transmitted to each end of the cell will depend not only on the particular rheological properties and deformation of the cell under stress but also on the adhesion bond distribution, the two models are coupled. Mathematically, the receptor distribution model describes events during the entire cycle whereas the viscoelastic model is used only for the translocation phase of the cycle. Then, the net speed determined from cell displacement during the translocation phase can be averaged over the entire movement cycle to obtain an observable movement speed.

In order to easily determine the distribution of adhesion receptors on a moving cell, the cell geometry is treated as two flat sheets – dorsal (top) and ventral (bottom) – with length L and width W, "sewn" together along all edges (see Figure 6–46). The dorsal sheet contains only free receptors, whereas the ventral has both free receptors and receptors bound to ligand immobilized on the underlying substratum. The distribution of adhesion receptors on the two faces of the cell surface depends on the parameters describing binding and trafficking events. Denote the number density of free receptors ($\#$/area) on the dorsal and ventral surfaces as n_d and n_v, respectively, and let n_b be the number density of bound receptors ($\#$/area) on the ventral surface; these quantities vary with respect to spatial locations x and y. The total number of adhesion receptors on the cell surface is R_T.

Free receptors on the ventral surface can bind reversibly with ligand on the underlying substratum, with bonds forming with rate constant k_f [$(\#$/area$)^{-1}$ time^{-1}] and dissociating with rate constant k_{r0} (time^{-1}). We also assume here that the ligand is present in excess at a constant and uniform number density, n_s ($\#$/area). Free receptors on the dorsal and ventral surfaces diffuse in the membrane with diffusion coefficient D_R. Because the dorsal and ventral surfaces are joined at all edges and receptors must be conserved, free receptors nearing the edge of the cell on one surface can diffuse over the edge onto the other surface. Further, free receptors may be internalized by endocytosis into the cell's interior with first-order rate constant k_e. Assuming the total number of adhesion receptors R_T is constant, the rates of insertion of receptors into the cell surface and internalization must be equal. Restricting the model's applicability to persistent, stable cell migration, we assume that the cell is directionally polarized and that fresh receptors (either recycled or newly synthesized) are inserted on both faces

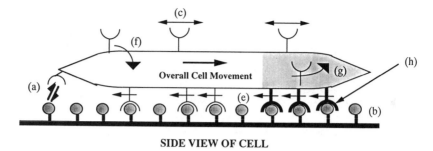

SIDE VIEW OF CELL

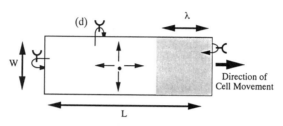

DORSAL VIEW OF CELL

Figure 6–46 Illustration of receptor dynamics portion of model by DiMilla *et al.* (1991). Key parameters include: (a) bond association and dissociation; (b) substratum ligand density; (c,d) receptor diffusivity; (e) apparent "convection" of bonds in reference frame of cell body; (f) receptor internalization; (g) receptor reinsertion into cell surface membrane (either recycled or newly-synthesized receptors); (h) relative strength of bonds between cell front and rear. Reproduced from the *Biophysical Journal*, 1991, Vol. 60, pp. 15–37 by copyright permission of the Biophysical Society.

of the cell's surface over a fraction λ of the cell length from the leading edge at a rate which balances internalization. These receptors will be inserted in equal distribution between the two faces of the cell over this forward region. Finally, choosing the frame of reference to be the cell body, bound receptor/ligand complexes will appear to "convect" backwards relative to the cell at the cell's overall forward velocity of v. It should be emphasized that no bulk membrane flow is assumed to occur, consistent with experimental observations (Kucik *et al.*, 1990; Lee *et al.*, 1990).

With these considerations, kinetic balance equations for free receptor densities, n_d and n_v, as functions of position x (length) and y (width) on the dorsal and ventral cell surfaces can be written at steady-state during the relaxation and extension stages:

$$0 = D_R\left(\frac{\partial^2 n_d}{\partial x^2} + \frac{\partial^2 n_d}{\partial y^2}\right) - k_e n_d + I_R \qquad (6\text{–}40a)$$

$$0 = D_R\left(\frac{\partial^2 n_v}{\partial x^2} + \frac{\partial^2 n_v}{\partial y^2}\right) - k_e n_v + I_R - k_f n_s n_v + k_{r0} n_b \qquad (6\text{–}40b)$$

The first term in each equation represents diffusion of free receptors in the membrane and the second term represents endocytic internalization of free receptors. The final two terms in Eqn. (6–40b) represent bond formation and dissociation on the ventral surface. The third term, I_R, in both equations represents insertion of fresh receptors (either recycled or newly synthesized) into the cell membrane, which may be restricted to taking place only over a fraction λ of the cell length from the front ($x = L$) so that

$$\left. \begin{array}{ll} 0 < x < (1 - \lambda)L, & I_R = 0 \\[2ex] (1 - \lambda)L < x < L, & I_R = \dfrac{k_e}{2\lambda LW}(R_T - B) \end{array} \right\} \qquad (6\text{–}40c)$$

Recall that the rate of receptor insertion is assumed to balance the rate of internalization, so that the total number of surface receptors, R_T, remains constant. B is the total number of adhesion bonds,

$$B = \int_0^L \int_{-W/2}^{-W/2} n_b(x, y)\, dx\, dy \qquad (6\text{–}41a)$$

Completing the set of receptor balances, n_b can be calculated from the steady-state kinetic balance for the bond density on the ventral cell surface:

$$0 = k_f n_s n_v - k_{ro} n_b + v \frac{\partial n_b}{\partial x} \qquad (6\text{–}41b)$$

In this equation the first two terms represent bond formation and dissociation, and the third term corresponds to the apparent "convection" resulting from choosing the cell body as a fixed frame of reference.

To solve Eqns. (6–40)–(6–41) nine boundary conditions are needed: four for each free receptor equation and one for the bound receptor equation. Matching free receptor numbers at the seams of adjoining surfaces provides four boundary conditions, and four more arise from similarly matching the free receptor fluxes. Conservation of total receptor number provides the final condition.

The remaining aspect of the model is *intracellular motile force generation*. Although in recent years many facets of cytoskeletal function have been identified, many of the mechanisms involved in force generation and transmission have yet to be definitively elucidated. Rather than develop a model for how cells generate force which depends on assumptions for detailed molecular events, we can represent cytoskeletal dynamics at a phenomenological level using a viscoelastic-solid model. During the contraction stage a *force* F_c (force/length) is produced within the cell. A portion of this force is transmitted to the substratum as traction while some is dissipated in cell deformation. Viscoelastic-solid models are useful for modeling the deformation of cells, which possess properties of both *stiffness* and *fluidity* (Elson, 1988) (see section 6.2.3.b). Stiffness can be represented by linear elastic springs, with *stress* (*i.e.*, force/area) proportional to *strain* (displacement) by the *elastic modulus*, E (force/area). Fluidity can be modeled by linear dashpots, or pistons, in which stress is proportional to the *rate of strain*, with *viscosity*, η, (force-time/area) serving as the proportionality constant.

A schematic of the viscoelastic-solid cell mechanics model used by DiMilla *et al.* is presented in Figure 6–47. The cell is divided into six compartments, each

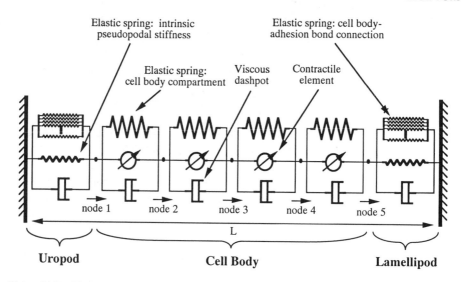

Figure 6–47 Illustration of cell mechanics portion of model by DiMilla *et al.* (1991). Elastic, viscous, and contractile elements are characterized by springs, dashpots, and vanes, respectively. Displacement of nodes represents movement of cell body. Reproduced from the *Biophysical Journal*, 1991, Vol. 60, pp. 15–37 by copyright permission of the Biophysical Society.

of length $L/6$. Six compartments were chosen so that all have length corresponding roughly to the lamellipodal and uropodal regions. The four interior compartments are identical and consist of a spring, dashpot, and contractile element in parallel. These compartments describe the mechanics of the cell body. The outer compartments, representing the uropod and lamellipod, also consist of dashpots and springs in parallel, but in these compartments we include two types of springs: a spring for the intrinsic stiffness of each pseudopod and springs representing connections between the cell body and adhesion bonds. These latter springs transmit the motile force to the adhesion bonds and the underlying substratum to provide the net traction necessary for movement in the presence of bond asymmetry. Mechanical properties and force generation of each compartment are assumed to be homogeneous for simplicity.

Locations of the compartment boundaries are represented by "nodes" denoted $i = 1$ through 5. Displacement of the position of each node, u_i, during the contraction stage is governed by a balance of the mechanical forces acting on it, as follows.

Node 1 (uropod/cell body):

$$\eta \frac{du_1}{dt} + Eu_1 + E_b B_u u_1 = \eta \frac{d(u_2 - u_1)}{dt} + E(u_2 - u_1) + F_2 \qquad (6\text{–}42a)$$

Nodes 2 through 4 (interior compartments):

$$\eta \frac{d(u_i - u_{i-1})}{dt} + E(u_i - u_{i-1}) + F_i = \eta \frac{d(u_{i+1} - u_i)}{dt} + E(u_{i+1} - u_i) + F_{i+1} \qquad (6\text{–}42b)$$

Node 5 (cell body/lamellipod):

$$\eta \frac{d(u_5 - u_4)}{dt} + E(u_5 - u_4) + F_5 = -\eta \frac{du_5}{dt} - Eu_5 - E_b B_1 u_5 \qquad (6\text{--}42c)$$

Each of these *force balances* contains three terms: (1) a viscous term involving the product of viscosity, η, and the rate of relative node displacement; (2) an elastic term involving the product of the cytoplasmic elastic modulus, E, and the relative node displacement; and (3) a contractile force term, F_i. The uropodal and lamellipodal compartments, which show up in the force balances for node 1 and node 5, respectively, additionally contain elastic terms representing the cytoskeleton/adhesion-bond linkages as springs (Dembo et al., 1988). These terms involve a bond elastic modulus E_b multiplied by the total number of bonds in these compartments, B_u and B_1, respectively.

It is important to recognize that these force balances do not follow the positions of the very ends of the lamellipod or uropod, but instead account only for "nodes" representing the interfaces of these extensions with the cell body. Thus, the lamellipod tip may stay fixed during the contraction stage whereas the position of node 5, u_5, is displaced. Also, although all bonds at the very end of the uropod must be disrupted for cell movement, there may remain a non-zero number of bonds, B_1, over the entire uropodal compartment.

Note that this model allows displacement of cell cytoplasm to take place by a combination of elastic deformation and viscous flow, consistent with observations by Heidemann et al. (1991). A new approach to studying strains and strain rates in moving cells by use of intracellular markers (Simon and Schmid-Schonbein, 1990) should soon provide more detailed information on cytoplasmic dynamics.

In order to solve the force balances comprising Eqns. (6–42a–c), we need to evaluate the *evolution in total uropodal and lamellipodal bond numbers*, B_u and B_1, during the contraction stage. Their values at the beginning of this stage are known from the solution of the receptor dynamics equations, Eqns. (6–40)–(6–41), integrating the bond number density distribution $n_b(x, y)$ over x and y within the uropodal and lamellipodal regions. But, given the stress acting on them during the contractile stage, the total bond numbers may be decreased. This process is modeled by treating the bond distributions in the uropodal and lamellipodal regions as uniform, thus governing changes in B_u and B_1 due to applied contractile stress. Assuming that receptor/ligand association and dissociation events occur on a faster time scale than does cell deformation, the bond numbers will quickly decrease to a new quasi-steady-state determined by the contractile stress. The equations governing this new steady state are simply kinetic balances for formation and dissociation of bonds in the uropodal and lamellipodal compartments, with the dissociation rate constants now reflecting the stress on the bonds:

$$\frac{dB_u}{dt} = 0 = k_f n_s R_u - k_{ru} B_u \qquad (6\text{--}43a)$$

$$\frac{dB_1}{dt} = 0 = k_f n_s R_1 - k_{r1} B_1 \qquad (6\text{--}43b)$$

R_u and R_l are the numbers of free receptors in the uropodal and lamellipodal compartments, respectively. R_{uT} $(= R_u + B_u)$ and R_{lT} $(= R_l + B_l)$ are the numbers of total receptors in the uropodal and lamellipodal regions; these quantities remain constant at their values obtained from the solution of Eqns. (6–40)–(6–41).

As originally suggested by Bell (1978) we assume that the bond association rate constant k_f is independent of the force acting on the bonds but the dissociation rate constant depends on applied forces. Hence, the values of k_{ru} and k_{rl} in the uropodal and lamellipodal regions, respectively, vary with the forces transmitted to bonds in each region. We also adopt Bell's expression for the role of force in the dissociation rate of a bond: the dissociation rate constants in the uropodal and lamellipodal compartments, k_{ru} and k_{rl}, respectively, become products of the intrinsic dissociation rate constants in these compartments absent stress, k_{r0}, and a Boltzmann-like exponential of mechanical energy applied per bond relative to the thermal energy (see Eqn. (6–9)). To allow the bonds in the rear of the cell to have a higher intrinsic dissociation rate than bonds in the front, we introduce ψ as the ratio of intrinsic dissociation rates in the absence of stress between front and back:

$$\psi = \frac{k_{rol}}{k_{rou}} \tag{6–44}$$

Allowing ψ to deviate from unity allows one means for providing receptor/ligand bond asymmetry across the cell length. $\psi < 1$, for instance, implies that the bond affinity is lower in the uropodal region than in the lamellipodal region, possibly due to covalent modification of the adhesion receptor or proteolysis of the ligand.

These considerations lead to expressions for the total number of bonds in each compartment during the contractile stage, derived from solutions of Eqns. (6–43a,b) using Eqns. (6–44) and (6–9):

$$B_u = \frac{R_{uT}}{1 + \dfrac{1}{\psi \kappa_A} \exp\{\varepsilon_b^2 u_1^2\}} \tag{6–45a}$$

$$B_l = \frac{R_{lT}}{1 + \dfrac{1}{\kappa_A} \exp\{\varepsilon_b^2 u_5^2\}} \tag{6–45b}$$

ε_b is the dimensionless bond elasticity, proportional to E_b. κ_A is the dimensionless intrinsic bond affinity, or *substratum adhesiveness*:

$$\kappa_A = \frac{k_f n_s}{k_{r0}} = \frac{n_s}{K_D} \tag{6–46}$$

κ_A is directly proportional to the surface ligand density and inversely proportional to the receptor/ligand equilibrium dissociation constant K_D, thus providing a measure of the strength of receptor/ligand association. Note in Eqns. (6–45a,b) that the effective affinity of the uropod is decreased by a factor of ψ compared to the lamellipod, according to Eqn. (6–44).

Substituting Eqns. (6–45a,b) into the force balances of Eqns. (6–42) allows determination of the node displacements following the contraction stage, applying a numerical integration algorithm to the resulting set of non-linear ordinary differential equations (see DiMilla *et al.*, 1991). The change of these displacements during a cycle represents the extent of cell body translocation during the cycle period. Thus, overall *cell speed* can then be calculated as the average velocity of the nodes over a full movement cycle of extension, contraction, and relaxation:

$$v = \frac{\sum_{i=1}^{5} u_i(t = t_c)}{5t_m} \qquad (6\text{–}47)$$

DiMilla *et al.* offer a spectrum of *model predictions* regarding the effect of key parameters on cell speed, based on estimates for the various parameter values. We will focus on a subset of these. Most important is the dependence of speed on the substratum adhesiveness, κ_A (defined in Eqn. (6–46)). Figure 6–48 shows a plot of cell speed as a function of κ_A for a series of values of the bond affinity asymmetry parameter ψ. A biphasic dependence is clear, with cell locomotion occurring over a limited range of adhesiveness. Hence, the model obtains a rigorous prediction corresponding to the intuitive argument outlined in Figure 6–45. Maximum speed is obtained when the adhesion bonds in the cell front are not disrupted by the motile force while the bonds at the cell rear are. It is thus dependent on the magnitude of the motile force and the fraction of the movement cycle over which it acts. The rising portion of the curve reflects an increasing ability of the bonds at the front to withstand the imposed stress, whereas the declining portion represents an increasing ability of the bonds at the rear to avoid disruption.

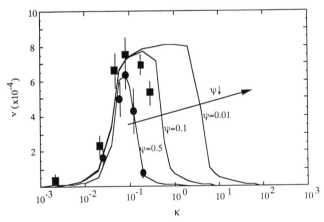

Figure 6–48 Calculated predictions from the model by DiMilla *et al.* (1991), for dimensionless cell speed, $v = v/k_{r0}L$, as a function of dimensionless cell/substratum adhesiveness $\kappa_A = n_s K_D$ with a series of values for the asymmetry in bond affinity ψ (the ratio of dissociation rate constants between cell front and rear, Eqn. (6–44)). Other parameter values used in the calculations are specified in the original reference. Superimposed are experimental data from Figure 6–44b normalized with respect to maximal speed value and location. The important result is that the range over which significant cell movement is observed can be accounted for by a modest asymmetry in bond affinity. Reproduced from the *Biophysical Journal*, 1991, Vol. 60, pp. 15–37 by copyright permission of the Biophysical Society.

It is significant to note that this model predicts that the extent of the range of adhesiveness over which cell locomotion can occur is primarily governed by the asymmetry in adhesiveness from the front to the rear of the cell. Figure 6–48 illustrates this for the case of a difference in bond affinity between the front and the rear. Polarized receptor trafficking, with insertion of fresh receptors into the membrane only near the front of the cell (*i.e.*, $\lambda \ll 1$ in Eqn. (6–40c)) provides similar though not identical predictions. In that case an increasing insertion rate, I_R of Eqns. (6–40a,b), yields a widening range of adhesiveness permissive for locomotion (see DiMilla *et al.*, 1991).

Superimposed on the model predictions in Figure 6–48 are the data from Figure 6–44b for smooth muscle cells migrating on plastic surfaces coated with fibronectin and collagen-IV. There is no attempt at a comprehensive quantitative match here, since most of the parameter values are merely estimates. But, the shapes of the theoretical and experimental curves are similar for ψ of the order of 0.1–0.5. That is, the extent of substratum adhesiveness allowing effective migration in this case can be accounted for by a ratio of bond affinities between lamellipod and uropod of the order of 2–10.

An important prediction that can be made from this plot concerns the dependence of cell speed on substratum ligand density, comparing ligands of different affinity for adhesion receptor binding. Recall that κ_A is the product of the ligand density, n_s, and the reciprocal of the binding dissociation constant, K_D. Thus, for a ligand of lower binding affinity (greater K_D), the biphasic curve will be shifted to larger values of n_s. Conversely, for a ligand of higher binding affinity (lower K_D) the curve should be shifted to smaller values of n_s. This prediction is illustrated qualitatively in Figure 6–49. Experimental data from Duband *et al.* (1991), for the migration of neutral crest cells, are shown for comparison in Figure 6–50, with the substratum-bound ligands being a panel of antibodies that bind to the fibronectin receptor. Using high affinity antibodies as the adhesion ligands, cell speed decreases with antibody density in the range studied, while when low affinity antibodies are used as the ligands, speed increases with antibody density. These observations may represent incomplete portions of biphasic curves, restricted to the particular ligand density ranges tested.

Another key parameter that shifts the location of the speed versus substratum adhesiveness curve is the total number of cell adhesion receptors, R_T. Increasing R_T moves this curve to lower adhesiveness, whereas decreasing R_T moves it to higher adhesiveness. This is a prediction that should be testable with cell lines exhibiting different levels of adhesion receptor expression.

Readers interested in further model predictions are referred to DiMilla *et al.* (1991). However, this model should be considered only as a starting point because of the need for more detailed information concerning central assumptions. For instance, one feature neglected in this model is any functional effect of adhesion-receptor organization into aggregates, such as focal contacts, by cytoskeletal interactions. It would be very interesting to examine the possible role of receptor organization in regulating cell migration, but there are few relevant data on this point so far. Also, while lamellipodal extension has been examined with theoretical models (*e.g.*, Oster and Perelson, 1987; Oster, 1988; Zhu and Skalak, 1988), until now these have not involved adhesion receptor interactions.

MODEL PREDICTIONS

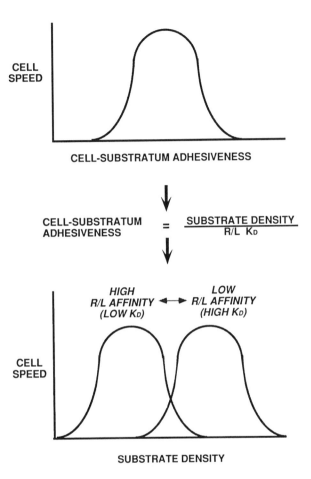

Figure 6–49 Illustration of predictions from model by DiMilla *et al.* (1991), for the effect of substratum ligand affinity for cell adhesion receptors on the dependence of cell speed on ligand density. Because the dimensionless cell/substratum adhesiveness κ_A is the product of ligand density and affinity, the curve shifts to lower ligand densities for higher affinities and *vice versa.*

6.3.3.b   Chemotaxic Receptors and Cell Movement Direction

We have earlier presented, in Figures 6–36 and 6–37, experimental data demonstrating the effect of chemoattractant ligand concentration and concentration gradient on movement direction, for neutrophil leukocytes and the slime mold amoeba *Dictyostelium discoideum.* Key features of these data are: (1) orientation bias increases asymptotically with the attractant concentration gradient steepness, ΔL, to a plateau value nearing perfect directional orientation bias; and (2) orientation bias has a biphasic dependence on absolute attractant concentration, L, (holding the relative gradient steepness $\varepsilon = \Delta L/L$ constant) with maximum response near the attractant/receptor binding dissociation constant K_D.

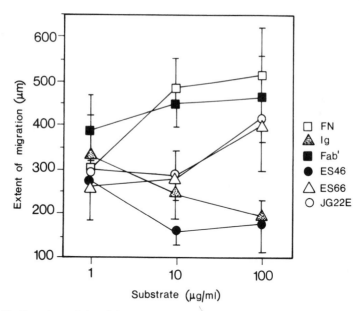

Figure 6–50 Experimental data from Duband *et al.* (1991) for migration by neural crest cells over ligand-coated surfaces, where the ligands are antibodies directed against cell adhesion receptors (along with fibronectin). Extent of migration was measured as distance travelled after 15 h. Increasing densities of low-affinity antibodies ES66 and JG22E enhance migration over the densities between 1 and 100 μg/ml, whereas the higher affinity antibodies ES46 and Ig2999 demonstrate the opposite trend. ES66 and JG22E have approximately the same affinity as fibronectin, Fn, while the others bind roughly two orders of magnitude more strongly. In comparison with the model predictions shown in Figure 6–49, these data may be interpreted as corresponding to rising and declining portions of curves for low- and high-affinity ligands, respectively.

Some conceptual underpinnings for these features are discussed by Devreotes and Zigmond (1988). A plausible picture is based on *modulation of endogenous cell polarity* by attractant/receptor binding (Figure 6–51). A locomoting cell is normally polarized, involving asymmetric cytoskeletal structure. In some as yet unknown manner this polarization helps inhibit lamellipod extension from the rear and sides of the cell. When a chemoattractant stimulus is present, receptor/ attractant complexes are formed on the cell membrane. Complexes generate intracellular signals leading to local increases in actin polymerization and weakening of the cortical cytoskeleton, both processes permitting lamellipodal extension in a region. Movement direction is then governed by competition among motile forces generated within localized regions of the cell front corresponding to areas of greatest attractant/receptor complex number. This picture seems to permit biased direction of movement toward higher attractant concentrations and to account – qualitatively, at least – for the features of Figures 6–36 and 6–37.

A fundamental issue that any successful model for cell migration direction must deal with is the *relationship between persistence time in uniform environments and directional orientation bias in attractant concentration gradients*. A model should explain why a cell moving in a given direction perceives a need to change

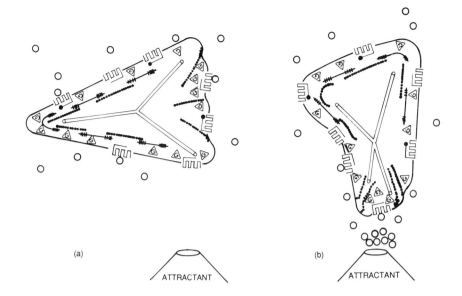

Figure 6–51 Schematic illustration of cell turning according to Devreotes and Zigmond (1988). (a) Locally high concentration of attractant molecules induces generation of actin filaments by G-protein interactions with receptor/attractant complexes. (b) Actin polymerization leads to turning impulse followed by orientation of polarity (signified by microtubules). Reproduced from the *Annual Review of Cell Biology*, *Vol.* 4, © 1988 by Annual Reviews Inc.

its direction, in the absence of any macroscopic attractant gradient; and, why at any point in time a certain fraction of the cells in a chemoattractant gradient are oriented in the "wrong" direction. These observations cannot be explained by heterogeneity in relevant cell properties. All cells undergo transient direction changes even in uniform attractant concentrations, and, in an attractant gradient, there are periods during which any given cell moves toward lower as well as higher chemoattractant concentrations.

A likely source of apparent cell "errors" in chemotactic orientation is that of *stochastic signal fluctuations* with magnitude comparable to the orientation signal. Consider a cell population with completely homogeneous properties – receptor number, receptor/ligand binding equilibrium and rate constants, signal transduction dynamics, motile force generation mechanics, etc. For such a population, the mean, or expected, orientation behavior would be uniform over the population at any point in time. If the external attractant gradient steepness is sufficient to induce correct orientation by one cell, then all other cells should exhibit precisely identical orientations. Plots like those of Figures 6–36 and 6–37 should demonstrate "all-or-none" behavior, with the fraction of correctly oriented cells being either 50% or 100% under any given condition. The intermediate behavior actually observed would, at least in part, arise from some source(s) of randomness occurring within the signal generation and/or response pathways.

Conceptually, the underlying premise is that of Figure 6–52. In macroscopically homogeneous environments fluctuations in the perceived attractant concentration at different regions of the cell can present to the cell a fluctuating apparent gradient signal leading to randomly-distributed orientation and movement direction changes. The time-frequency of these fluctuations would underlie the frequency of turns in a cell path, and thus be inversely related to the persistence time. In

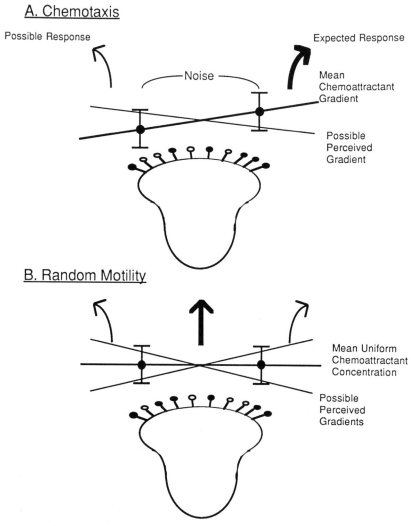

Figure 6–52 Receptor/ligand binding fluctuations as an underlying determinant of directional persistence in random motility (b) and limited directional bias in chemotaxis (a) (from Tranquillo *et al.*, 1988b). In uniform environments, fluctuations give rise to apparent gradients causing cell turning. In gradients, fluctuations give rise to false gradients causing incorrect orientation. Reproduced from the *Journal of Cell Biology*, 1988, Vol. 106, pp. 303–309 by copyright permission of the Rockerfeller University Press.

macroscopic concentration gradients, similar fluctuations could occasionally make the cell perceive an apparent reverse gradient signal leading to incorrect orientation and movement direction changes. Intuitively, the greater the ratio of fluctuations to mean signal, the smaller the persistence time and the lesser the chemotactic orientation bias. A valid mechanistic model for cell directional orientation, combined with a realistic estimate of the source and level of signal fluctuations, should simultaneously account for the observed values of persistence time and chemotactic sensitivity.

Signal noise resulting from the probabilistic kinetics of receptor/ligand binding has been calculated to be much greater than that from thermal fluctuations in local ligand concentration (Tranquillo, 1990) (see section 4.1). Since there is likely to be numerical amplification of the receptor binding signal along the signal transduction pathway (Devreotes and Zigmond, 1988), and consequently greater numbers of molecules involved in these subsequent steps, there is good reason to believe that the receptor-binding step produces the major share of the overall signal noise. It is this stage at which the minimum number of molecules of a given species (*i.e.*, receptors act). In section 4.1.2, we found that the *relative root-mean-square fluctuation* in receptor/ligand complex numbers at equilibrium, $(\delta C_{eq}/C_{eq})$, due to stochastic binding was given by (see Eqn. (4–15)):

$$\frac{\delta C_{eq}}{C_{eq}} = \left(\frac{K_D}{LR_T}\right)^{1/2} \tag{6–48}$$

Hence, at $L = K_D$ with $R_T = 10^4$ #/cell, $(\delta C_{eq}/C_{eq}) \sim 1\%$. Since this is the same magnitude as the attractant concentration gradients cells can effectively orient in at this concentration, the idea illustrated in Figure 6–52 is plausible.

Eqn. (6–48) was obtained for the case of instantaneous perception of attractant concentration via cell receptor binding. It neglects any time-averaging, or integrated memory, of earlier binding events. Since signal transduction resulting from receptor binding occurs through succeeding biochemical reactions possessing finite rates, attractant/receptor signaling events are effectively integrated over a *time-averaging period*, t_{avg}. DeLisi and Marchetti (1983) derived a more complicated expression for the relative root-mean-square fluctuation in receptor/ligand complexes, $(\delta C_{eq}/C_{eq})$, appropriate for a system exhibiting integrand memory. In this case, signal noise decreases as t_{avg} increases, because a greater number of binding events are sampled by the cell. One especially important prediction from their expression is that signal noise also decreases as the complex dissociation rate constant, k_r, increases. An explanation is that a greater rate of complex turnover, or new binding events, allows statistical sampling of more attractant molecules in the given time-averaging period, t_{avg}. Experimental testing of this prediction, for instance by using ligands possessing binding affinities for the receptor, is made difficult by the presence of endocytic trafficking (see Chapter 3). Internalization rate constants for receptor/ligand complexes are typically as great or greater than dissociation rate constants, so any effects of changes in k_r could be masked.

Given this foundation, equilibrium (Tranquillo and Lauffenburger, 1986) and dynamic models (Tranquillo and Lauffenburger, 1987) have been developed to explain cell directional movement properties. By rigorously quantifying known sources of randomness and incorporating these into simple hypotheses for signal generation and response, these models offer theoretical predictions that can be tested by comparison to quantitative experimental results like the data in Figures 6–36 and 6–37 and to individual cell paths obtained from analysis with the phenomenological models of the previous section. The equilibrium model is concerned only with cell orientation data, whereas the dynamic model additionally deals with directional persistence data during random motility. Another distinction is that the equilibrium model explicitly imposes a time-averaging period as a model parameter, whereas the dynamic model includes signal transduction and cell turning response kinetics so that signal integration arises implicitly from underlying rate constants. An excellent discussion of these models, plus the beginnings of a next-generation effort, is given by Tranquillo (1990).

Analysis of the relationship between persistence time in uniform environments and directional orientation bias in attractant gradients is crucial in our opinion, because cell turning mechanisms must be general – independent of whether or not a gradient exists macroscopically. Since the equilibrium model cannot treat this issue, we will focus here on a detailed discussion of the *dynamic model*. Substantial mathematical treatment of this model is provided by Tranquillo and Lauffenburger (1987) and Tranquillo (1990), and conceptual discussions and comparison to experimental data are given by Tranquillo et al. (1988a,b).

Consider a cell moving at constant speed, v, and with orientation angle θ relative to a fixed axis, such as the direction of a chemoattractant gradient (Figure 6–53). The cell is divided into two compartments along its polarization axis. The local extracellular attractant concentrations near each compartment are L_1 and L_2, leading to instantaneous cell surface receptor/attractant complex numbers C_1 and C_2, respectively. These complex numbers are dynamically governed by stochastic differential equations, derived from probabilistic balance equations for the probabilities that a given number of complexes are present (see section 4.1.2):

$$dC_i = \{[k_f L_i (R_T - C_{im}) - k_r C_{im}] - [(k_f L_i + k_r)(C_i - C_{im})]\}\, dt$$

$$+ \frac{1}{R_T^{1/2}} \{k_f L_i (R_T - C_{im}) + k_r C_{im}\}^{1/2}\, dW_i \qquad i = 1, 2 \qquad (6\text{--}49a,b))$$

The first bracketed set of terms is the *deterministic part* of these equations governing the mean, or expected, changes in complex number C_i during a time interval dt. Indeed, C_{im}, $i = 1, 2$, are the expected, or mean, values of C_i corresponding to attractant concentrations L_i, governed by deterministic species balance equations:

$$\frac{dC_{im}}{dt} = k_f L_i (R_T - C_{im}) - k_r C_{im} \qquad i = 1, 2 \qquad (6\text{--}50a,b)$$

The *probabilistic part* of Eqns. (6–49a,b) is the last term, the one involving dW_i. The instantaneous number of complexes in each compartment thus fluctuates around the mean, expected value according to the level of noise due to kinetic binding probabilistic effects, as in the plots of Figure 4–1. dW_i represents a random number taken from a Gaussian distribution possessing a mean of zero and standard deviation of unity. Hence, the magnitude of the fluctuations is

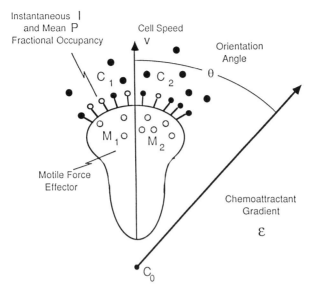

Figure 6–53 Schematic illustration for model of chemosensory cell movement direction by Tranquillo and Lauffenburger (1987). The cell translocates along its polarity axis with constant speed, v, and changes its movement direction angle θ with respect to a chemoattractant concentration gradient at a rate proportional to the difference in motile effector concentration, M, between the two compartments of an idealized lamellipodium. Signal is generated within each compartment with rate proportional to the instantaneous number of receptor/attractant complexes C on the surface of that compartment. C in each compartment is governed by stochastic binding kinetics (Eqns. (6–49a,b)).

determined by the coefficient terms multiplying dW_i. Solutions to Eqns. (6–49) and (6–50) can be obtained by numerical integration (using an explicit method such as Cauchy–Euler in order to prevent false averaging of fluctuations). Random noise inputs corresponding to dW_i are multiplied by the probabilistic part, then added to the deterministic part at each time interval (see Tranquillo and Lauffenburger, 1987, for details). This is how the plots in Figure 4–1 were, in fact, generated from Eqns. (6–49) and (6–50).

Note that as the total receptor number, R_T, increases the relative magnitude of the random term decreases; this corresponds to sampling a greater number of events as in the simple fluctuation expression of Eqn. (6–48). The relative significance of fluctuations also decreases as the association and dissociation rate constants k_f and k_r increase, because they appear in the probabilistic part to only the $1/2$ power compared to first-order in the deterministic part.

With transient receptor binding events specified, kinetic balance equations for the concentrations, M_i, of the critical intracellular *motile effector* in the two compartments can be written:

$$\frac{dM_1}{dt} = k_{gen} G(C_1) - k_{deg} M_1 - k_{diff}(M_1 - M_2) \tag{6–51a}$$

$$\frac{dM_2}{dt} = k_{gen} G(C_2) - k_{deg} M_2 - k_{diff}(M_2 - M_1) \tag{6–51b}$$

This motile effector could be a second messenger generated by the receptor/attractant complexes, *e.g.* Ca^{2+} or a lipid metabolite. Alternatively, it could be a component further along the signal transduction and response pathway, such as polymerized actin. At the present time, it would be premature to specify a particular candidate, especially since the primary goal of the model is to explore the hypothesis that receptor binding fluctuations are responsible for cell turning behavior. k_{gen} and k_{deg} are generation and degradation rate constants for this signal species, respectively, and k_{diff} is a rate constant for diffusive molecular transport between the two compartments. k_{diff} is basically equal to the motile effector diffusion coefficient divided by the square of the compartment length scale, so it has units of time^{-1}. Signal degradation is assumed to occur as a first-order process in signal concentration, perhaps by sequestration, enzymatic reaction, or protein-binding. The signal generation rate, $G(C_i)$, may take on a variety of forms, the simplest corresponding to an absolute complex number signal: $G(C_i) = C_i$; an alternative could represent a relative complex number signal: $G(C_i) = C_i/(C_1 + C_2)$.

Given intracellular motility effector levels, the model postulates that the cell changes direction with an angular rate proportional to the imbalance between the motile effects levels in the two compartments:

$$\frac{d\theta}{dt} = \omega(M_1 - M_2) \tag{6-52}$$

Recall that θ is the angle of current cell movement with respect to fixed reference axis. ω characterizes the *turning sensitivity* of the cell to an intracellular signal imbalance. One way to interpret Eqn. (6–52) is as a phenomenological *force balance* normal to the cell polarity axis: the net motile force induced by the imbalance in motile effector, transmitted to the substratum via cell/substratum adhesion bonds, is balanced by the resistive force opposing turning – likely due to passive cytoplasmic rheology and active cytoskeletal mechanics.

Eqns. (6–49)–(6–52) make up the stimulus–response portion of this model. In order to complete the system, one must close the loop by expressing the changes in the stimulus environment encountered as the cell continues to move. Equations for the local chemoattractant concentrations external to the two cell compartments can be written as:

$$\frac{dL_1}{dt} = \left[v \cos\theta - r_c \frac{d\theta}{dt} \sin(\theta + \theta_c) \right] \frac{\Delta L}{\Delta x} \tag{6-53a}$$

$$\frac{dL_2}{dt} = \left[v \cos\theta - r_c \frac{d\theta}{dt} \sin(\theta - \theta_c) \right] \frac{\Delta L}{\Delta x} \tag{6-53b}$$

$(\Delta L/\Delta x)$ is the chemoattractant gradient along the axis $\theta = 0$, and r_c and θ_c together define characteristic locations of the two cell compartments.

As the cell moves with speed v, its two lamellipodial compartments generate intracellular signal levels in accord with receptor/attractant binding events governed by stochastic binding kinetics and local attractant concentrations. These intracellular signal levels induce a smooth turning behavior superimposed on the linear cell speed. In this manner, *individual cell paths* can be simulated by numerical

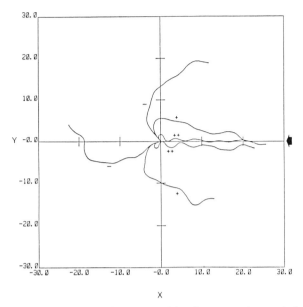

Figure 6–54 Sample computed paths for two model cells over a time period of 7.5 min (from Tranquillo and Lauffenburger, 1987) in chemoattractant gradients of relative steepness ε directed along the x-axis from the right. Axes denote position in μm. For each cell, the curve labelled '–' represents movement in the absence of a chemoattractant gradient, *i.e.*, $\varepsilon = 0$. '+' and '++' curves represent movement in a moderate ($\varepsilon = 0.008$) and large ($\varepsilon = 0.08$) gradient steepness, respectively. The set of random numbers characterizing stochastic binding fluctuations in Eqns. (6–49a,b) remain the same for a particular cell as the gradient steepness is changed.

solution of Eqns. (6–49)–(6–53), for individual realizations of random number sets, dW_i, causing receptor binding fluctuations in Eqns. (6–49a,b).

Sample cell paths are illustrated in Figure 6–54. This plot shows a set of three paths for each of two cells. ε is the fractional gradient, $\varepsilon = (1/2)\Delta L/L$, where ΔL is the concentration difference across the cell length. Each cell path is distinguished by its own distinct random number set, and the three paths are results for three different attractant concentration gradients oriented along the x-axis from the right-hand side: ($-$) $\varepsilon = 0$ (random motility), ($+$) $\varepsilon = 0.008$ (a moderate chemotactic gradient similar to that found in the Zigmond bridge assay), and ($++$) $\varepsilon = 0.08$ (a huge chemotactic gradient, greater than those typically found in experimental assays). Notice how, for $\varepsilon = 0$, qualitatively reasonable cell paths are obtained with short-term persistence and longer-term randomness. As the attractant gradient increases, direction changes become biased toward higher concentrations. Figure 6–55 gives "scatter diagrams", showing locations for a population of 100 model cells after 2.5 min and 7.5 min of migration, for the same three values of ε. Again, persistent random walk behavior, biased in attractant gradients, is seen.

Quantitative comparison to experimental data is accomplished by estimating as many model parameters as possible from independent sources, then fixing all others to give agreement with experimental persistence time measurements in uniform environments. R_T, k_f, and k_r are known for the particular system of

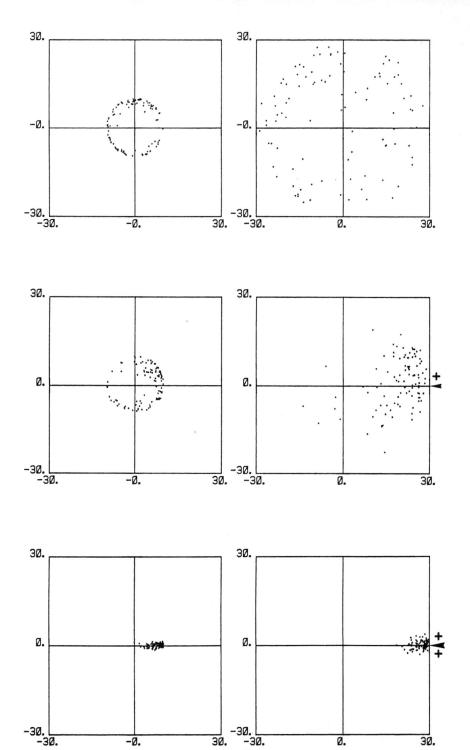

Figure 6–55 Scatter diagrams for a population of 100 model cells (from Tranquillo and Lauffenburger, 1987). Axes denote position in μm. Coordinates for cell location are plotted at 2.5 min (left-hand column) and 7.5 min (right-hand column) in chemoattractant gradients of zero (top, ε = 0), moderate (middle, ε = 0.008), and large (bottom, ε = 0.08) relative steepness from the right.

focus, neutrophil leukocytes. For *random motility* in the absence of an attractant gradient, analysis yields an explicit expression for the persistence time, P (Tranquillo and Lauffenburger, 1987):

$$P = \frac{k_r}{R_T} \left(\frac{\tau_R^2}{\tau_D} \right)^2 \left[\frac{\left(1 + \dfrac{L}{K_D} \right)^3}{\left(\dfrac{L}{K_D} \right)} \right] \tag{6–54}$$

where τ_R and τ_D are time constants characterizing cell turning response and turning signal decay, respectively:

$$\tau_R = \frac{1}{(\omega k_{gen})^{1/2}} \tag{6–55a}$$

$$\tau_D = \frac{1}{2k_{diff} + k_{deg}} \tag{6–55b}$$

The *response time constant* τ_R decreases for larger turning sensitivity, ω, and motile effector generation rate constant, k_{gen}. Similarly, the *decay time constant* τ_D decreases for larger motile effector diffusion rate constant, k_{diff}, and motile effector degradation rate constant, k_{deg}. Note that all the unknown model parameters – ω, k_{gen}, k_{diff}, and k_{deg} – affect the model results only through the lumped time constants τ_R and τ_D.

An important prediction implied by Eqn. (6–54) is that the persistence time, P, should increase as the receptor/attractant dissociation rate constant, k_r, increases. Increasing P is associated with a smaller likelihood of false gradients (Figure 6–52). This behavior arises from the role of k_r in determining the relative magnitude of binding fluctuations: greater values of k_r result in a greater number of binding events available to the cell for statistical sampling for a given integrated memory period. Note that this is consistent with the theoretical findings of DeLisi and Marchetti (1983) mentioned previously, that ligands with lower affinity for receptor binding should lead to smaller relative fluctuations in binding. Again, experimental test of this prediction using ligands possessing different affinities might be interesting, except for the anticipated masking effects of endocytic receptor trafficking which would take place on the same time scale as complex dissociation.

Eqn. (6–54) was obtained under the assumption that signal generation is proportional to the absolute complex number, *i.e.*, $G(C_i) = C_i$. Under an alternative assumption, that signal generation is proportional to the relative complex number, $G(C_i) = C_i/R_T$, the quantity R_T would be in the numerator instead of the denominator of the expression for P. An interesting test of these two alternatives would be to study persistence of cells possessing differing receptor numbers. If cells with greater R_T exhibited greater persistence times than those with fewer R_T a model with signal generation proportional to the relative complex number would be favored; on the other hand, if smaller persistence times were found for cells with greater R_T a model with signal generation proportional to the absolute complex number would be more consistent.

Simulations of *chemotactic motility* in attractant gradients reveal that only the ratio (τ_R^2/τ_D) matters for directional orientation, as well as for directional persistence (Eqn. (6–54)), so the quantities necessary to compute chemotactic movement paths are completely determined for this system once the persistence time P is measured. That is, predictions of directional orientation in a specified chemoattractant gradient of relative steepness ε are *a priori*. Figure 6–56 presents a comparison of model predictions and experimental data for directional orientation as a function of gradient magnitude. The data represent the percentage of cells oriented correctly up the gradient as a function of the fractional gradient, ε, with mean attractant concentration equal to the binding equilibrium dissociation constant, K_D (Zigmond, 1977). Model predictions are cell orientation percentages calculated from 100 simulated cell paths at each of the fractional gradients at this concentration, using known parameter values $R_T = 1 \times 10^4$ and $k_r = 0.4 \, \text{min}^{-1}$ along with the ratio $\tau_R^2/\tau_D = 10^2 \, \text{min}$ determined from fitting to $P = 4 \, \text{min}$. Agreement is from fair to excellent, depending on the value used for the cell radius. Under normal conditions, $r_c \sim 5 \, \mu\text{m}$ for neutrophil leukocytes, but in the bridge assay the cells are squeezed substantially between the cover slip and microscope slide bridge. Squeezing them to roughly twice their normal radius, so that $\tau_c \sim 10 \, \mu\text{m}$, gives an effective fractional gradient twice that previously estimated (Zigmond *et al.*, 1985). With this gradient, the *a priori* model predictions match the experimental measurements very well, clearly exhibiting the linear dependence of orientation bias on gradient size for small gradients and the asymptotic approach to a plateau at "large" gradients. Simulations aimed at the dependence of orientation bias on attractant concentration were not attempted, because the

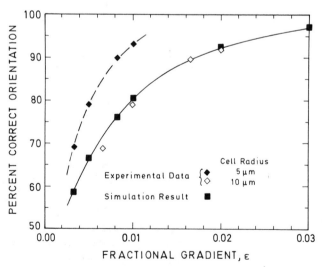

Figure 6–56 Directional orientation accuracy predicted for model cells (solid curve) and observed experimentally for neutrophil leukocytes (diamond symbols) from Tranquillo *et al.* (1988b). Two cell radius values are shown for calculation of fractional chemoattractant gradient across a cell length for the experimental orientation data; normally the cell radius is 5 μm, but when squeezed under a cover slip for the Zigmond visual assay used here (Zigmond, 1977) it is increased to about 10 μm.

degree of cell polarization (corresponding to the turning response term) does not appear to be constant as attractant concentration varies. It is also unlikely that the first-order approximations for signal transduction processes maintain constant parameter values across a wide range of ligand concentration (see Chapter 5).

A question to consider at this point is how the directional orientation accuracy might depend on receptor number, R_T, and complex dissociation rate constant, k_r. Since the relative level of binding fluctuations should decrease as k_r increases, orientation accuracy is predicted to be improved for lower affinity ligands. (This presumes that there is no lower limit for mean receptor occupancy time required to generate an effective signal.) As with the predicted effect on persistence time, an experimental test of this theoretical result would be very interesting if the interference of endocytic trafficking could be avoided. Recalling that the effect of receptor number on the relative level of binding fluctuations could be in either direction, depending on whether the motile effector is produced in proportion to the absolute complex number or the relative complex number, its consequent influence on orientation accuracy should be directly correlated with that on persistence time in either case.

The central finding of this work is that stochastic receptor binding fluctuations, with statistical magnitude known from measured kinetic binding and dissociation rate constants, coupled with a simple, linear intracellular turning signal response and decay mechanism, simultaneously provide explanation for both the frequency of turning in random motility and the bias in turning in chemotaxis.

Given some confidence in this model, an additional intriguing prediction follows. Figure 6–57 plots lines of constant persistence time, P, along with degree of orientation bias (quantified as percentage of cells correctly oriented), as functions of response and decay time constants, τ_R and τ_D. Values of total receptor number, $R_T = 1 \times 10^4$, and receptor/attractant complex dissociation rate constant, $k_r = 0.4 \text{ min}^{-1}$, are set at their known values, and $L/K_D = 1$ as always in this work. It is clear that both persistence and orientation bias depend primarily on the ratio τ_R^2/τ_D, as discussed earlier. The new prediction is that there is a special value of the persistence time that yields *maximal orientation bias*. That is, when the quantity τ_R^2/τ_D, and hence P, is too small, orientation is poor. As τ_R^2/τ_D, and P, increase orientation improves – until they become too great, at which time orientation once again becomes poor. Apparently, there is some intermediate, optimal level of signal "noise" coming out of the response/decay "filter". It is noteworthy that the predicted value of P for maximal orientation bias, 3 min, is close to that measured experimentally, 4 min, for this cell type (Zigmond et al., 1985).

The value of the quantity τ_R^2/τ_D giving this near-maximal, and apparently observed, orientation behavior is approximately 10^2 min. A crude estimate of τ_R of about 10–30 s for neutrophil leukocytes can be gained from experiments in which cell turning behavior was observed following local stimulation by attractant molecules released from a micropipette (Gerisch and Keller, 1981). Using $\tau_R = 20$ s yields $\tau_D \sim 0.1$ s as a signal decay time constant due to intracellular messenger diffusion or degradation. This is certainly plausible given that we might estimate the transport rate constant k_{diff}, to be $D/r_c^2 = (10^{-6} \text{ cm}^2/\text{s})/(5 \text{ }\mu\text{m}^2) \sim 4 \text{ s}^{-1}$ for a small messenger molecule. Substituting into Eqn. (6–55b) along with $k_{\text{deg}} = 0$, we obtain $\tau_D = 0.12$ s.

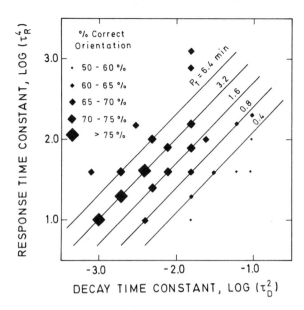

Figure 6–57 Model predictions for dependence of directional orientation accuracy on time constants (Eqns. (6–55a,b)) for system response, τ_R, and signal decay, τ_D, for $R_T = 10^4$ #/cell and $k_r = 0.4$ min^{-1}. Computed results for the percentage of cells oriented toward higher chemoattractant concentrations in a gradient of moderate relative steepness ($\varepsilon = 0.008$) is denoted by diamond symbols with size increasing as accuracy increases. Lines represent constant persistence time; as expressed in Eqn. (6–54) these follow a constant ratio of τ_R^2/τ_D. Notice that accuracy is greatest for values of τ_R^2/τ_D corresponding to a persistence time of about 3 min, which compares favorably with experimentally measured values. From Tranquillo and Lauffenburger (1987).

An additional set of relevant experimental data has been presented for alveolar macrophages, another type of phagocytic white blood cell. In response to the same chemoattractant, fNLLP, these cells exhibit orientation bias much poorer than do neutrophil leukocytes, with CI in the range 0.2–0.3 (Tranquillo et al., 1988a, Farrell et al., 1990). Although the receptor number and binding kinetics are similar between these cell types, the macrophages also move more slowly, by a factor of about three, and possess a greater persistence time by the same factor (Farrell et al., 1990). On Figure 6–57, this could correspond to an increase in the response time constant, τ_R, by approximately 3^4, or roughly 80. The figure predicts that increased persistence time and decreased orientation bias should be simultaneous consequences, both of which are observed as mentioned. This result is also consistent with the data plotted in Figure 6–33, comparing paired speed and persistence time measurements for neutrophil leukocytes and alveolar macrophages as well as a few other cell types.

6.3.3.c Combined Model for Movement Speed and Direction

Dickinson and Tranquillo (1993a) have recently presented a model that basically combines the elements of the previous two subsections, with the goal of accounting for the influence of adhesion receptors and substratum-bound ligand properties on both speed and direction of cell movement. An extension of the stochastic

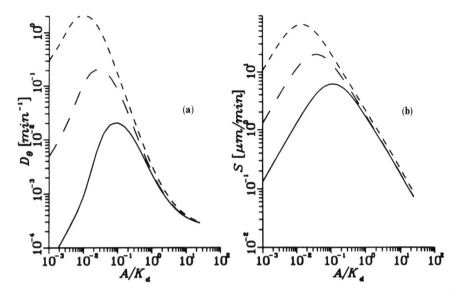

Figure 6–58 Predictions for simultaneous variation of speed and persistence time with effective motile force, f_r, according to the model of Dickinson and Tranquillo (1993a). (a) D_θ, the rotational diffusion coefficient, which is inversely proportional to the persistence time, P; (b) S, the cell speed. Both are plotted as a function of substratum ligand density (A in this figure) divided by the receptor/ligand equilibrium dissociation constant, K_D. Three values of the effective motile force are considered: $f_r = 5 \times 10^{-11}$ dynes (solid curve), $f_r = 5 \times 10^{-10}$ dynes (dashed curve), and $f_r = 5 \times 10^{-9}$ dynes (dotted curve). With all other model parameters held constant (see original reference), speed increases while persistence time decreases (i.e., D_θ increases) as f_r is increased. Compare to the experimental measurements of speed and persistence time shown in Figure 6–33, which appear to demonstrate similar inverse correlation between speed and persistence time.

framework underlying the dynamic model by Tranquillo and Lauffenburger (1987) is applied to interactions between adhesion receptors and substratum-bound ligands instead of to those between chemosensory receptors and soluble chemoattractant ligands. A third compartment representing the cell uropod is included. This new model is capable of predicting persistence time for cells moving randomly over uniform substrata and directional migration for cells moving on substrata possessing gradients in immobilized ligand. It uses a mechanical force balance involving transmission of intracellular motile force to traction on the substratum via adhesion bonds, in a fashion similar to that in the models by Lauffenburger (1989) and DiMilla et al. (1991), to predict cell speed as a function of ligand density. Additionally it uses a torque balance analogous to Eqn. (6–52) to predict cell turning behavior.

Although the fundamental concepts are therefore reminiscent of those discussed in the previous subsections, the model of Dickinson and Tranquillo is an extremely important advance because it has the potential for simulating cell movement paths comprehensively based on receptor/ligand properties and phenomenological cell mechanical features. It holds promise for rigorous analysis of hypotheses concerning haptotaxis, for instance.

More immediately, this model permits simultaneous prediction of cell speed and persistence time as they depend on underlying mechanistic parameters. As an example, Figure 6–58 shows plots of speed and a rotational diffusion coefficient, D_θ (which is the inverse of persistence time) as they vary with the effective cell motile force, essentially a ratio of motile force to cell/substratum adhesiveness. The graphs in this figure show S and D_θ as a function of substratum ligand density. For a given ligand density, cell speed increases and persistence time decreases (*i.e.*, D_θ increases) together as the effective motile force is increased. Thus, the model of Dickinson and Tranquillo predicts – at least under certain conditions – an inverse correlation between speed and persistence time.

Comparing this result with the experimental measurements of speed and persistence time pairs shown for various cell types in Figure 6–33, the model predictions are certainly in qualitative agreement with the data obtained to date. An essential point to reiterate about the speed and persistence time values plotted in Figure 6–33 is that they are merely representative measurements under particular conditions of substratum ligand density and chemoattractant concentration. Hence, for all cell types considered on that plot both speed and persistence time will exhibit a distribution of values depending on the measurement environment. Figure 6–58 emphasizes this point by elucidating predicted dependence of these quantities on substratum ligand density.

NOMENCLATURE

Symbol	Definition	Typical Units
a	contact area radius	cm
A	area	cm^2
A_c	contact area	μm^2
A_i	total surface area of cell i	μm^2
A_{max}	maximum cell–cell contact area	μm^2
B	membrane bending elastic modulus	dyn-cm
B	number of receptor/ligand complexes acting as bridges	#/cell
B_u	number of bridges in uropodal compartment	#/cell
B_l	number of bridges in lamellipodal compartment	#/cell
c	cell number density	cells/area or cells/volume[a]
CI	chemotactic index	
C	number of receptor/ligand complexes	#/cell
C_i	number of intracellular receptor/ligand complexes (section 6.1)	#/cell
C_i	number of receptor/ligand complexes in compartment i ($i = 1, 2$) (section 6.3)	#/cell
C_{im}	mean number of receptor/ligand complexes in compartment i ($i = 1, 2$)	#/cell
$C_{maximal}$	number of receptor/ligand complexes for maximal cell proliferation response	#/cell
C_s	number of receptor/ligand complexes on the surface	#/cell
$C_{threshold}$	number of receptor/ligand complexes for threshold cell proliferation response	#/cell
C_T	total number of receptor/ligand complexes	#/cell
dW_i	random number	
$\langle d \rangle$	mean cell displacement	μm
$\langle d^2 \rangle$	mean-square cell displacement	μm^2
D_R	receptor diffusion coefficient	cm^2/s

D_θ	rotational diffusion coefficient	s^{-1}
e	exponential constant ($\cong 2.7$)	
E	cell elastic modulus	dyn/cm^2
E_b	bond elastic modulus	dyn/cm^2
f_L	fraction of endocytosed ligands degraded	
f_R	fraction of endocytosed receptors degraded	
f	force applied to or by a bond	$dyn/bond$
f_c	force that will break a bond	$dyn/bond$
F	force	dyn
F_c	force per unit length acting on a cell during contraction	dyn/cm
F_i	force per unit length acting on a cell at node i	dyn/cm
G	Gibbs free energy (section 6.2)	$kcal/mole$
G	signal generation rate (section 6.3)	
I_R	rate of receptor insertion	$\#/cm^2$-s
$\mathbf{J}_c$	cell-population migration flux	$\#/cm$-s or $\#/cm^2$-s[a]
k_{deg}	overall rate constant for degradation (section 6.1)	min^{-1}
k_{deg}	rate constant for degradation of motile effector (section 6.3)	min^{-1}
k_{diff}	rate constant for diffusive transport of motile effector	min^{-1}
k_e	rate constant for internalization of free receptors	min^{-1}
k_{eC}	rate constant for internalization of complexes	min^{-1}
k_f	rate constant for association of receptor/ligand complexes	$M^{-1} min^{-1}$ or $cm^2/\#$-s[b]
k_g	specific cell growth rate constant	min^{-1}
k_{gen}	rate constant for generation of motile effector (for case of $G(C_i) = C_i$)	$M/(\#/cell)$-min
k_{gmax}	maximum specific cell growth rate constant	min^{-1}
k_{on}	intrinsic receptor/ligand association rate constant	$M^{-1} min^{-1}$
k_r	rate constant for dissociation of receptor/ligand complexes	min^{-1}
k_{rl}	rate constant for dissociation of receptor/ligand complexes in lamellipodal region	min^{-1}
k_{r0}	unstressed rate constant for dissociation of receptor/ligand complexes	min^{-1}
k_{r0l}	unstressed rate constant for dissociation of receptor/ligand complexes in lamellipodal region	min^{-1}
k_{ru}	rate constant for dissociation of receptor/ligand complexes in uropodal region	min^{-1}
k_{r0u}	unstressed rate constant for dissociation of receptor/ligand complexes in uropodal region	min^{-1}
k_t	transport rate constant for receptor movement	min^{-1}
k_u	rate constant for receptor/ligand uncoupling or dissociation	min^{-1}
K	membrane elastic area compressibility modulus	dyn/cm
K_A	binding equilibrium constant	M^{-1} or $cm^2/\#$ [b]
K_B	Boltzmann constant ($= 1.38 \times 10^{-16}$ g-cm^2/s^2-degree Kelvin)	
K_{cp}	coated pit affinity constant	
K_D	equilibrium dissociation constant	M or $\#/cm^2$ [b]
K_M	Monod model saturation constant	M
K_{ss}	apparent cellular affinity constant	
l_b	receptor/ligand bond length	cm
l_c	critical receptor/ligand bond length	cm
L	equilibrium bond length (section 6.2)	μm
L	cell length (section 6.3)	μm
L_{path}	cell path length	μm
L	ligand concentration	M
L_i	ligand concentration near compartment i ($i = 1, 2$)	M
M	bending moment	dyn

M_i	concentration of intracellular motile effector in compartment i ($i = 1, 2$)	M
n	number of dimensions for cell movement	
n_b	bridge density	$\#/\mu m^2$
n_{bc}	threshold value of the bond density	$\#/\mu m^2$
n_d	density of free receptors on dorsal surface	$\#/\mu m^2$
n_{ic}	density of free receptors within the contact area on cell i ($i = 1, 2$)	$\#/\mu m^2$
n_{iT}	density of total receptors on cell i ($i = 1, 2$)	$\#/\mu m^2$
n_{io}	density of free receptors outside the contact area on cell i ($i = 1, 2$)	$\#/\mu m^2$
n_j	density of receptors in jth state	$\#/\mu m^2$
n_s	ligand density on a surface	$\#/\mu m^2$
n_v	density of free receptors on ventral surface	$\#/\mu m^2$
N_j	number of receptors in jth state	$\#/cell$
p	probability density function	
p_a	probability that at least one bond exists	
p_{exp}	experimental sticking probability	
P	cell movement persistence time	min
P_B	probability that B bonds exist	
q	autocrine ligand secretion rate	moles/cell-min
Q	dimensionless autocrine ligand secretion rate	
Q_m	transverse shear resultant	dyn/cm
r_c	characteristic radius of cell lamellipod compartment	μm
r_0	range of bond potential energy minimum	nm
R	cell radius	cm
R_1	radius of curvature	cm
R_2	radius of curvature	cm
R_c	total number of receptors in the contact area	$\#/cell$
R_i	number of intracellular free receptors	$\#/cell$
R_{ic}	total number of free receptors in the contact area on cell i ($i = 1, 2$)	$\#/cell$
R_{io}	total number of free receptors outside the contact area on cell i ($i = 1, 2$)	$\#/cell$
R_{iT}	total surface receptor number on cell i ($i = 1, 2$)	$\#/cell$
R_l	number of free receptors in lamellipodal compartment	$\#/cell$
R_{lT}	total number of receptors in lamellipodal compartment	$\#/cell$
R_s	number of free receptors on the cell surface	$\#/cell$
R_T	total surface receptor number	$\#/cell$
R_u	number of free receptors in uropodal compartment	$\#/cell$
R_{uT}	total number of receptors in uropodal compartment	$\#/cell$
R_{iT}^c	critical total surface receptor number on cell	$\#/cell$
s	arc length	cm
s_c	value of arc length at edge of contact zone	cm
s'	arc length in moving reference frame	cm
s	cell–cell or cell–substrate separation distance	μm
S	fluid shear rate (section 6.2)	cm/min
S	cell movement speed (section 6.3)	$\mu m/min$
t	time	min
t_{avg}	time period for averaging of signaling events	min
t_c	contact time (section 6.2)	min
t_c	time period for force transmission (section 6.3)	min
t_m	time period entire cycle of cell movement	min
t_r	time period for relaxation	min
t_1	time period for lamellipod extension	min
T	absolute temperature	degrees Kelvin
T_f	tension	dyn/cm
T_m^0	applied membrane tension	dyn/cm
T_m	membrane tension	dyn/cm

T_f^c	critical tension	dyn/cm
u	scaled bond number	
u_i	position of node i	cm
v	fluid velocity (section 6.2)	cm/s
v	cell velocity (section 6.3)	cm/s
$\mathbf{v}$	vector of fluid flow rates	cm/s
v_c	velocity at edge of contact zone	cm/s
V_s	rate of new receptor synthesis and expression	(#/cell)/min
W	interaction potential per unit area (section 6.2)	J/m²
W	cell width (section 6.3)	μm
α	ratio of cell volume to ligand diffusion volume (section 6.1)	
α	fractional change in area (section 6.2)	
α_b	bond breakage energy	
β	geometric factor for cell growth substratum (section 6.1)	
β	relative contact area (section 6.2)	
γ	complex mitogenic signal generating activity (section 6.1)	% max DNA synthesis rate/#
γ	bond interaction range (section 6.2)	nm
δ	scaled receptor accumulation rate (section 6.2)	
δC	standard deviation in the value of C	#/cell
δ_g	thickness of the glycocalyx	μm
Δt	time step	min
ε	attractant gradient steepness	
ε_b	scaled bond elasticity	
κ	scaled equilibrium dissociation constant	
κ_A	scaled intrinsic bond affinity	
κ_b	spring constant for a bond	dyn/cm
κ_{ts}	spring constant for a bond transition state	dyn/cm
λ	fraction of cell length over which receptors are inserted	
λ_1	ratio of current length of element side to initial length	
λ_2	ratio of current length of element side to initial length	
η	fractional cell surface coverage of growth substratum (section 6.1)	
η	fluid shear viscosity (section 6.2 and 6.3)	dyn-s/cm²
μ	membrane elastic shear modulus (section 6.2)	dyn/cm
μ	cell random motility coefficient (section 6.3)	cm²/s
μ_b	bond or bridge chemical potential	dyn-cm
μ_b^0	standard chemical potential of a bond or bridge	dyn-cm
μ_j	chemical potential of the jth receptor state	dyn-cm
μ_j^0	standard chemical potential of the jth receptor state	dyn-cm
Γ	non-specific interaction energy per area of contact	dyn/cm
ζ	ratio of non-specific repulsion to specific binding	
ρ_{cell}	dimensionless association rate constant for ligand binding to cell surface receptors	
θ	angle between membrane and surface (section 6.2)	radians
θ	angle between cell direction and direction of attractant gradient (section 6.3)	radians
θ	scaled bond formation rate (section 6.2)	
θ_c	angle between cell axis and chemoattractant gradient axis	radians
θ_f	angle	radians
θ_0	macroscopic contact angle	radians
ψ	adhesion energy per unit area of contact	dyn/cm
τ	scaled time	
τ	isotropic tension	dyn/cm
τ_D	decay time constant	min
τ_R	response time constant	min

τ_s	surface shear resultant	dyn/cm
τ_s	shear stress	dyn/cm^2
σ	compressibility coefficient of the glycocalyx	dyn
σ_n	local normal stress due to receptor/ligand bonds	dyn/cm^2
Ω	angular velocity of the cell	radians/s
∇L	gradient in ligand concentration	M/cm
∇v	rate of fluid strain	s^{-1}
Δ	change in a quantity	
χ	chemotaxis coefficient	cm^2/s-M
χ_0	chemotactic sensitivity	cm/#
ψ	ratio of intrinsic dissociation rates	
ω	turning sensitivity	radians/M-min

[a] First set of units applies to the case of cell movement in two dimensions. Second set of units applies to the case of all movement in three dimensions.

[b] First set of units applies to the case of solution ligand binding to cell surface receptors. Second set of units applies to the case of surface-bound ligand binding to cell surface receptors, a two-dimensional reaction.

REFERENCES

Aaronson, S. A. (1991). Growth factors and cancer. *Science*, **254**:1146–1153.

Adamson, A. W. (1976). *Physical Chemistry of Surfaces*. New York: Wiley.

Aharanov, A. M., Pruss, R. M. and Herschman, H. R. (1978). Epidermal growth factor: relationship between receptor regulation and mitogenesis in 3T3 cells. *J. Biol. Chem.*, **253**:3970–3977.

Albelda, S. M. and Buck, C. A. (1990). Integrins and other cell adhesion molecules. *FASEB J.*, **4**:2868–2880.

Alt, W. (1980). Biased random walk models for chemotaxis and related diffusion approximations. *J. Math. Biol.*, **9**:147–177.

Aznavoorian, S., Stracke, M. L., Krutzsch, H., Schiffmann, E. and Liotta, L. A. (1990). Signal transduction for chemotaxis and haptotaxis by matrix molecules in tumor cells. *J. Cell Biol.*, **110**:1427–1438.

Bailey, J. and Ollis, D. (1986). *Biochemical Engineering Fundamentals*, 2nd edition. New York: McGraw-Hill, 984 pp.

Baserga, R. (1985). *The Biology of Cell Reproduction*. Cambridge, MA: Harvard University Press.

Bell, G. I. (1978). Models for the specific adhesion of cells to cells. *Science*, **200**:618–627.

Bell, G. I. (1988). Models of cell adhesion involving specific binding. *In* P. Bongrand (Ed.), *Physical Basis of Cell–Cell Adhesion*, pp. 227–258. Boca Raton, FL: CRC Press.

Bell, G. I., Dembo, M. and Bongrand, P. (1984). Cell adhesion: competition between nonspecific repulsion and specific bonding. *Biophys. J.*, **45**:1051–1064.

Berenson, R. J., Bensinger, W. I. and Kalamasz, D. (1986). Positive selection of viable cell populations using avidin-biotin immunoadsorption. *J. Immunol. Meth.*, **91**:11–19.

Blood, C. H., Sasse, J., Brodt, P. and Zetter, B. R. (1988). Identification of a tumor cell receptor for VGVAPG, an elastin-derived chemotactic peptide. *J. Cell Biol.*, **107**:1987–1993.

Bohmer, R. M. and Beattie, L. D. (1988). Probability of transition into cell cycle does not depend on growth factor concentration. *J. Cell. Physiol.*, **136**:194–197.

Bongrand, P. and Bell, G. I. (1984). Cell–cell adhesion: parameters and possible mechanisms. *In* A. S. Perelson, C. DeLisi and F. W. Wiegel (Ed.), *Cell Surface Dynamics: Concepts and Models*, pp. 459–493. New York: Marcel Dekker.

Brachmann, R., Lindquist, P. B., Nagashima, M., Kohr, W., Lipari, T., Napier, M. and Derynck, R. (1989). Transmembrane TGF-alpha precursors activate EGF/TGF-alpha receptors. *Cell*, **56**:691–700.

Brandley, B. K., Swiedler, S. J. and Robbins, P. W. (1990). Carbohydrate ligands of the LEC cell adhesion molecules. *Cell*, **63**:861–863.

Bray, D. (1992). *Cell Movement*. New York: Garland Publishing. 406 pp.

Bretscher, M. (1984). Endocytosis: relation to capping and locomotion. *Science*, **224**:681–686.

Buck, C. A. and Horwitz, A. F. (1987). Cell surface receptors for extracellular matrix molecules. *Annu. Rev. Cell Biol.*, **3**:179–205.

Buettner, H. M., Zigmond, S. H. and Lauffenburger, D. A. (1989). Cell transport in the millipore filter assay. *AICHE J.*, **35**:459–465.

Burck, K. B., Liu, E. T. and Larrick, J. W. (1988). *Oncogenes: An Introduction to the Concept of Cancer Genes*. New York: Springer-Verlag, 300 pp.

Burridge, K., Fath, K., Kelly, T., Nuckolls, G. and Turner, C. (1988). Focal adhesions: transmembrane junctions between the extracellular matrix and the cytoskeleton. *Annu. Rev. Cell Biol.*, **4**:487–525.

Campion, S. R., Matsunami, R. K., Engler, D. A. and Niyogi, S. K. (1990). Biochemical properties of site-directed mutants of human epidermal growth factor. *Biochemistry*, **29**:9988–9993.

Cantley, L. C., Auger, K. R., Carpenter, C., Duckworth, B., Graziani, A., Kapeller, R. and Soltoff, S. (1991). Oncogenes and signal transduction. *Cell*, **64**:281–302.

Capo, C., Garrouste, F., Benoliel, A. M., Bongrand, P., Ryter, A. and Bell, G. (1982). Concanavalin-A-mediated thymocyte agglutation: a model for quantitative study of cell adhesion. *J. Cell Sci.*, **56**:21–48.

Carpenter, G. (1987). Receptors for epidermal growth factor and other polypeptide mitogens. *Annu. Rev. Biochem.*, **56**:881–914.

Carpenter, G. and Wahl, M. I. (1990). The epidermal growth factor family. *In* M. B. Sporn and A. B. Roberts (Eds), *Peptide Growth Factors and Their Receptors I*, pp. 69–171. Berlin: Springer-Verlag.

Chan, P., Lawrence, M. B., Dustin, M. L., Ferguson, L. M., Golan, D. E. and Springer, T. A. (1991). Influence of receptor lateral mobility on adhesion strengthening between membranes containing LFA-3 and CD2. *J. Cell Biol.*, **115**:245–255.

Chen, W. S., Lazar, C. S., Lund, K. A., Welsh, J. B., Chang, C. P., Walton, G. M., Der, C. J., Wiley, H. S., Gill, G. N. and Rosenfeld, M. G. (1989). Functional independence of the epidermal growth factor receptor from a domain required for ligand-induced internalization and calcium regulation. *Cell*, **59**:33–43.

Cozens-Roberts, C., Lauffenburger, D. A. and Quinn, J. A. (1990a). Receptor-mediated cell attachment and detachment kinetics: I. Probabilistic model and analysis. *Biophys. J.*, **58**:841–856.

Cozens-Roberts, C., Quinn, J. A. and Lauffenburger, D. A. (1990b). Receptor-mediated adhesion phenomena: model studies with the radial-flow detachment assay. *Biophys. J.*, **58**:107–125.

Cozens-Roberts, C., Quinn, J. A. and Lauffenburger, D. A. (1990c). Receptor-mediated cell attachment and detachment kinetics: II. Experimental model studies with the radial-flow detachment assay. *Biophys. J.*, **58**:857–872.

Cross, M. and Dexter, T. M. (1991). Growth factors in development, transformation, and tumorigenesis. *Cell*, **64**:271–280.

Curtis, A. S. G. and Lackie, J. M. (Eds.) (1991). *Measuring Cell Adhesion*. New York: John Wiley.

Curtis, A. and Pitts, J. (1980). *Cell Adhesion and Motility*. Cambridge: Cambridge University Press.

Czech, M. P., Clairmont, K. B., Yagaloff, K. A. and Corvera, S. (1990). Properties and regulation of receptors for growth factors. *In* M. B. Sporn and A. B. Roberts (Ed.), *Peptide Growth Factors and Their Receptors I*, pp. 37–65. Berlin: Springer-Verlag.

DeLisi, C. and Marchetti, F. (1983). A theory of measurement error and its implications for spatial and temporal gradient sensing during chemotaxis. *Cell Biophys.*, **5**:237–253.

Dembo, M., Harlow, F. H. and Alt, W. (1984). The biophysics of cell surface motility. *In* A. S. Perelson, C. DeLisi and F. W. Wiegel (Ed.), *Cell Surface Dynamics: Concepts and Models*, pp. 495–542. New York: Marcel Dekker.

Dembo, M., Torney, D. C., Saxman, K. and Hammer, D. (1988). The reaction-limited kinetics of membrane-to-surface adhesion and detachment. *Proc. R. Soc. Lond. B*, **234**:55–83.

Devreotes, P. N. and Zigmond, S. H. (1988). Chemotaxis in eukaryotic cells: a focus on leukocytes and *Dictyostelium*. *Annu. Rev. Cell Biol.*, **4**:649–686.

Dickinson, R. B. and Tranquillo, R. T. (1933a). A stochastic model for adhesion-mediated cell random motility and haptotaxis. *J. Math. Biol.*, in press.

Dickinson, R. B. and Tranquillo, R. T. (1993b). Quantitative characterization of cell invasion *in vitro*. *Ann. Biomed. Eng.*, in press.

DiMilla, P. A., Barbee, K. and Lauffenburger, D. A. (1991). Mathematical model for the effects of adhesion and mechanics on cell migration speed. *Biophys. J.*, **60**:15–37.

DiMilla, P. A., Quinn, J. A., Albelda, S. A. and Lauffenburger, D. A. (1992). Measurement of cell migration parameters for human smooth muscle cells. *AICHE J.*, **38**:1092–1104.

DiMilla, P. A., Stone, J., Quinn, J. A., Albelda, S. M. and Lauffenburger, D. A. (1993). An optimal adhesiveness exists for smooth muscle cell migration on type-IV collagen and fibronectin. Submitted for publication.

Dong, C., Skalak, R., Sung, K. P., Schmid-Schonbein, G. W. and Chien, S. (1988). Passive deformation analysis of human leukocytes. *Journal of Biomechanical Engineering*, **110**:27–36.

Draetta, G. (1990). Cell cycle control in eukaryotes: molecular mechanisms of cdc2 activation. *Trends Biochem. Sci.*, **15**:378–383.

Duband, J., Dufour, S., Yamada, S. S., Yamada, K. M. and Thiery, J. P. (1991). Neural crest cell locomotion induced by antibodies to $\beta 1$ integrins: a tool for studying the roles of substratum molecular avidity and density in migration. *J. Cell Sci.*, **98**:517–532.

Dunn, G. A. (1981). Chemotaxis as a form of directed cell behavior: some theoretical considerations. *In* J. M. Lackie and P. C. Wilkinson (Eds.), *Biology of the Chemotactic Response*, pp. 247–280. Cambridge: Cambridge University Press.

Dunn, G. A. (1983). Characterising a kinesis response: time averaged measures of cell speed and directional persistence. *Agents and Actions [Suppl.]*, **22**:14–33.

Dustin, M. L. (1990). Two-way signalling through the LFA-1 lymphocyte adhesion receptor. *BioEssays*, **12**:421–427.

Dustin, M. L. and Springer, T. A. (1989). T-cell receptor cross-linking transiently stimulates adhesiveness through LFA-1. *Nature*, **341**:619–624.

Ebner, R. and Derynck, R. (1991). Epidermal growth factor and transforming growth factor alpha: differential intracellular routing and processing of ligand-receptor complexes. *Cell Reg.*, **2**:599–612.

Egelhoff, T. T. and Spudich, J. A. (1991). Molecular genetics of cell migration: *Dictyostelium* as a model system. *Trends Genet.*, **7**:161–166.

Elson, E. L. (1988). Cellular mechanics as an indicator of cytoskeletal structure and function. *Annu. Rev. Biophys. Chem.*, **17**:397–430.

Evans, E., Berk, D. and Leung, A. (1991a). Detachment of agglutinin-bonded red blood cells: I. Forces to rupture molecular-point attachments. *Biophys. J.*, **59**:838–848.

Evans, E., Berk, D., Leung, A. and Mohandas, N. (1991b). Detachment of agglutinin-bonded red blood cells: II. Mechanical energies to moderate large contact areas. *Biophys. J.*, **59**:849–860.

Evans, E. and Needham, D. (1987). Physical properties of surfactant bilayer membranes: thermal transitions, elasticity, rigidity, cohesion, and colloidal interactions. *J. Phys. Chem.*, **91**:4219–4228.

Evans, E. and Yeung, A. (1989). Apparent viscosity and cortical tension of blood granulocytes determined by micropipet aspiration. *Biophys. J.*, **56**:151–160.

Evans, E. A. (1985a). Detailed mechanics of membrane–membrane adhesion and separation: I. Continuum of molecular cross-bridges. *Biophys. J.*, **48**:175–183.

Evans, E. A. (1985b). Detailed mechanics of membrane–membrane adhesion and separation: II. Discrete kinetically trapped molecular cross-bridges. *Biophys. J.*, **48**:185–192.

Evans, E. A. (1988). Mechanics of cell deformation and cell-surface adhesion. *In* P. Bongrand (Ed.), *Physical Basis of Cell–Cell Adhesion*, pp. 91–123. Boca Raton, FL: CRC Press.

Evans, E. A. (1989). Structure and deformation properties of red blood cells: concepts and quantitative methods. *Methods Enzymol.*, **173**:3–35.

Evans, E. A. and Skalak, R. (1980). *Mechanics and Thermodynamics of Biomembranes*. Boca Raton, FL: CRC Press.

Farmer, S. R. and Dike, L. E. (1989). Cell shape and growth control: role of cytoskeleton–extracellular matrix interactions. *In* W. D. Stein and F. Bronner (Eds.), *Cell Shape: Determinants, Regulation, and Regulatory Control*, pp. 173–202. San Diego, CA: Academic Press.

Farrell, B. E., Daniele, R. P. and Lauffenburger, D. A. (1990). Quantitative relationships between single-cell and cell-population model parameters for chemosensory migration responses of alveolar macrophages to C5a. *Cell Motil. Cytoskeleton*, **16**:279–293.

Felder, S. and Elson, E. L. (1990). Mechanics of fibroblast locomotion: quantitative analysis of forces and motions at the leading lamellas of fibroblasts. *J. Cell Biol.*, **111**:2513–2526.

Fischer, E. H., Charbonneau, H. and Tonks, N. K. (1991). Protein tyrosine phosphatases: a diverse family of intracellular and transmembrane enzymes. *Science*, **253**:401–406.

Fisher, P. R., Merkl, R. and Gerisch, G. (1989). Quantitative analysis of cell motility and chemotaxis in *Dictyostelium discoideum* by using an image processing system and a novel chemotaxis chamber providing stationary chemical gradients. *J. Cell Biol.*, **108**:973–984.

Foxall, C., Watson, S. R., Dowbenko, D., Fennie, C., Lasky, L. A., Kiso, M., Hasegawa, A., Asa, D. and Brandley, B. K. (1992). The three members of the selectin receptor family recognize a common carbohydrate epitope, the sialyl Lewis$z$ oligosaccharide. *J. Cell Biol.*, **117**:895–902.

Gail, M. H. and Boone, C. W. (1970). The locomotion of mouse fibroblasts in tissue culture. *Biophys. J.*, **10**:980–993.

Gardiner, C. W. (1983). *Handbook of Stochastic Methods for Physics, Chemistry, and the Natural Sciences*. New York: Springer-Verlag.

Gerard, N. P. and Gerard, C. (1991). The chemotactic receptor for human C5a anaphylatoxin. *Nature*, **349**:614–620.

Gerisch, G. and Keller, H.-U. (1981). Chemotactic reorientation of granulocytes stimulated with micropipettes containing f-Met-Leu-Phe. *J. Cell Sci.*, **52**:1–10.

Gill, G. N. (1989). Growth factors and their receptors. *In* R. Weinberg (Ed.), *Oncogenes and the Molecular Origins of Cancer*, pp. 67–96. Cold Spring Harbor, NY: Cold Spring Harbor Laboratory Press.

Gill, G. N., Bertics, P. J. and Santon, J. B. (1987). Epidermal growth factor and its receptor. *Molec. Cell Endocrinol.*, **51**:169–186.

Glacken, M. W., Adema, E. and Sinskey, A. J. (1988). Mathematical descriptions of hybridoma culture kinetics: I. Initial metabolic rates. *Biotech. Bioeng.*, **32**:491–506.

Goldman, A. J. Cox, R. G. and Brenner, H. (1967a). Slow viscous motion of a sphere parallel to a plane wall: I. Motion through a quiescent fluid. *Chem. Eng. Sci.*, **22**:637–652.

Goldman, A. J., Cox, R. G. and Brenner, H. (1967b). Slow viscous motion of a sphere parallel to a plane wall: II. Couette flow. *Chem. Eng. Sci.*, **22**:653–659.

Goodman, S. L., Deutzmann, R. and von der Mark, K. (1987). Two distinct cell-binding domains in laminin can independently promote nonneuronal cell adhesion and spreading. *J. Cell Biol.*, **105**:595–610.

Goodman, S. L., Risse, G. and von der Mark, K. (1989). The E8 subfragment of laminin promotes locomotion of myoblasts over extracellular matrix. *J. Cell Biol.*, **109**:799–809.

Gospodarowicz, D., Massaglia, S., Cheng, J. and Fujii, D. (1989). Fibroblast growth factor. *CRC Crit. Rev. Oncol.*, **1**:1–26.

Grinnell, F. (1986). Focal adhesion sites and the removal of substratum-bound fibronectin. *J. Cell Biol.*, **103**:2697–2706.

Hammer, D. A. and Apte, S. A. (1992). Simulation of cell rolling and adhesion on surfaces in shear flow: general results and analysis of selectin-mediated neutrophil adhesion. *Biophys. J.*, **63**:35–57.

Hammer, D. A. and Lauffenburger, D. A. (1987). A dynamical model for receptor-mediated cell adhesion to surfaces. *Biophys. J.*, **52**:475–487.

Hammer, D. A., Linderman, J. J., Graves, D. J. and Lauffenburger, D. A. (1987). Affinity chromatography for cell separation: mathematical model and experimental analysis. *Biotech. Progress*, **3**:189–204.

Harris, A. K. (1989). Protrusive activity of the cell surface and the movements of tissue cells. *In* N. Akkas (Ed.), *Biomechanics of Active Movement and Deformation of Cells*, pp. 249–294. Berlin: Springer-Verlag.

Harris, A. K., Wild, P. and Stopak, D. (1980). Silicone rubber substrata: a new wrinkle in the study of cell locomotion. *Science*, **208**:177–179.

Hartmann, N. R., Gilbert, C. W., Jansson, B., MacDonald, P. D. M., Steel, G. G. and Valleron, A. J. (1975). A comparison of computer methods for the analysis of fraction labelled mitosis curves. *Cell Tiss. Kinet.*, **8**:119–124.

Heath, J. P. and Holifield, B. F. (1991). Cell locomotion: new research tests old ideas on membrane and cytoskeletal flow. *Cell Motil. Cytoskeleton*, **18**:245–257.

Heidemann, S. R., Lamoureux, P. and Buxbaum, R. E. (1991). On the cytomechanics and fluid dynamics of growth cone motility. *J. Cell Sci. Suppl.*, **15**:35–44.

Helm, C. A., Knoll, W. and Israelachvili, J. N. (1991). Measurement of ligand–receptor interactions. *Proc. Natl. Acad. Sci. USA*, **8**:8169–8173.

Hemler, M. E. (1990). VLA proteins in the integrin family: structures, functions, and their role on leukocytes. *Annu. Rev. Immunol.*, **8**:365–400.

Holmes, W. E., Lee, J., Kuang, W. J., Rice, G. C. and Wood, W. I. (1991). Structure and functional expression of a human interleukin-8 receptor. *Science*, **253**:1278–1280.

Hubbe, M. A. (1981). Adhesion and detachment of biological cells in vitro. *Prog. Surface Sci.*, **111**:65–138.

Hudlicka, O. and Tyler, K. R. (1986). *Angiogenesis: The Growth of the Vascular System.* London: Academic Press, 221 pp.

Hunkapiller, T. and Hood, L. (1986). The growing immunoglobulin gene superfamily. *Nature*, **323**:15–16.

Hunter, T. (1991). Cooperation between oncogenes. *Cell*, **64**:249–270.

Hunter, T. and Pines, J. (1991). Cyclins and cancer. *Cell*, **66**:1071–1074.

Hynes, R. O. (1992). Integrins: versatility, modulation, and signalling in cell adhesion. *Cell*, **69**:11–25.

Israelachvili, J. N. and McGuiggan, P. M. (1988). Forces between surfaces in liquids. *Science*, **241**:795–800.

Kater, S. K. and Letourneau, P. (1985). *Biology of the Nerve Growth Cone*. New York: Liss, 351 pp.

Keller, E. F. and Segel, L. A. (1971). Model for chemotaxis. *J. Theor. Biol.*, **30**:225–234.

Klagsbrun, M. and Baird, A. (1991). A dual receptor system is required for basic fibroblast growth factor activity. *Cell*, **67**:229–231.

Knauer, D. J., Wiley, H. S. and Cunningham, D. D. (1984). Relationship between epidermal growth factor receptor occupancy and mitogenic response: quantitative analysis using a steady-state model system. *J. Biol. Chem.*, **259**:5623–5631.

Kolodney, M. S. and Wysolmerski, R. B. (1992). Isometric contraction by fibroblasts and endothelial cells in tissue culture: a quantitative study. *J. Cell Biol.*, **117**:73–82.

Kucik, D. F., Elson, E. and Sheetz, M. (1990). Cell migration does not produce membrane flow. *J. Cell Biol.*, **111**:1617–1622.

Kuo, S. C. and Lauffenburger, D. (1993). Relationship between receptor/ligand bond affinity and specific adhesion strength. *Biophys. J.*, **65**:2191–2200.

Kupfer, A. and Singer, S. J. (1989). Cell biology of cytotoxic and helper T cell functions: immunofluorescence microscopic studies of single cells and cell couples. *Annu. Rev. Immunol.*, **7**:309–337.

Lackie, J. M. (1986). *Cell Movement and Cell Behavior*. London: Allen & Unwin, 316 pp.

Lackie, J. M., Chaabane, N. and Croket, K. V. (1987). A critique of the methods used to assess leukocyte behavior. *Biomed. Pharmacother*, **41**:265–278.

Lackie, J. M. and Wilkinson, P. C. (1984). Adhesion and locomotion of neutrophil leukocytes on 2-D substrata and in 3-D matrices. *In* H. J. Meiselman, M. A. Lichtman and P. L. LaCelle (Eds.), *White Cell Mechanics: Basic Science and Clinical Aspects*, pp. 237–254. New York: Liss.

Lauffenburger, D. A. (1983). Measurement of phenomenological parameters for leukocyte motility and chemotaxis. *Agents and Actions* [*Suppl.*], **12**:34–53.

Lauffenburger, D. A. (1989). A simple model for the effects of receptor-mediated cell–substratum adhesion on cell migration. *Chem. Eng. Sci.*, **44**:1903–1914.

Lauffenburger, D. A. and Cozens, C. (1989). Regulation of mammalian cell growth by autocrine growth factors: analysis of consequences for inoculum cell density effects. *Biotech. Bioeng.*, **33**:1365–1378.

Lauffenburger, D. A., Farrell, B., Tranquillo, R., Kistler, A. and Zigmond, S. (1987). Gradient perception by neutrophil leucocytes, continued. *J. Cell Sci.*, **88**:415–416.

Lauffenburger, D. A., Linderman, J. and Berkowitz, L. (1987). Analysis of mammalian cell growth factor receptor dynamics. *Ann. NY Acad. Sci.*, **506**:147–162.

Lauffenburger, D. A., Rothman, C. and Zigmond, S. J. (1983). Measurement of leukocyte motility and chemotaxis parameters with a linear under-agarose assay. *J. Immunol.*, **131**:940–947.

Lauffenburger, D. A., Tranquillo, R. T. and Zigmond, S. H. (1988). Concentration gradients of chemotactic factors in chemotaxis assays. *Methods Enzymol.*, **162**:85–101.

Lawrence, M. B., Smith, C. W., Eskin, S. G. and McIntire, L. V. (1990). Effect of venous shear stress on CD18-mediated neutrophil adhesion to cultured endothelium. *Blood*, **75**:227–237.

Lawrence, M. B. and Springer, T. A. (1991). Leukocytes roll on a selectin at physiologic flow rates: distinction from and prerequisite for adhesion through integrins. *Cell*, **65**:859–873.

Lee, J., Gustafsson, M., Magnusson, K. and Jacobson, K. (1990). The direction of membrane lipid flow in locomoting polymorphonuclear leukocytes. *Science,* **247**:1229–1233.

Lund, K. A., Opresko, L. K., Starbuck, C., Walsh, B. J. and Wiley, H. S. (1990). Quantitative analysis of the endocytic system involved in hormone-induced receptor internalization. *J. Biol. Chem.,* **265**:15713–15723.

McClay, D. R. and Ettensohn, C. A. (1987). Cell adhesion in morphogenesis. *Annu. Rev. Cell Biol.,* **3**:319–345.

McClay, D. R., Wessel, G. M. and Marchase, R. B. (1981). Intercellular recognition: quantitation of initial binding events. *Proc. Natl. Acad. Sci. USA,* **78**:4975–4979.

McCutcheon, M. (1946). Chemotaxis in leukocytes. *Physiol., Rev.,* **26**:319–336.

McKeehan, W. L. and McKeehan, K. A. (1981). Extracellular regulation of fibroblast multiplication: a direct kinetic approach to analysis of role of low molecular weight nutrients and serum growth factors. *J. Supramol. Struct. Cell. Biochem.,* **15**:83–110.

Mege, J. L., Capo, C., Benoliel, A. and Bongrand, P. (1987). Use of cell contour analysis to evaluate the affinity between macrophages and glutaraldehyde-treated erythrocytes. *Biophys. J.,* **52**:177–186.

Monod, J. (1949). The growth of bacterial cultures. *Annu. Rev. Microbiol.,* **3**:371–394.

Mueller, S. C., Kelly, T., Dai, M., Dai, H. and Chen, W. (1989). Dynamic cytoskeleton–integrin associations induced by cell binding to immobilized fibronectin. *J. Cell Biol.,* **109**:3455–3464.

Murray, A. W. and Kirschner, M. W. (1989). Dominoes and clocks: the union of two views of the cell cycle. *Science,* **246**:614–621.

Murthy, U., Basu, M., Sen-Majumdar, A. and Das, M. (1986). Perinuclear location and recycling of epidermal growth factor receptor kinase. *J. Cell Biol.,* **103**:333–342.

Nossal, R. (1988). On the elasticity of cytoskeletal networks. *Biophys. J.,* **53**:349–359.

Oppenheim, J. J., Zachariae, C. O., Mukaida, N. and Matsushima, N. (1991). Properties of the novel proinflammatory supergene "intercrine" cytokine family. *Annu. Rev. Immunol.,* **9**:617–648.

Oster, G. (1988). Biophysics of the leading lamella. *Cell Motil. Cytoskeleton,* **10**:164–171.

Oster, G. F. and Perelson, A. S. (1987). The physics of cell motility. *J. Cell Sci. Suppl.,* **8**:35–54.

Othmer, H. G., Dunbar, S. R. and Alt, W. (1988). Models of dispersal in biological systems. *J. Math. Biol.,* **26**:263–298.

Pardee, A. B. (1989). G_1 events and regulation of cell proliferation. *Science,* **246**:603–608.

Parkhurst, M. R. and Saltzmann, W. M. (1992). Quantification of human neutrophil motility in three-dimensional collagen gels: effect of collagen concentration. *Biophys. J.,* **61**:306–315.

Pines, J. and Hunter, T. (1990). p34[cdc2]: The S and M kinase? *New Biol.,* **2**:389–401.

Pinsky, M. A. (1991). *Partial Differential Equations and Boundary-Value Problems with Applications.* Second edition. New York: McGraw-Hill, 461 pp.

Regen, C. and Horwitz, A. F. (1992). Dynamics of $\beta1$ integrin-mediated adhesive contacts in motile fibroblasts. *J. Cell Biol.,* **119**:1347–1359.

Rein, A. and Rubin, H. (1968). Effects of local cell concentrations upon the growth of chick embryo cells in tissue culture. *Exp. Cell Res.,* **49**:666–678.

Rivero, M. A., Tranquillo, R. T., Buettner, H. M. and Lauffenburger, D. A. (1989). Transport models for chemotactic cell populations based on individual cell behavior. *Chem. Eng., Sci.,* **44**:2881–2897.

Rothman, C. and Lauffenburger, D. A. (1983). Analysis of the linear under-agarose leukocyte chemotaxis assay. *Ann. Biomed. Eng.,* **11**:451–460.

Rutishauser, U. and Sachs, L. (1975). Receptor mobility and the binding of cells to lectin-coated fibers. *J. Cell Biol.,* **66**:76–85.

Ryan, T. A., Myers, J., Holowka, D., Baird, B. and Webb, W. W. (1988). Molecular crowding on the cell surface. *Science*, **239**:61–64.

Sager, R. (1989). Tumor suppressor genes: the puzzle and the promise. *Science*, **246**:1406–1412.

Saterbak, E. A., Kuo, S. C. and Lauffenburger, D. A. (1993). Heterogeneity and probabilistic binding contributions to receptor-mediated cell detachment kinetics. *Biophys. J.*, **65**:243–252.

Schmid-Schonbein, G. W., Sung, K. P., Tozeren, H., Skalak, R. and Chien, S. (1981). Passive mechanical properties of human leukocytes. *Biophys. J.*, **36**:243–256.

Sharma, S. K. and Mahendroo, P. P. (1980). Affinity chromatography of cells and cell membranes. *J. Chromatogr.*, **184**:471–499.

Shields, E. D. and Noble, P. B. (1987). Methodology for detection of heterogeneity of cell locomotory phenotypes in three-dimensional gels. *Exp. Cell Biol.*, **55**:250–256.

Simon, S. I. and Schmid-Schonbein, G. W. (1990). Cytoplasmic strains and strain rates in motile polymorphonuclear leukocytes. *Biophys. J.*, **58**:319–332.

Singer, J. S. (1992). Intercellular communication and cell–cell adhesion. *Science*, **255**:1671–1677.

Singer, S. J. and Kupfer, A. (1986). The directed migration of eukaryotic cells. *Annu. Rev. Cell Biol.*, **2**:337–365.

Skalak, R., Dong, C. and Zhu, C. (1990). Passive deformations and active motions of leukocytes. *J. Biomech. Eng.*, **112**:295–302.

Sorkin, A. and Carpenter, G. (1991). Dimerization of internalized epidermal growth factor receptors. *J. Biol. Chem.*, **266**:23453–23460.

Sporn, M. B. and Roberts, A. B. (1990). *Peptide Growth Factors and Their Receptors, vol. I*. Berlin: Springer-Verlag, 794 pp.

Springer, T. A. (1990). Adhesion receptors of the immune system. *Nature*, **346**:425–434.

Starbuck, C. and Lauffenburger, D. A. (1992). Mathematical model for the effects of epidermal growth factor receptor trafficking dynamics on fibroblast proliferation responses. *Biotech. Prog.*, **8**:132–143.

Starbuck, C., Wiley, H. S. and Lauffenburger, D. A. (1990). Epidermal growth factor binding and trafficking dynamics in fibroblasts: relationship to cell proliferation. *Chem. Eng. Sci.*, **45**:2367–2373.

Stokes, C. L., Williams, S. K. and Lauffenburger, D. A. (1991). Migration of individual microvessel endothelial cells: stochastic model and parameter measurement. *J. Cell Sci.*, **99**:419–430.

Stopak, D. and Harris, A. K. (1982). Connective tissue morphogenesis by fibroblast traction: I. tissue culture observations. *Dev. Biol.*, **90**:383–398.

Swan, G. W. (1977). *Some Current Mathematical Topics in Cancer Research*. Ann Arbor, MI: University Microfilms International.

Takeichi, M. (1990). Cadherins: a molecular family important in selective cell–cell adhesion. *Annu. Rev. Biochem.*, **59**:237–252.

Tapley, P., Horwitz, A. F., Buck, C., Duggan, K. and Rohrschneider, L. (1989). Integrins isolated from Rous sarcoma virus-transformed chicken embryo fibroblasts. *Oncogene*, **4**:325–333.

Thomas, K. M., Pyun, H. Y. and Navarro, J. (1990). Molecular cloning of the fMet-Leu-Phe receptor from neutrophils. *J. Biol. Chem.*, **265**:20061–20064.

Torney, D. C., Dembo, M. and Bell, G. I. (1986). Thermodynamics of cell adhesion II: freely mobile repellers. *Biophys. J.*, **49**:501–507.

Tozeren, A., Sung, K. P. and Chien, S. (1989). Theoretical and experimental studies on cross-bridge migration during cell disaggregation. *Biophys. J.*, **55**:479–487.

Tranquillo, R. T. (1990). Theories and models of gradient perception. *In* J. P. Armitage and J. M. Lackie (Eds.), *Biology of the Chemotactic Response*, pp. 35–75. Cambridge: Cambridge University Press.

Tranquillo, R. T. and Alt, W. (1990). Glossary of terms concerning oriented movement. *In* W. Alt and G. Hoffman (Eds.), *Biological Motion*, pp. 510–517. Berlin: Springer-Verlag.

Tranquillo, R. T., Fisher, E. S., Farrell, B. E. and Lauffenburger, D. A. (1988a). A stochastic model for chemosensory cell movement: application to neutrophil and macrophage persistence and orientation. *Math. Biosci.*, **90**:287–303.

Tranquillo, R. T. and Lauffenburger, D. A. (1986). Consequences of chemosensory phenomena for leukocyte chemotactic orientation. *Cell Biophysics*, **8**:1–46.

Tranquillo, R. T. and Lauffenburger, D. A. (1987). Stochastic model of leukocyte chemosensory movement. *J. Math. Biol.*, **25**:229–262.

Tranquillo, R. T., Lauffenburger, D. A. and Zigmond, S. J. (1988b). A stochastic model for leukocyte random motility and chemotaxis based on receptor binding fluctuations. *J. Cell Biol.*, **106**:303–309.

Tranquillo, R. T., Zigmond, S. H. and Lauffenburger, D. A. (1988c). Measurement of the chemotaxis coefficient for human neutrophils in the under-agarose migration assay. *Cell Motil. Cytoskeleton*, **11**:1–15.

Trinkaus, J. P. (1984). *Cells Into Organs. The Forces That Shape the Embryo*. Englewood Cliffs, NJ: Prentice-Hall.

Ullrich, A. and Schlessinger, J. (1990). Signal transduction by receptors with tyrosine kinase activity. *Cell*, **61**:203–212.

Van de Vijver, M. J., Kumar, R. and Mendelsohn, J. (1991). Ligand-induced activation of A431 cell epidermal growth factor receptors occurs primarily by an autocrine pathway that acts upon receptors on the surface rather than intracellularly. *J. Biol. Chem.*, **266**:7503–7508.

Ward, M. D. and Hammer, D. A. (1993). A theoretical analysis for the effect of focal contact formation on cell–substrate attachment strength. *Biophys. J.*, in press.

Wattenbarger, M. R., Graves, D. J. and Lauffenburger, D. A. (1990). Specific adhesion of glycophorin liposomes to a lectin surface in shear flow. *Biophys. J.*, **57**:765–777.

Weigel, P. H., Schnaar, R. L., Kuhlenschmidt, M. S., Schmell, E., Lee, R. T., Lee, Y. C. and Roseman, S. (1979). Adhesion of hepatocytes to immobilized sugars: a threshold phenomenon. *J. Biol. Chem.*, **254**:10830–10838.

Wells, A., Welsh, J. B., Lazar, C. S., Wiley, S., Gill, G. N. and Rosenfeld, M. G. (1990). Ligand-induced transformation by a noninternalizing epidermal growth factor receptor. *Science*, **247**:962–964.

Wiley, H. S. and Cunningham, D. D. (1981). A steady state model for analyzing the cellular binding, internalization and degradation of polypeptide ligands. *Cell*, **25**:433–440.

Wiley, H. S., Herbst, J. J., Walsh, B. J., Lauffenburger, D. A., Rosenfeld, M. G. and Gill, G. N. (1991). The role of tyrosine kinase activity in endocytotic compartmentation and down-regulation of the epidermal growth factor receptor. *J. Biol. Chem.*, **266**:11083–11094.

Wilkinson, P. C. (1982). *Chemotaxis and Inflammation*, second edition. Edinburgh: Churchill-Livingstone, 249 pp.

Yarden, Y. and Ullrich, A. (1988). Growth factor receptor tyrosine kinases. *Annu. Rev. Biochem.*, **57**:443–478.

Yeung, A. and Evans, E. (1989). Cortical shell–liquid core model for passive flow of liquid-like spherical cells into micropipets. *Biophys. J.*, **56**:139–149.

Zhu, C. (1991). A thermodynamic and biomechanical theory of cell adhesion. Part I: General Formalism. *J. Theor. Biol.*, **150**:27–50.

Zhu, C. and Skalak, R. (1988). A continuum model of protrusion of pseudopod in leukocytes. *Biophys. J.*, **54**:1115–1137.

Zhurkov, S. N. (1965). Kinetic concept of the strength of solids. *Int. J. Fract. Mech.*, **1**:311–323.

Zicha, D., Dunn, G. A. and Brown, A. F. (1991). A new direct-viewing chemotaxis chamber. *J. Cell Sci.*, **99**:769–775.

Zietz, S. and Nicolini, C. (1978). Flow microfluorometry and cell kinetics: a review. *In* A. J. Valleron and P. B. M. MacDonald (Eds.), *Biomathematics and Cell Kinetics*, pp. 357–394. Amsterdam: Elsevier/North-Holland.

Zigmond, S. H. (1974). Mechanisms of sensing chemical gradients by polymorphonuclear leukocytes. *Nature*, **249**:450–452.

Zigmond, S. H. (1977). Ability of polymorphonuclear leukocytes to orient in gradients of chemotactic factors. *J. Cell Biol.*, **75**:606–616.

Zigmond, S. H. (1981). Consequences of chemotactic peptide receptor modulation for leukocyte orientation. *J. Cell Biol.*, **88**:644–647.

Zigmond, S. H. (1982). Polymorphonuclear leukocyte response to chemotactic gradients. *In* R. Bellairs, A. Curtis and G. Dunn (Eds.), *Cell Behavior*, pp. 183–202. Cambridge: Cambridge University Press.

Zigmond, S. H. (1989). Cell locomotion and chemotaxis. *Curr. Opin. Cell Biol.*, **1**:80–86.

Zigmond, S. S., Klausner, R., Tranquillo, R. T. and Lauffenburger, D. A. (1985). Analysis of the requirements for time-averaging of receptor occupancy for gradient detection by polymorphonuclear leukocytes. *In* M. Czech and C. R. Kahn (Eds.), *Membrane Receptors and Cellular Regulation*, pp. 347–356. New York: Alan R. Liss.

Zigmond, S. H. and Lauffenburger, D. A. (1986). Assays of leukocyte chemotaxis. *Annu. Rev. Med.*, **37**:149–155.

Zigmond, S. H., Sullivan, S. J. and Lauffenburger, D. A. (1982). Kinetic analysis of chemotactic peptide receptor modulation. *J. Cell Biol.*, **92**:34–43.

7

Future Directions

Recalling the intent we outlined in Chapter 1, to build a bridge between cell biology and engineering in the area of cell receptor phenomena, we recognize that this book provides only a beginning. In offering a framework for analysis of the effects of receptor and ligand properties on cell responses, we have tried to integrate the influence of these properties through the various levels involved in behavioral responses: binding, trafficking, signaling, and interaction with other cell components. Yet, for the most part, the specific modeling efforts we have discussed have been limited in the number of levels combined in any given example. None really put together detailed description of events in all these four areas to generate predictions of resulting cell function. Further, we have not identified any body of work to date on relating receptor signaling to gene regulation processes, so this important level of response has been present only implicitly at best.

Hence, our first comment on directions for future work is that integration of more levels of response should be given great attention. In each of the cell functions considered in this book, possibilities for this sort of effort beckon. In cell proliferation, receptor/growth factor binding and trafficking properties have been related to an eventual mitogenic response, but intermediate steps of signal transduction and propagation (kinase and phosphatase activity, interactions with cyclins and growth suppressor proteins) should soon be accessible to modeling to fill in at least a portion of the connection. In cell adhesion, regulation of adhesion receptor binding and trafficking processes, as well as detailed interactions with cytoskeletal elements, should be incorporated into the model structures. Cell migration may be closest to a state of integrating many response levels, because key signal transduction events appear to occur within a comparatively short time scale. In all cases, however, we appeal to judicious disregard of detail merely for its own sake. An ability to choose the most economical characterization of a complicated system, though an art, is still essential for seeing the forest despite the trees.

A different type of integration will also be needed, that of the cells with a more complex environment. Obviously, multiple ligands of diverse sorts are typically present in a physiological system, so a significant next step will be to analyze how complementary, competing, or irrelevant signals are combined to elicit an overall cell behavior. At the same time, tissues (whether *in vivo* or *in vitro*) are composed of numerous cell types that interact by means of receptor/ligand interactions. These must be integrated as well when the ultimate objective is understanding or predicting cell function in viable tissue.

At the heart of all future modeling work, though, must be model validation through experiment. To us this is perhaps the most exciting aspect of this approach. Given the capabilities of molecular biologists to alter molecular structures of receptors, ligand, and accessory components, there is no excuse for avoiding the direct testing of models with intentional modification of system parameter values. The coming years should be a golden age for molecular-based modeling of cell receptor phenomena, because models can be formulated and tested using manipulation of key molecular species in terms of their important quantitative properties.

Finally, a brief word on the future of mathematical models themselves is in order. A familiar habitat of engineers, applied mathematicians, physicists, and others trained in the use of modeling tools has traditionally been differential equations, which characterize changes in system variables. With the utility of molecular simulations, in which many individual entities are followed to assess population or system behavior, this has begun to change. Given the dramatically increasing power of computational facilities available to researchers, we anticipate widespread application of direct simulation methods to model situations of great complexity. Just as a single example, one can envision simulations of cell adhesion processes in which hundreds of receptors, ligands, cytoskeletal elements, and signal transduction components are all followed individually through their dynamic play governed by intermolecular forces. We currently capture these dynamics with rate constants – unquestionably a convenient symbolism – but a visual realization of the phenomena may bring additional insight. Of course, simulations too must be subject to experimental validation. Here the proper experimental view may be direct visualization also, using microscopic image analysis procedures competent to identify molecular events. Molecular properties must still be manipulated, nonetheless, in order to verify that a simulation can predict an outcome as with more traditional modeling approaches.

Finally, the most certain statement we can make about future directions of work in this area is that they will arise from innovations that have not even occurred to us. We will be eager to see what those are.

Index